Changing Climates in North American Politics

American and Comparative Environmental Policy
Sheldon Kamieniecki and Michael E. Kraft, series editors

Russell J. Dalton, Paula Garb, Nicholas P. Lovrich, John C. Pierce, and John M. Whiteley, *Critical Masses: Citizens, Nuclear Weapons Production, and Environmental Destruction in the United States and Russia*

Daniel A. Mazmanian and Michael E. Kraft, editors, *Toward Sustainable Communities: Transition and Transformations in Environmental Policy*

Elizabeth R. DeSombre, *Domestic Sources of International Environmental Policy: Industry, Environmentalists, and U.S. Power*

Kate O'Neill, *Waste Trading among Rich Nations: Building a New Theory of Environmental Regulation*

Joachim Blatter and Helen Ingram, editors, *Reflections on Water: New Approaches to Transboundary Conflicts and Cooperation*

Paul F. Steinberg, *Environmental Leadership in Developing Countries: Transnational Relations and Biodiversity Policy in Costa Rica and Bolivia*

Uday Desai, editor, *Environmental Politics and Policy in Industrialized Countries*

Kent Portney, *Taking Sustainable Cities Seriously: Economic Development, the Environment, and Quality of Life in American Cities*

Edward P. Weber, *Bringing Society Back In: Grassroots Ecosystem Management, Accountability, and Sustainable Communities*

Norman J. Vig and Michael G. Faure, eds., *Green Giants? Environmental Policies of the United States and the European Union*

Robert F. Durant, Daniel J. Fiorino, and Rosemary O'Leary, eds., *Environmental Governance Reconsidered: Challenges, Choices, and Opportunities*

Paul A. Sabatier, Will Focht, Mark Lubell, Zev Trachtenberg, Arnold Vedlitz, and Marty Matlock, eds., Sw*imming Upstream: Collaborative Approaches to Watershed Management*

Sally K. Fairfax, Lauren Gwin, Mary Ann King, Leigh S. Raymond, and Laura Watt, *Buying Nature: The Limits of Land Acquisition as a Conservation Strategy, 1780–2004*

Steven Cohen, Sheldon Kamieniecki, and Matthew A. Cahn, *Strategic Planning in Environmental Regulation: A Policy Approach That Works*

Michael E. Kraft and Sheldon Kamieniecki, eds., *Business and Environmental Policy: Corporate Interests in the American Political System*

Joseph F. C. DiMento and Pamela Doughman, eds., *Climate Change: What It Means for Us, Our Children, and Our Grandchildren*

Christopher McGrory Klyza and David J. Sousa, *American Environmental Policy, 1990–2006: Beyond Gridlock*

John M. Whiteley, Helen Ingram, and Richard Perry, eds., W*ater, Place, and Equity*

Judith A. Layzer, *Natural Experiments: Ecosystem-Based Management and the Environment*

Daniel A. Mazmanian and Michael E. Kraft, editors, *Toward Sustainable Communities: Transition and Transformations in Environmental Policy*, 2nd edition

Henrik Selin and Stacy D. VanDeveer, eds., *Changing Climates in North American Politics: Institutions, Policymaking, and Multilevel Governance*

Changing Climates in North American Politics

Institutions, Policymaking, and Multilevel Governance

edited by Henrik Selin and Stacy D. VanDeveer

The MIT Press
Cambridge, Massachusetts
London, England

MIT Press books may be purchased at special quantity discounts for business or sales promotional use. For information, please email special_sales@mitpress.mit.edu or write to Special Sales Department, The MIT Press, 55 Hayward Street, Cambridge, MA 02142.

This book was set in Sabon on 3B2 by Asco Typesetters, Hong Kong. Printed and bound in the United States of America. Printed on recycled paper.

Library of Congress Cataloging-in-Publication Data

Changing climates in North American politics : institutions, policymaking, and multilevel governance / edited by Henrik Selin and Stacy D. VanDeveer.
p. cm. — (American and comparative environmental policy)
Includes bibliographical references and index.
ISBN 978-0-262-01299-7 (hardcover : alk. paper) — ISBN 978-0-262-51286-2 (pbk. : alk. paper)
1. Climatic changes—Political aspects—North America. 2. Climatic changes—Government policy—North America. 3. Environmental policy—Political aspects—North America.
4. Global warming—Political aspects—North America. I. Selin, Henrik, 1971–
II. VanDeveer, Stacy D.
QC982.8.C53 2009
363.738′7456097—dc22 2008042153

10 9 8 7 6 5 4 3 2 1

This book is dedicated to the life and memory of Alex Farrell. He is greatly missed.

Contents

Series Foreword

Climate change is arguably the most important environmental policy challenge of the twenty-first century. Yet governments around the world have struggled to devise politically acceptable solutions. This is really not surprising. Climate change is the poster child for what many call third-generational environmental problems. They are global in origins and effects, their most serious impacts may come well into the future, they are technically complex and beset by scientific uncertainties, they can be costly to address, and solutions may require major changes in personal lifestyles, corporate behavior, consumerism, and government policies. Moreover, the benefits of acting on climate change are uncertain over the long term and broadly distributed to the public, whereas the costs tend to be more certain, imposed in the short term, and concentrated on particular groups or economic sectors eager to avoid them. This is not a recipe for successful policymaking.

As of this writing, the European Union is experiencing considerable problems with its effort to implement a market-based permit trading system to control carbon dioxide emissions in its member nations. Many government leaders outside of Europe, including those in North America and Asia, have been hoping that this new market system for carbon dioxide permit trading might provide a model for reducing greenhouse gas emissions in their countries. If proposed reforms do not correct the present difficulties with the market-based system in Europe, public officials will have to look elsewhere for new and innovative approaches to control greenhouse gas emissions.

Under these circumstances, scholars and policymakers would benefit from any assessment that speaks to how political systems can respond effectively to climate change. This volume provides just this kind of analysis, and it does so in a way that should appeal to a broad audience. The focus is on policymaking institutions and policy actors in the North American setting, with special consideration given to the multiple levels of government involved; emerging institutions, policies, and practices; the roles for public, private, and civic actors and their interaction; and processes of policy innovation and diffusion across various governmental jurisdictions. The fifteen chapters offer a wealth of descriptive detail on the evolution of climate change policies and the conditions that facilitate policymaking. The authors cover a breathtaking array of policy developments from local to national and international levels, including the roles of businesses, college campuses, and communications media. However, they wisely concentrate on actions within Canada, the United States, and Mexico to provide a unique geopolitical perspective. Many readers will be familiar with the broad outline of the policy developments, but nonetheless they will find the full accounts here to be a fascinating review of how far climate change policies have evolved in recent years. The individual chapters provide well-told and captivating stories of policy initiatives, plentiful insights into policymaking processes, and valuable lessons for future policy actions.

Of special interest is the authors' collective assessment of multilevel governance, where policy development occurs simultaneously and somewhat awkwardly in multiple jurisdictions, and where political authority is widely dispersed among an exceptionally diverse set of policy actors operating across local, state, regional, national, continental, and international venues. In the North American case, they find that climate change governance has been much more of a bottom-up than a top-down phenomenon, in contrast to the European Union, where the top-down approach has been dominant. Although the North American history reflects, in part, the reluctance of the national governments in the United States and Canada to advance climate change policy, the chapters also speak to the manifold opportunities afforded policy actors in the North American context to further their cause. The book's findings on the rich institutional and civil society capabilities in the North American political systems are encouraging. They also add much to our understanding of how complex policymaking dynamics affect the search for solutions to climate change as well as other environmental challenges.

The book illustrates well the goals of the MIT Press series in American and Comparative Environmental Policy. We encourage work that examines a broad range of environmental policy issues. We are particularly interested in volumes that incorporate interdisciplinary research and focus on the linkages between public policy and environmental problems and issues both within the United States and in cross-national settings. We welcome contributions that analyze the policy dimensions of relationships between humans and the environment from either a theoretical or empirical perspective. At a time when environmental policies are increasingly seen as controversial and new approaches are being implemented widely, we especially encourage studies that assess policy successes and failures, evaluate new institutional arrangements and policy tools, and clarify new directions for environmental politics and policy. The books in this series are written for a wide audience that includes academics, policymakers, environmental scientists and professionals, business and labor leaders, environmental activists, and students concerned with environmental issues. We hope they contribute to public understanding of environmental problems, issues, and policies of concern today and also suggest promising actions for the future.

Sheldon Kamieniecki, *University California, Santa Cruz*

Michael Kraft, *University of Wisconsin–Green Bay*

American and Comparative Environmental Policy Series Editors

Acknowledgments

This volume has been over three years in the making. It is the product of iterated interaction among the book's editors and participating authors, and it has benefited from a host of other commentators, discussants, and peer reviewers. In May 2006 we organized the public conference and authors' workshop Climate Change Politics in North America in Washington, D.C., at the Woodrow Wilson International Center for Scholars (WWICS). At this event, authors presented first drafts of their chapters. We are grateful for conference funding from the Embassy of Canada (Washington, D.C.), the Energy Foundation, and WWICS's Canada Institute and the Environmental Change and Security Project.

The authors and editors received valuable feedback from a series of commentators (academics, practitioners, and policy advocates). We would like to thank all who attended the conference for their participation and engaging questions, comments, and critiques. The participants, who contributed much to our discussions, included Tim Kennedy (Global Public Affairs), Julie Anderson (Union of Concerned Scientists), Truman Semans (Pew Climate Center), Andrew Aulisi (World Resources Institute), and Joseph Dukert (independent energy analyst). We are extremely grateful to the staff of the Canada Institute and the Environmental Change and Security Project for their hard work and invaluable input, especially David Biette, Geoff Dabelko, Katherine Ostrye, and Christophe Leroy. Also, our thinking on this volume and every contribution benefited from the careful attention and stewardship of Clay Morgan of the MIT Press, the series editors Sheldon Kamieniecki and Michael Kraft, and three extremely detailed, informative, and challenging anonymous peer reviews.

Henrik Selin would like to thank the Department of International Relations at Boston University and the European Commission's Jean Monnet program for financial support. Stacy D. VanDeveer is grateful for funding from the Department of Political Science and the College of Liberal Arts at the University of New Hampshire, the University of New Hampshire's Ronald H. O'Neal professorship, and the

European Commission's Jean Monnet program. We are both grateful to the Carbon Solutions New England project based at the University of New Hampshire and the many individuals in the United States, Canada, and Mexico who made time to be interviewed during the course of the research.

Introduction

1

Changing Climates and Institution Building across the Continent

Henrik Selin and Stacy D. VanDeveer

Introduction

Global climate change engenders complex and seemingly contradictory policy responses and outcomes. Whereas the U.S. federal government of George W. Bush rejected the Kyoto Protocol and opposed mandatory greenhouse gas (GHG) restrictions, California simultaneously developed some of the world's most comprehensive climate change laws and regulations. While Canada's federal government ratified the Kyoto Protocol in 2002, Canadian GHG emissions increased faster than U.S. national emissions over the past decade. Many major North American firms fiercely opposed climate change policy in the 1990s and early 2000s, but recently a growing number of corporate leaders have joined governors, premiers, mayors, and civil society representatives from all over the continent to call on federal policymakers to establish national cap-and-trade schemes to regulate GHG emissions. In Mexico, the state-owned oil company was among the first to engage constructively in climate change issues in the 1990s, in stark contrast to leading U.S. and Canadian oil firms.

The causes of climate change play out at multiple governance scales as GHG emissions result from actions and choices of individuals influenced by a range of institutions, from the local to the global. Impacts of climate change are also experienced at global, regional, national, and local levels (DiMento and Doughman 2007). As such, climate change governance involves and links multiple levels of political authority. Much has been written on governance of climate change at the global level (Dessler and Parson 2006; Luterbacher and Sprinz 2001; Miller and Edwards 2001). Recently, scholars have paid more attention to expanding regional, national, and local climate change mitigation (e.g., GHG reduction) and adaptation (e.g., making societies less vulnerable to climatic changes) efforts developing in parallel to global politics and policymaking (Bulkeley and Moser 2007; Harrison and Sundstrom 2007; Selin and VanDeveer 2007). This volume examines causes and implications of climate change–related political action in North America, from continental to local governance levels, involving a wide range of public, private, and civil society actors.

Climate change politics and policymaking have grown increasingly complex and dynamic across North America during the early years of the twenty-first century. Many of these actions are developing against the backdrop of growing scientific certainty about human influence on the global climate system (Cowie 2007), following trends in climate change–related policy developments in many other regions of the world (Fisher 2004; Harrison 2004; Low 2005; Schreurs 2002; Weart 2003). In its latest set of reports, the Intergovernmental Panel on Climate Change (IPCC) (2007a; 2007b; 2007c) stated that the current warming trend is primarily caused by anthropogenic releases of GHGs and that political and social actions are necessary to avert potentially disastrous effects. Continuing warming may, for example, affect precipitation and storm patterns, alter seasonal patterns, and accelerate melting of the polar ice caps thereby causing rising sea levels. Collectively, these and other climate-related changes would have considerable impacts on human societies and ecological systems worldwide.

Concern among prominent climate scientists about human-induced climate change dates back several decades, but most international and domestic policy debates began in the 1980s (Weart 2003). By 1992, when the world's governments gathered at the United Nations Conference on Environment and Development in Rio de Janeiro, Brazil, climate change was firmly established on the global political agenda. There, governments adopted the United Nations Framework Convention on Climate Change (UNFCCC), which sets the goal of "stabilization of greenhouse gas concentrations in the atmosphere at a level that would prevent dangerous anthropogenic interference with the climate system" (UNFCC 1992, article 2). Five years later, the much-debated Kyoto Protocol was adopted under the legal framework of the UNFCCC, setting mandatory individual-emission targets on carbon dioxide (CO_2) and five other GHGs for thirty-nine industrialized countries and countries with economies in transition, but not for developing countries.

Countries with GHG emission targets that have ratified the Kyoto Protocol are obliged to meet these no later than 2012. Of the three North American countries, only Canada and Mexico ratified the Kyoto Protocol (see table 1.1). Under the Kyoto Protocol, Canada has taken on a legally binding commitment to reduce emissions by 6 percent below its 1990 levels. Mexico, as a developing country, is exempted from mandatory emission reductions. One thing that all three countries have in common, however, is that their national emissions have grown since 1990 (see table 1.2).[1] In general, North American GHG emissions grew about 1 percent per year after 1990 (U.S. Climate Change Science Program 2007). Although Mexican and Canadian emissions have grown faster than those of the United States, the aggregate growth in U.S. emissions exceeds the combined increase of Canada and Mexico.

Table 1.1
North American signatures and ratifications of global climate change agreements

	Canada	United States	Mexico
1992 Framework Convention Adopted: June 1992 Entry into force: March 1993	S: 6/12/1992 R: 12/4/1992	S: 6/12/1992 R: 10/5/1992	S: 6/13/1992 R: 3/11/1993
1997 Kyoto Protocol Adopted: December 1997 Entry into force: February 2005	S: 4/29/1998 R: 12/17/2002	S: 12/11/1998	S: 6/9/1998 R: 9/7/2000

Notes: S = signed; R = ratified. Source: http://unfccc.int.

Table 1.2
North American and global GHG emissions

	Canada	United States	Mexico
Total emissions (CO_2 equivalent) 1990	599 megatons	6109 megatons	383 megatons
Total emissions 2004	758 megatons	7074 megatons	643 megatons*
Total emissions increase since 1990	26.5%	15.8%	67.9%
Per capita emissions (tons CO_2 equivalent) (2000)	22.1	24.5	5.2
Global ranking of per capita emissions (2000)	7th	6th	76th
Population (2000)	31 million	280 million	100 million
Percent of total global emissions (2000)	2.0%	20.6%	1.5%

Notes: * Mexico figure is for 2002. Sources: U.S. Environmental Protection Agency 2006; Environment Canada 2004; UNFCCC 2005; and Baumert, Herzog, and Pershing 2005.

The United States is the world's largest GHG emitter in absolute, cumulative terms. China recently surpassed the United States as the leading global emitter on a yearly basis, but U.S. annual GHG emissions remain high compared to other countries. Furthermore, Canada and the United States have some of the highest per capita emissions in the world. It should also be noted that the U.S. Energy Information Administration (2007) projects that U.S. and Canadian CO_2 emissions will grow by about 1 percent annually through 2030 (absent policy interventions or other changes), while Mexico's emissions are projected to grow by an average of 2.3 percent during the same period. These projections would mean an increase in the three countries' total CO_2 emissions by over 36 percent by 2030.

At a meeting in Bali in December 2007, almost all of the world's countries—including all three North American ones—launched the difficult political process of negotiating a new climate change agreement intended to take effect in 2012, when the Kyoto Protocol expires. Much of North American politics is influenced by the fact that Canada, the United States, and Mexico are pursuing deeper economic integration and higher economic growth under the North American Free Trade Agreement (NAFTA). While there are distinct differences among the North American countries in national attitudes toward the future of global climate change policy and characteristics of federal climate change action, complexities and divergences of climate change policymaking and action are even greater at the subnational level. Climate change policy initiatives are discussed and developed in a multitude of states, provinces, municipalities, and firms.

These many initiatives shape ongoing political debate and policy change at local, federal, and international levels. To better understand North American climate politics and policymaking, it is necessary to move the focus of debate to issues beyond the Kyoto Protocol. The changes in climate policy in North America are driven by continental, national, and local political developments and ongoing climate change and energy initiatives within civil society and private sector organizations. This volume examines four broad questions critical to understanding dynamic and continuing developments in North American climate change politics and policymaking. Some of the volume's chapters speak to all four questions, while others focus primarily on one or two.

1. What are the new or emerging institutions, policies, and practices in the area of climate change governance under development in North America?
2. What roles do major public, private, and civil society actors play, and how do they interact to shape policy and governance?
3. Through which pathways are climate change policies and initiatives diffused across jurisdictions in North America?
4. To what extent can North American climate change action be characterized as existing or emerging multilevel governance, and are local and federal institutions across the continent facilitating or impeding such developments?

North American GHG emissions are quite significant in both absolute and per capita terms (see table 1.2). Quickly stabilizing and then reducing North American emissions is necessary (but not sufficient) to prevent a doubling in atmospheric concentrations of CO_2 from preindustrial levels (which many scientists believe is important). As such, analysts and policymakers interested in short-term political actions and long-term solutions to the climate challenge should pay close attention to the plethora of efforts associated with climate change mitigation across North America and critically assess the potential accomplishments and limitations of these efforts. North American countries, firms, and citizens are often identified as laggards on GHG reduction efforts. This volume demonstrates, however, that more climate change mitigation politics and policymaking are taking place across the continent than is frequently believed. A rapidly growing number of public- and private-sector actors in North America are preparing for, and often trying to help create, a future of higher costs of carbon emissions where more stringent policies limit the release of GHG emissions.

North American Climate Change Politics in Brief

Multilevel governance across different levels of social organization has attracted growing analytical attention (Bache and Flinders 2004; Finger, Tamiotti, and Allouche 2006; Hooghe and Marks 2003; Young 2002). Multilevel climate change governance is developing simultaneously in multiple jurisdictions as policymaking and regulatory authority are dispersed among actors operating across global, continental, national, and local levels (Betsill and Bulkeley 2006; Bulkeley and Moser 2007; Harrison and Sundstrom 2007). North American climate change governance includes extensive horizontal and vertical interaction among federal, state, provincial, and municipal policymakers, private sector leaders, and civil society representatives. In contrast to European Union (EU) multilevel governance, North American climate change politics has not included strong national government leadership or much effort to coordinate policies across jurisdictions. As such, North American climate change governance has seen much more of bottom-up dynamics and noticeably less of top-down policymaking than has the EU.

There is a century-long history of intense and complex bilateral ecopolitics between U.S. and Canadian federal and state/provincial authorities covering a wide range of pollution and natural resource issues (Dorsey 1998; Le Prestre and Stoett 2006). The U.S.-Mexican border region has also played host to transnational and interstate environmental politics for several decades (Mumme 2003). In addition, Canadian, U.S., and Mexican trade and environmental issues—especially in the context of impacts of NAFTA—have received much scholarly attention (Audley 1997; Deere and Esty 2002; Gallagher 2004; Markell and Knox 2003). However, developing North American multilevel climate change governance appears to involve a greater number of actors, jurisdictions, and institutions than most other areas of transcontinental environmental policymaking and cooperation.

All three North American countries have federal structures. Climate change policies are under debate, enactment, and implementation at the federal level, in states and provinces, in municipalities, and in many firms across the continent. North

American civil society has also become more active and engaged in climate change mitigation and advocacy efforts since the 1990s. However, decision-making authority is divided differently in each of the three federal systems. Furthermore, many federal divisions of authority remain unsettled, as demonstrated by a series of ongoing lawsuits between U.S. states and the federal government and the spirited debates between Canadian provinces and federal authorities around climate change policymaking. As such, issues of federal- and local-level relations must be taken into consideration when examining North American climate change politics. This section briefly summarizes such developments, setting the context for subsequent chapters.

Federal Politics and Policymaking

In the United States, the Bush administration consistently opposed mandatory GHG emission reductions; the previous Clinton administration favored climate change policy action but enacted very little of it. Instead, federal policy focused on voluntary programs with a goal of reducing the GHG intensity of the U.S. economy as measured by national emissions/gross domestic product. However, this policy did not prevent absolute increases in GHG emissions. U.S. federal policy has also favored scientific study of climate change and the development of emissions-reducing technologies. One of the great ironies of climate change politics is that the Bush administration funded more climate change research than any other country—only to ignore most of its findings. The Climate Change Science Program (CCSP) and the Climate Change Technology Program (CCTP) were created in 2002 (Victor 2004). The CCSP was established to support climate monitoring and research on causes of climate change and outlines a plan for development of tools to aid policymaking, while the CCTP is tasked with creating and implementing a research and development program for climate change–related technology.

Throughout the 1990s and the early years of the twenty-first century, the U.S. Congress consistently opposed mandatory GHG controls with strong opposition from post-1994 Republican leaders. In July 1997, the Senate passed the Byrd-Hagel resolution by 95–0, stating its opposition to an international climate change treaty that did not include meaningful participation and commitments from developing countries. According to most senators, the Kyoto Protocol did not pass this test. Senator James Inhofe, former chair of the Senate Committee on Environment and Public Works, famously described global warming as "the greatest hoax ever perpetrated on the American people" (Kolbert 2005). In the early 2000s, the Senate voted down several proposals to establish mandatory national GHG regulations. Proposed GHG emission reductions typically faced even stronger opposition in the House of Representatives.

There are, however, signs of political change in Washington. The 2006 congressional elections resulted in narrow Democratic majorities in both the House of Representatives and the Senate, paving the way for several new initiatives in the 110th Congress (2007–2009). The ruling by the U.S. Supreme Court in April 2007 that the Clean Air Act gives the U.S. Environmental Protection Agency the authority to regulate CO_2 emissions from vehicles was a critical event acting as an important driver of federal policy change. In December 2007, Congress passed a bill increasing Corporate Average Fuel Economy (CAFE) standards for vehicles. While this increase was rather modest, setting the target of thirty-five miles per gallon by 2020, it was the first increase in CAFE standards by Congress in over thirty years. The higher CAFE standards were signed into law by President Bush in the Energy Independence and Security Act of 2007, which also included new efficiency standards for light bulbs and appliances along with substantial subsidies and mandates designed to increase the use of corn-based ethanol and biofuels.

With enhanced majorities of the Democratic Party, leaders in the Senate and the House of Representatives vowed that climate change will receive sustained attention in the 111th Congress (2009–2011), organizing multiple committee hearings on scientific and political aspects of climate change. A host of legislative proposals targeting GHG emissions are under development in both the House of Representatives and the Senate. Several of these bills build on proposals defeated by earlier Congresses and draw on state initiatives. Furthermore, the arrival of President Barack Obama's administration in January 2009 means that supporters of more aggressive climate change policy and action (both domestically and internationally) currently control both ends of Pennsylvania Avenue (i.e., the executive and legislative branches of the federal government).

Canada's national parliament, after lengthy debate, ratified the Kyoto Protocol in 2002. Canada was an early supporter of global climate policy. For example, Canada convened the World Conference on the Changing Atmosphere in Toronto in 1988, which set the "Toronto Target" calling on states to reduce their CO_2 emissions by 20 percent below 1988 levels by 2005 (a goal widely missed by all North American countries). In 1998, four years before ratifying the Kyoto Protocol, the Canadian federal government launched a national process to address climate change. In October 2000, the ministers of Energy and Environment Canada jointly announced a National Implementation Strategy on Climate Change. This Strategy created a framework for Canada's federal, provincial, and territorial governments to collaborate and develop a series of action plans and initiatives to be taken individually and collectively.

As a part of the Strategy, the first National Climate Change Business Plan, released in 2000, set general objectives for research on alternative energy, adaptation, and education. In 2002, the federal government released a Climate Change Plan for Canada, which contained more specific information including a goal of reducing annual GHG emissions by 240 megatons. In addition, Canada's federal government concluded a series of voluntary initiatives. This action plan was updated in 2005, outlining a series of regulatory and incentive-based efforts to reduce GHG emissions (Pew Center on Global Climate Change 2005). Nevertheless, Canada has steadily increased its GHG emissions since 1990, and Canadian government officials acknowledged in 2006 (and repeatedly thereafter) that Canada will not meet its Kyoto commitments. Several additional federal-level proposals to reduce national GHG emissions were developed in 2007 and 2008, but it remains unclear what will eventually be enacted by Parliament and implemented by the federal government.

Compared to Canada and the United States, climate change has been subject to less political debate in Mexico. Mexico's federal government became engaged on climate change issues largely as a result of the entry into force of the UNFCCC in 1993 and the start of the Kyoto Protocol negotiations in 1995. Since then, successive Mexican governments have publicly recognized the threat of climate change but formulated little federal policy designed to reduce GHG emissions. However, Mexico is increasingly developing projects under the Kyoto Protocol's Clean Development Mechanism (CDM).[2] Mexico is one of only two members of the Organisation for Economic Co-operation and Development (OECD) that did not accept mandatory GHG emission reduction goals under the Kyoto Protocol (the other is South Korea). Mexico is likely to come under sustained pressure to accept future international obligations under the global climate regime during the post-Kyoto negotiations. As such, the role of Mexico in North American and global climate change policy and GHG reduction is likely to increase in the future.

States, Provinces, and Municipalities

U.S. states and Canadian provinces are forging ahead with climate change plans and policymaking beyond what is mandated by federal authorities. Many states and provinces have done so precisely because of what local political leaders, policymakers, and officials see as federal inaction on climate change (Rabe 2004; Rabe 2008; Selin and VanDeveer 2007). These local and regional policy developments are significant as many states and provinces emit GHG emissions at the level of industrialized countries and large developing countries. Together, state and provincial climate change policies and support for renewable energy add up to serious GHG reductions if fully implemented (Byrne, Hughes, Rickerson, et al. 2007; Lutsey and Sperling 2008). Furthermore, just as state and provincial GHG emission levels and trends vary substantially, so too do the enacted policies and political debates across these many jurisdictions (Jiusto 2008; Rabe 2008).

Several states and provinces have adopted GHG reduction targets. For example, the six New England states (Connecticut, Maine, Massachusetts, New Hampshire, Rhode Island, and Vermont) and five eastern Canadian provinces (New Brunswick, Newfoundland and Labrador, Nova Scotia, Prince Edward Island, and Québec) in 2001 committed to reduce GHG emissions to 1990 levels by 2010 and to 10 percent below 1990 levels by 2020. They also pledged to ultimately decrease emissions to levels that do not pose a threat to the climate, which according to an official estimate would require a 75 to 85 percent reduction from 2001 emission levels. In 2005, Governor Schwarzenegger committed to reduce California's GHG emissions to 2000 levels by 2010 and to reach 1990 levels by 2020, with the long-term goal of reducing emissions to 80 percent below 1990 levels by 2050. In 2006, the state passed AB32, putting into state law the target of reaching 1990 levels by 2020. Other states—including Connecticut, Hawaii, New Jersey, and Washington—followed suit, writing similar GHG reduction goals into state law.

A majority of U.S. states have initiated a series of policy initiatives designed to reduce their GHG emissions. Both the number of states developing climate change policies and the stringency of their actions have increased over time. Specific actions include the issuing of statewide climate change action plans, mandating that electric utilities generate a specific minimum amount of power from renewable energy sources (so-called renewable portfolio standards) to be increased over time, and establishing public funds to support energy efficiency and/or renewable energy development. In fact, by 2008 over half of U.S. states had enacted renewable portfolio standards and a host of other incentives for renewable energy, helping states like Texas and California to become wind and solar energy leaders (Krauss 2008; Rabe 2004; Richtel and Markoff 2008). Many states have also developed building and product standards and issued rules for public sector purchasing designed to reduce energy use and CO_2 emissions.[3] In addition, California has formulated controls on CO_2 emissions from vehicles and launched a regulatory process designed to produce standards that lower the carbon content of gasoline and other fuels. Several other states are poised to adopt these standards if they survive ongoing legal challenges.

While some Canadian provinces also have stated intentions to reduce GHG emissions, their efforts have typically been fewer in number and more modest in scope than those in many U.S. states. For example, the five eastern Canadian provinces that signed onto a regional action plan with the six New England states in 2001 have taken less action to implement this plan compared to their U.S. counterparts (Selin and VanDeveer 2005). Since 2003, however, Ontario has pursued a series of policies designed to expand renewable energy generation (Rowlands 2007). In 2007, British Columbia, Ontario, New Brunswick, Alberta, Saskatchewan, and Manitoba announced new climate change initiatives and GHG reduction targets. Canada's western provinces have also accelerated climate change cooperation with states in the western United States (Point Carbon 2007b). By 2008, British Columbia had established itself as the provincial climate change leader with enactment of aggressive GHG reduction goals and a broad-based carbon taxation scheme.

Much public sector debate involving extensive state and provincial participation focuses on the establishment of GHG emissions trading schemes in North America. In 2003, states in the U.S. northeast launched a policy process, which created a regional CO_2 trading scheme involving ten states under the Regional Greenhouse Gas Initiative (RGGI). Joint structures for GHG emission trading are also under development among western U.S. states and some Canadian provinces. Finally, ideas about a possible CO_2 trading system under NAFTA have been raised by a number of actors. These various developments and proposals are drawing technical and policy lessons from existing North American regional and national trading schemes for SO_2 and NO_x (Aulisi, Farrell, Pershing, et al. 2005) as well as experiences with the EU Emissions Trading Scheme, which was formally launched in 2005 covering over 11,500 different installations in all twenty-seven EU member states (Skjærseth and Wettestad 2008).

A growing number of North American municipalities are also initiating climate change action. Many are members of the International Council for Local Environmental Initiatives (ICLEI) and its Cities for Climate Protection (CCP) program. By 2007, the CCP program had over 260 members from the United States, Canada, and Mexico. In joining the CCP program, municipalities commit to, among other things, establishing local GHG emission reduction targets and working toward their implementation. In addition, by 2008, over 800 U.S. mayors from all fifty states and representing approximately 80 million Americans had signed a declaration stating their goal to meet or exceed the U.S. emissions reductions called for in the Kyoto Protocol (7 percent reduction below 1990 emissions levels by 2012). In Canada, the Federation of Canadian Municipalities, with over 1,400 members from all ten provinces and three territories, plays a similar role.

Private Sector Initiatives and Civil Society Engagement

In the 1990s, many North American corporations and trade associations led the opposition against mandatory GHG emission reductions nationally and internationally (Levy 2005; Skjærseth and Skodvin 2001). For example, the Coal Association of Canada ran full-page ads in major newspapers during the final stages of Kyoto Protocol negotiations that read: "Some Japanese terms Canadians ought to know: Seppuku: Ritual suicide with honor. Kyoto: Economic suicide by ignorance" (Macdonald and Smith 1999–2000, 107). Similarly, an industry-funded campaign in the United States portrayed Kyoto as unfair to the United States because developing countries were not required to make GHG reductions.

During the 2000s, however, a growing number of North American firms are taking significant measures to reduce GHG emissions. In doing so, many corporate leaders are discovering that they can save substantial amounts of money in the process. Corporate executives also prepare for a future carbon-constrained economy by, for example, increasing their investments in the development of more energy efficient products and technologies that reduce GHG emissions. More aggressive climate change policy, moreover, creates opportunities for companies in renewable energy generation. The market for consultancy and accounting firms offering their services to private and public organizations that want to participate in credit and/or offset schemes for CO_2 reductions is also growing sharply. In addition, climate change offers both financial challenges and business possibilities to the insurance and reinsurance sector. Thus, in contrast to the 1990s, many large corporations (outside of the oil and automobile sectors) have added their voices and lobbyists to those advocating more serious climate change mitigation and adaptation polices.

Another North American private sector initiative is the creation of a voluntary market for CO_2 emissions permits, the Chicago Climate Exchange (CCX), which opened in December 2003. Membership, which increased from twenty-three firms in late 2003 to seventy by mid-2008, includes U.S. and Canadian corporate giants such as DuPont, Motorola, and Manitoba Hydro. Members commit to reducing their North American emissions by one percent every year for four years, but face no penalties if they do not meet their targets. Some have already made considerable progress, with DuPont reducing its emissions by over 70 percent below 1990 levels by the early 2000s (Goodell 2006). Members join CCX because they recognize the climate change problem, but they also participate for strategic reasons. By joining, companies gain valuable experience in managing GHG emissions, they position themselves at the frontline for a future mandatory trading system they believe is likely, and they hope to reap public relations benefits. The number of traded permits doubled between 2006 and 2007 to 23 million, but prices have remained low (Point Carbon 2008, 4). Further, CCX launched a California division designed to develop financial instruments to serve the California market.

Yet, there remains considerable North American private and public sector opposition and ambivalence to more stringent GHG policy as firms can exercise considerable influence over environmental policymaking (Kraft and Kamieniecki 2007). Although a few continue to question the science behind human-induced climate change, the larger group of skeptics argues that the costs of regulating CO_2 emissions will be too high. Opposition strategies include a host of measures, as evident in the political and legal challenges by the U.S. automotive industry to California's state law to regulate CO_2 emissions from vehicles, which would be copied by at least sixteen other states. The lawsuit by the auto industry was rejected by a federal judge in December 2007, but enforcement was delayed by the U.S. Environmental

Protection Agency's procrastination in granting California the necessary waiver (Egelko 2007). The Competitive Enterprise Institute launched an aggressive media campaign against efforts to control CO_2 emissions with the slogan "They call it pollution; we call it life." Many U.S. states have also yet to commit to long-term measures to reduce GHG emissions. In Canada, Alberta has voiced strong opposition to Canada's Kyoto commitments and at the same time invested heavily in tar sands oil extraction. In the United States, environmental advocacy groups, despite growing membership, have had relatively little impact on federal climate change policy and the national debate even as opponents of climate change policy stepped up their activities in Washington, D.C.

Nevertheless, recent increases in civil society activity around climate change issues include multiple initiatives by nongovernmental organizations (NGOs), and a growing social movement around climate change issues can be identified (Moser 2007). Private foundations and universities also show growing support for North American climate change initiatives. Such support has helped to sponsor state, provincial, and municipal efforts to address climate change. In addition, the sharp increase in local-level climate change action is receiving greatly expanded media coverage and editorial support. In fact, by 2008, about 50 percent of the U.S. population lived in a state or municipality with stated GHG reduction goals. Furthermore, local U.S. governments have taken a "prescribed pattern of inventorying their emissions, establishing climate change action plans, setting emission reduction targets similar to those of the Kyoto Protocol, enacting state-level regulations and standards explicitly targeting GHGs, and forging multi-government alliances to reinforce and support their actions" (Lutsey and Sperling 2008, 673). Much the same can be said of growing Canadian provincial and municipal action.

The Way Forward

This volume analyzes the dynamism and innovation of contemporary climate change policies across North America, including those involving many U.S. states and Canadian provinces, large corporations, NAFTA bodies, universities, NGOs and private firms. The chapters examine issues critical to our understanding of climate change politics in North America, focusing on a multitude of existing and potential policy developments at continental, national, regional, and local governance levels in the public sector, in the private sector, and in civil society. Taken as a whole, the subsequent chapters provide an analysis of multilevel climate change and energy debates and policymaking efforts across Canada, the United States, and Mexico. Several chapters also examine major transnational and international issues and policy efforts involving two or all three North American countries.

Chapters are divided into four thematic sections. The first, *Between Kyoto and Washington*, focuses on climate change politics and policymaking in Mexico and Canada. Simone Pulver's chapter—*Climate Change Politics in Mexico*—examines how global and U.S. federal climate change politics have influenced Mexican climate change action. Pulver argues that the initial Mexican agenda for climate change action was set by climate scientists in the national university and by bureaucrats in the environment ministry. With the rise in international attention, a wider array of government ministries began to engage in the climate policy process and bureaucratic politics impeded forward action in the 1990s. In contrast to the United States and Canada, industry actors including Mexico's state-owned oil company Petróleos Mexicanos have advocated precautionary climate change action. Pulver notes that Mexican environmental NGOs have been largely absent from the climate debate.

Peter J. Stoett's chapter—*Looking for Leadership: Canada and Climate Change Policy*—examines major Canadian developments and challenges. Disharmony between the federal and provincial governments has been a constant factor in Canadian climate change policy. The election of a minority Conservative government in 2006 further distanced Ottawa from the Kyoto process as Canadian climate change debate and action are significantly influenced by U.S. policymaking. Stoett suggests that, despite the need for more effective federal-provincial cooperation, strong Canadian leadership on climate change will not emanate from the federal level but will more likely reflect local and nongovernmental initiatives. While Canada is very unlikely to meet its Kyoto Protocol targets by 2012, the chapter discusses the fact that there are other approaches to climate change mitigation and adaptation that Canadians could pursue.

The second thematic section, *States and Cities Out Front*, focuses on the plethora of local-level policy developments in North America. Barry G. Rabe's chapter—*Second-Generation Climate Policies in the States: Proliferation, Diffusion, and Regionalization*—details rapidly expanding policymaking efforts on climate change in U.S. states. The chapter examines trends in states' climate policy formation and implementation, which includes continuing proliferation of a diverse array of GHG reduction policy tools and multistate collaboration that brings a regional dimension to these state efforts. The chapter also examines alternative venues for state climate policy development, including direct democracy and litigation through elected attorneys general. It concludes with a comparison of the evolving American system with other multilevel governance systems and a discussion of potential stumbling blocks facing its bottom-up approach to policy development.

Alexander E. Farrell and W. Michael Hanemann, in *Field Notes on the Political Economy of California Climate Policy*, focus attention on recent developments in climate policy in California. The state is in the midst of implementing a broad set of sectoral policies to reduce GHG emissions. Their chapter examines major events since 2000 when the California Global Warming Solutions Act (AB32) became law. The authors demonstrate how California's history of leadership in air quality, energy efficiency, and other aspects of energy policy were, and remain, tightly linked to climate change policy in the state, and how the development of climate policy was linked to broader political trends. The chapter discusses critical policy choices, such as the role of market-based mechanisms, and points out some of the difficult decisions that lie ahead in the implementation phase of California's climate policy development.

The chapter by Henrik Selin and Stacy D. VanDeveer—*Climate Leadership in Northeast North America*—examines regional and local-level policy developments in the continent's northeast. In particular, it focuses on two major regional state-led policy developments: the 2001 Climate Change Action Plan of the New England Governors Conference and the Eastern Canadian Premiers and its implementation, and the creation of a regional cap-and-trade scheme for CO_2 emissions from power plants under RGGI between 2003 and 2009. In addition, the chapter examines growing municipal and civil society engagement on climate change and GHG mitigation in the region, arguing that regional networks of policy advocates channel influence through overlapping pathways of policy change. The chapter concludes with a discussion of the potential and limitations of developing climate change policymaking in the Northeast.

Christopher Gore and Pamela Robinson, in *Local Government Response to Climate Change: Our Last, Best Hope?*, examine municipal climate change action. A growing number of North American cities have formally committed to GHG emission reductions and are members of domestic and/or international associations of municipalities that work together on climate change issues. The authors discuss the central role that such transnational and national networks of municipalities play in promoting climate responses. Experiences of two cities that are leaders in municipal action are explored in greater detail: Toronto and Portland, Oregon. The authors argue that local governments in North America should be recognized as leaders in climate change response and that collectively they have real and potential power to drive further municipal action and to shape future provincial, state, and national climate change action.

The third thematic section, *Continental Politics,* examines issues of continental integration and collaboration on climate change science and politics. Michele M. Betsill's chapter—*NAFTA as a Forum for CO_2 Permit Trading?*—takes discussions within the North American Commission for Environmental Cooperation (CEC), NAFTA's environmental organ, about establishing a continental CO_2 trading system to mitigate the environmental impacts of electricity generation as a starting point. Following a brief discussion of emissions trading as a mechanism for addressing climate change and an overview of the CEC discussion on climate change, the chapter addresses three sets of issues related to establishing a CEC-based CO_2 permit trading system with particular focus on its implications for climate protection: the institutional context, design elements, and interplay with other trading systems. Based on this analysis, Betsill questions the wisdom of establishing a CO_2 permit trading system under NAFTA.

Ian H. Rowlands' chapter—*Renewable Electricity Politics across Borders*—analyzes the ways in which cross-border relations between Canada and the United States have affected the development of renewable electricity in each country. The chapter argues that the most significant issue that has arisen to date revolves around the Canadian export of electricity generated by large-scale hydropower facilities, which has been resisted by some in the United States who argue that Canadian hydropower should not be given privileged access. The chapter anticipates further debates regarding the definition of renewable or green energy; issues related to cross-border investment, green procurement, subsidies, and tradable certificates are also identified.

In chapter 10, *Arctic Climate Change: North American Actors in Circumpolar Knowledge Production and Policymaking*, Annika E. Nilsson examines issues of scientific assessment and policy making in an Arctic context with a particular Canadian and U.S. focus. Through a case study of the Arctic Climate Impact Assessment (ACIA), the chapter analyzes how the Arctic Council helped bring indigenous peoples' perspectives to the fore, making the assessment scientifically credible, politically legitimate, and salient to new actors in Arctic climate politics and policymaking. The chapter also discusses political differences among Arctic states concerning assessment-related issues and the limitation of regional efforts in changing established power dynamics in the global climate arena.

The final thematic section, *Climate Action among Firms, Campuses, and Individuals*, examines major developments in North American private and civil society sectors. The chapter by Charles A. Jones and David L. Levy—*Business Strategies and Climate Change*—analyzes reactions and strategies of private sector actors to the climate change issue and expanding climate policy. Voluntary measures taken by business to reduce GHG emissions represent substantial investments, but North American business responses to climate change are often ambiguous and tentative. The authors argue that business is prepared to take action consistent with a fragmented and voluntary regime, while simultaneously opposing any policy that would mandate a more robust response. In addition, the connection between tentative corporate response and weak government policies reinforces the inertia in the current energy system.

Virginia Haufler's chapter *Insurance and Reinsurance in a Changing Climate* focuses on the insurance and reinsurance sector of the economies of North America. The chapter explores the insurance industry's responses to changing weather risks today and how it may react in the future. The North American insurance industry has been slow to recognize the potential threat of climate change. This is changing under the impact of environmental activism; pressure from European reinsurers, who have been more knowledgeable about the issues; and the apparent crisis posed by successive major weather disasters. Instead of the traditional industry responses to disaster—raising prices and/or withdrawing from markets—insurers are beginning to recognize their leverage over the behavior of customers. By redesigning contracts and pricing structures, the insurance industry may become a significant source of new incentives for improved environmental performance.

Next, Dovev Levine's chapter, *Campus Climate Action*, examines characters and drivers of the rapid expansion of climate change action on a host of university campuses in North America. Levine notes that university climate change action remains little studied, and he argues that campus action can have important consequences for GHG mitigation efforts and climate change policymaking. Specifically, the chapter discusses how and why university climate change action is developing in areas of curriculum designs, university operations, research, and outreach activities with local communities. Levine argues that campus action holds much potential as a source of political and economic influence, shaping more innovative and stringent climate change and renewable energy policies in North America.

Susanne C. Moser, in chapter 14, *Communicating Climate Change and Motivating Civic Action: Renewing, Activating, and Building Democracies*, discusses the role of communication in motivating citizen action and support for more aggressive climate change policy. In the absence of federal leadership, bottom-up pressure is building to force national policy changes. This chapter focuses on how civic mobilization and engagement on climate change can be fostered through effective communication. It lays out why effective communication is essential to bringing about different types of civic engagement, offers specific communication strategies that can increase civic engagement, and illustrates these with best practices and examples from the current North American context.

The final chapter, *North American Climate Governance: Policymaking and Institutions in the Multilevel Greenhouse*, returns to the four broad questions outlined at the beginning of this introduction. The chapter's authors, Henrik Selin and Stacy D. VanDeveer, draw on insights and arguments from the volume's other chapters to address each of the four questions. The chapter also identifies four possible scenarios for the future of continental climate change politics based on combinations of high and low federal and subnational involvement, paying particular attention to opportunities and challenges of complex multilevel governance. The volume concludes with a few remarks on continuing governance issues and challenges in the North American greenhouse.

Notes

1. Note that GHG emissions data and estimates can and do vary somewhat across chapters in the volume due to differences in estimation methods, dates, and other technical factors.

2. The CDM was established under the Kyoto Protocol as one of five options for Annex I parties (i.e., industrialized countries and countries with economies in transition) to meet their mandatory emission reduction obligations. Under the CDM, Annex I parties can earn credits for lowering GHG emissions in non–Annex I countries (i.e., developing countries without mandatory Kyoto targets). As such, Annex I countries can pursue projects that reduce GHG emissions in developing countries if they believe that these are more cost-effective than reducing domestic emissions. This will reduce emissions in developing countries that participate in CDM projects, and could also serve to diffuse technology to developing countries. Critics have argued, however, that the CDM allows high-emitting countries to buy themselves free from the responsibility to reduce domestic emissions.

3. See the Pew Center's database of state and local climate change initiatives at www.pewclimate.org.

References

Audley, John Joseph. 1997. *Green Politics and Global Trade: NAFTA and the Future of Environmental Politics*. Washington, DC: Georgetown University Press.

Aulisi, Andrew, Alexander F. Farrell, Jonathan Pershing, and Stacy D. VanDeveer. 2005. *Greenhouse Gas Emissions Trading in U.S. States: Observations and Lessons from the OTC NO_x Budget Program*. Washington, DC: World Resources Institute.

Bache, Ian, and Mathew Flinders, eds. 2004. *Multi-level Governance*. Oxford: Oxford University Press.

Baumert, Kevin, Timothy Herzog, and Jonathan Pershing. 2005. *Navigating the Numbers: Greenhouse Gas Data and International Climate Policy*. Washington, DC: World Resources Institute.

Betsill, Michele M., and Harriet Bulkeley. 2006. Cities and the Multilevel Governance of Global Climate Change. *Global Governance* 12(2): 141–159.

Bulkeley, Harriet, and Susanne C. Moser, eds. 2007. Responding to Climate Change: Governance and Social Action beyond Kyoto. *Global Environmental Politics* 7(2): 1–144.

Byrne, John, Kristen Hughes, Wilson Rickerson, and Lado Kurdgelashvili. 2007. American Policy Conflict in the Greenhouse: Divergent Trends in Federal, Regional, State, and Local Green Energy and Climate Change Policy. *Energy Policy* 35(9): 4555–4573.

Cowie, Jonathan. 2007. *Climate Change: Biological and Human Aspects*. Cambridge: Cambridge University Press.

Deere, Carolyn L., and Daniel C. Esty, eds. 2002. *Greening the Americas: NAFTA's Lessons for Hemispheric Trade*. Cambridge, MA: MIT Press.

Dessler, Andrew E., and Edward A. Parson. 2006. *The Science and Politics of Global Climate Change: A Guide to the Debate*. Cambridge: Cambridge University Press.

DiMento, Joseph F. C., and Pamela Doughman, eds. 2007. *Climate Change: What It Means for Us, Our Children, and Our Grandchildren*. Cambridge, MA: MIT Press.

Dorsey, Kurk. 1998. *The Dawn of Conservation Diplomacy: U.S.-Canadian Wildlife Protection Treaties in the Progressive Era*. Seattle: University of Washington Press.

Egelko, Bob. 2007. California's Emission-Control Law Upheld on 1st Test in U.S. Court. *San Francisco Chronicle*, December 13.

Environment Canada. 2004. *Greenhouse Gas Inventory*. Ottawa: Environment Canada.

Finger, Matthias, Ludvine Tamiotti, and Jeremy Allouche. 2006. *The Multi-Governance of Water: Four Case Studies*. Albany: State University of New York Press.

Fisher, Dana R. 2004. *National Governance and the Global Climate Change Regime*. New York: Rowman and Littlefield.

Gallagher, Kevin P. 2004. *Free Trade and the Environment: Mexico, NAFTA, and Beyond*. Stanford: Stanford University Press.

Goodell, Jeff. 2006. Capital Pollution Solution? *New York Times Magazine*, July 30.

Harrison, Kathryn, and Lisa McIntosh Sundstrom, eds. 2007. The Comparative Politics of Climate Change. *Global Environmental Politics* 7(4): 1–139.

Harrison, Neil E. 2004. Political Responses to Changing Uncertainty in Climate Science. In *Science and Politics in the International Environment*, edited by Neil E. Harrison and Gary C. Bryner. New York: Rowman and Littlefield.

Hooghe, Liesbet, and Gary Marks. 2003. Unraveling the Central State, but How? Types of Multi-Level Governance. *American Political Science Review* 97(2): 233–243.

Intergovernmental Panel on Climate Change. 2007a. *Climate Change 2007—Impacts, Adaptation and Vulnerability*. Cambridge: Cambridge University Press.

Intergovernmental Panel on Climate Change. 2007b. *Climate Change 2007—Mitigation of Climate Change*. Cambridge: Cambridge University Press.

Intergovernmental Panel on Climate Change. 2007c. *Climate Change 2007—The Physical Science Basis*. Cambridge: Cambridge University Press.

Jiusto, Scott. 2008. An Indicator Framework for Assessing U.S. State Carbon Emissions Reduction Efforts (with Baseline Trends from 1990 to 2001). *Energy Policy* 36(6): 2234–2252.

Kolbert, Elizabeth. 2005. Annals of Science: The Climate of Man—III. *New Yorker* 81(12): 52–63.

Kraft, Michael E., and Sheldon Kamieniecki, eds. 2007. *Business and Environmental Policy: Corporate Interests in the American Political System*. Cambridge, MA: MIT Press.

Krauss, Clifford. 2008. Move Over, Oil, There's Money in Texas Wind. *New York Times*, February 23.

Le Prestre, Philippe, and Peter Stoett, eds. 2006. *Bilateral Ecopolitics: Continuity and Change in Canadian-American Environmental Relations*. Aldershot, UK: Ashgate.

Levy, David L. 2005. Business and the Evolution of the Climate Regime: The Dynamics of Corporate Strategies. In *The Business of Global Environmental Governance*, edited by David L. Levy and Peter J. Newell. Cambridge, MA: MIT Press.

Low, Pak Sum, ed. 2005. *Climate Change and Africa*. Cambridge: Cambridge University Press.

Luterbacher, Urs, and Detlef F. Sprinz. 2001. *International Relations and Global Climate Change*. Cambridge, MA: MIT Press.

Lutsey, Nicholas, and Daniel Sperling, eds. 2008. America's Bottom-Up Climate Change Mitigation Policy. *Energy Policy* 36(2): 673–685.

Macdonald, Douglas, and Heather A. Smith. 1999–2000. Promises Made, Promises Broken: Questioning Canada's Commitment to Climate Change. *International Journal* 55 (Winter 1999–2000): 107–124.

Markell, David L., and John H. Knox, eds. 2003. *Greening NAFTA: The North American Commission for Environmental Cooperation*. Stanford: Stanford University Press.

Miller, Clark, and Paul A. Edwards, eds. 2001. *Changing the Atmosphere: Expert Knowledge and Environmental Governance*. Cambridge, MA: MIT Press.

Moser, Susanne C. 2007. In the Long Shadows of Inaction: The Quiet Building of a Climate Protection Movement in the United States. *Global Environmental Politics* 7(2): 124–144.

Mumme, Stephen P. 2003. Environmental Politics and Policy in U.S.-Mexican Border Studies: Developments, Achievements, and Trends. *Social Science Journal* 40(4): 593–606.

Pew Center on Global Climate Change. 2005. "Canada's Climate Change Plan"(summarizing the Canadian government's "Moving Forward on Climate Change: A Plan for Honouring our Kyoto Commitment"). Available at http://www.pewclimate.org/policy_center/international_policy/canada_climate_plan.cfm.

Point Carbon. 2007a. *Carbon Market North America*, July 18.

Point Carbon. 2007b. *Carbon Market North America*, June 20.

Point Carbon. 2008. *Carbon Market North America* 3(1): 1–8, January 16.

Rabe, Barry G. 2004. *Statehouse and Greenhouse: The Emerging Politics of American Climate Change Policy*. Washington, DC: Brookings Institution Press.

Rabe, Barry G. 2008. States on Steroids: The Intergovernmental Odyssey of American Climate Policy. *Review of Policy Research* 25(2): 105–128.

Richtel, Matt, and John Markoff. 2008. A Green Energy Industry Takes Root in California. *New York Times*, February 1.

Rowlands, Ian H. 2007. The Development of Renewable Electricity Policy in the Province of Ontario: The Influence of Ideas and Timing. *Review of Policy Research* 24(3): 185–207.

Schreurs, Miranda. 2002. *Environmental Politics in Japan, Germany and the United States*. Cambridge: Cambridge University Press.

Selin, Henrik, and Stacy D. VanDeveer. 2005. Canadian-U.S. Environmental Cooperation: Climate Change Networks and Regional Action. *American Review of Canadian Studies* 35(2): 353–378.

Selin, Henrik, and Stacy D. VanDeveer. 2007. Political Science and Prediction: What's Next for U.S. Climate Change Policy? *Review of Policy Research* 24(1): 1–27.

Skjærseth, Jon Birger, and Tora Skodvin. 2001. Climate Change and the Oil Industry: Common Problems, Different Strategies. *Global Environmental Politics* 1(4): 43–64.

Skjærseth, Jon Birger, and Jørgen Wettestad. 2008. *EU Emissions Trading: Initiating, Decision-Making and Implementation*. Aldershot, UK: Ashgate.

United Nations Framework Convention on Climate Change (UNFCCC). 2005. *Sixth Compilation and Synthesis of Initial National Communications from Parties Not Included in Annex I to the Convention*. Bonn, Germany: Secretariat of the UNFCC.

United Nations Framework Convention on Climate Change (UNFCCC). 1992. Article 2 of the United Nations Framework Convention on Climate Change. Available at http://unfccc.int/essential_background/convention/background/items/1349.php.

U.S. Climate Change Science Program. 2007. *The First State of the Carbon Cycle Report (SOCCR): The North American Carbon Budget and Implications for the Global Carbon Cycle*. Washington, DC: U.S. Climate Change Science Program.

U.S. Energy Information Administration. 2007. *Emissions of Greenhouse Gases in the United States 2006*. Washington, DC: U.S. Energy Information Administration.

U.S. Environmental Protection Agency. 2006. *U.S. Emissions Inventory 2006*. Washington, DC: U.S. EPA.

Victor, David G. 2004. *Climate Change: Debating America's Policy Options*. New York: Council on Foreign Relations.

Weart, Spencer R. 2003. *The Discovery of Global Warming*. Cambridge, MA: Harvard University Press.

Young, Oran R. 2002. *The Institutional Dimensions of Environmental Change: Fit, Interplay, and Scale*. Cambridge, MA: MIT Press.

I

Between Kyoto and Washington

2

Climate Change Politics in Mexico

Simone Pulver

To date, Mexico has been a small player in the global climate change arena. When ranking countries based on annual greenhouse gas (GHG) emissions, Mexico falls twelfth among the top twenty emitters. In 2002, Mexico's annual emissions of approximately 640 million metric tons (Mt) of carbon dioxide (CO_2) equivalent represented less than 2 percent of global GHG emissions (Government of Mexico 2006). Likewise, in a North American context, both Mexico and Canada are dwarfed by the United States, accounting for only about 6 percent and 8 percent of North American GHG emissions, respectively (WRI 2007). From a political perspective, Mexico has also been a minor player, especially when compared to its northern neighbors. Federal and state climate policy decisions made in the United States, the largest historical and per capita emitter of GHGs, have significant biophysical and political ramifications. Likewise Canada, though a relatively small GHG emitter, made international headlines in the climate and energy domains through its 2002 decision to ratify the Kyoto Protocol and its more recent large-scale investment in nonconventional fossil fuels (Harrison 2007).

However, Mexico's political prominence in the climate arena will increase over the coming decades. First, regardless of the future of the Kyoto regime, international climate policy is likely to move toward binding GHG emissions reduction commitments for developing countries. Mexico and South Korea, as the only two members of the Organization for Economic Cooperation and Development (OECD) that did not take on targets under the Kyoto Protocol, will be at the forefront of the negotiations. Second, given the extensive economic integration and institutional bases for cooperation, the North American region could become a promising site for piloting a climate regime that integrates the economies of developed and developing countries.

This chapter examines the evolution of climate change politics in Mexico, focusing on the underlying structural conditions and the short-term policy events that drive Mexican climate politics past, present, and future. Three key features emerge. First, Mexico's GHG emissions trajectory is driven by energy-intensive economic growth. Approximately 60 percent of Mexico's GHG emissions are attributable to the energy sector (Government of Mexico 2006). This sector is already the target of much policy activity, with the primary goal of improving local air quality (Di Sbroiavacca and Girardin 2000). One primary task of advocates for action on climate change in Mexico is to integrate climate change concerns into ongoing energy-sector policymaking.

Second, key actors in academia, the federal government and the private sector have been the most active proponents of action on climate change in Mexico. They have succeeded in building awareness about climate change across a broad range of constituencies (Martinez and Fernandez Bremauntz 2004). However, this generally supportive stance has not translated into a consistent push for effective climate policy because of conflicts between federal agencies and fluctuations over time in federal government and private sector interest in and support for action on climate change.

Third, mostly absent from the climate change debates in Mexico are environmental groups and state and local governments. Mexico has not yet seen the emergence of vibrant NGO, state- and city-level climate politics, which are the focus of much action in the United States and Canada. None of Mexico's thirty-one states has taken independent action on climate change. Likewise, only three Mexican cities are members of the International Council for Local Environmental Initiatives' climate protection program. While there have been some initiatives to reduce Mexico City's CO_2 emissions, which account for 20 percent of the country's total emissions (Climate Group 2004; Gore and Robinson, this volume), yet city-level action on climate change in the capitol and largest city in the country has been blocked by problems of institutional fit and capacity (Romero Lankao 2007).

The chapter begins with an introduction to the climate and energy context in Mexico. It then explores five key periods in the evolution of Mexican climate politics. Finally, the chapter turns to the respective roles of civil society and private-sector groups in influencing federal government action on climate change.

The Climate and Energy Context

In facing climate change, two issues are of primary concern to Mexican policymakers. First, Mexico is vulnerable to climate change impacts. In particular, drought and desertification pose threats to food production and livelihoods in the northern and central regions of the country. Coastal areas are threatened by rising sea levels and increased tropical storms. Both threats are of particular concern to the oil-producing region in the Gulf of Mexico (Government of Mexico 2006). Second, in considering GHG mitigation options, Mexican policymakers are focused on initiatives that will not hamper the country's continued economic growth. Mexico has

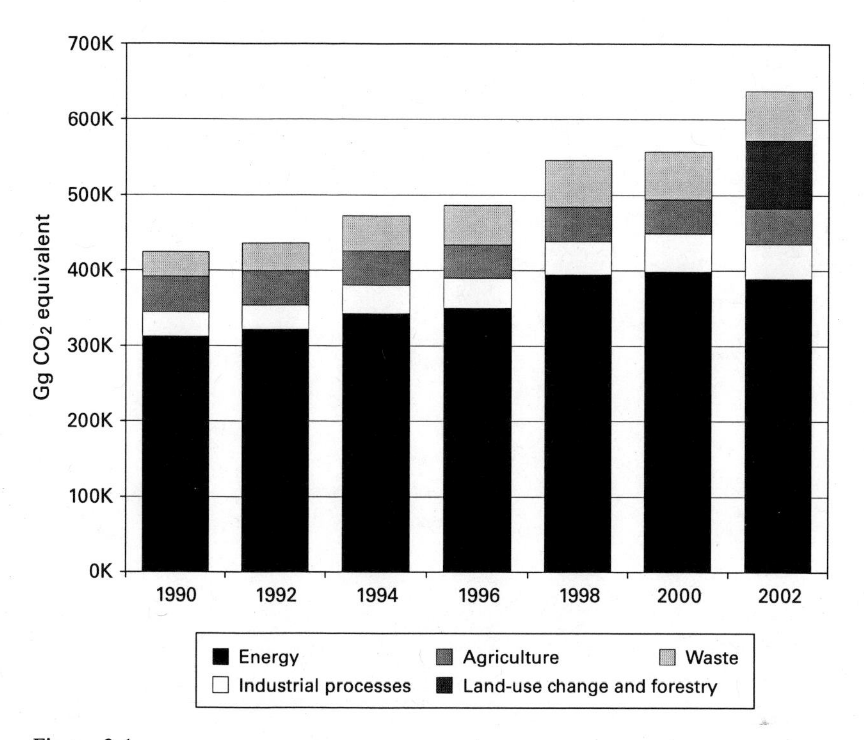

Figure 2.1
Annual greenhouse gas emissions for Mexico (Gg CO_2e). Source: CCS 2008.

a dual economy, characterized by both modern industry and traditional agriculture. Over the past decades, the economy has been restructured to emphasize product specialization and exports and has maintained an average GDP growth rate of about 3 percent. Per capita GDP has increased from $4,456 in 1990 to $9,158 in 2003 (Government of Mexico 2006).

Between 1990 and 2002, Mexico's GHG emissions increased from 420 to 640 Mt of CO_2 equivalent, primarily due to population growth and economic development (figure 2.1). The major emissions sectors are energy production and transformation (43 percent), transportation (18 percent), and land-use change and agriculture (21 percent) (Government of Mexico 2006). The energy and forestry sectors are the most promising sites for GHG mitigation opportunities (Government of Mexico 2006). In the forestry sector, mitigation activities have focused on forest protection and sustainable forest management (Di Sbroiavacca and Girardin 2000).

Changes in the energy sector are also creating opportunities for large-scale reduction in Mexico's GHG emissions. In particular, the fuel oil percentage of Mexico's energy consumption is expected to decline, with a shift toward cleaner fuels, primarily natural gas. There are several drivers of this shift. First, some analysts contend that Mexican oil production peaked in the mid-2000s and is now declining (EIA 2007). Second, high world oil prices favor export over the use of oil in domestic power generation. The United States is the primary importer of Mexican oil (Belausteguigoitia, Carlos, and Lopez-Bassols 1999; EIA 2007). Finally, government policy initiatives have incentivized the shift to cleaner fuels and to increased efficiency in Mexico's energy system (Di Sbroiavacca and Girardin 2000; Government of Mexico 2006).

The role of government policy in shaping priorities and investment in Mexico's energy infrastructure provides an opening for climate policy advocates. However, any restructuring of Mexico's energy system is hampered by the importance of oil to the economy. Fuel oil, combusted in power plants and car engines, accounts for approximately 60 percent of Mexico's energy consumption (EIA 2007). Mexico is the tenth-largest oil producer in the world (PIW 2006), and oil currently accounts for 30 percent of government revenue and 7 percent of Mexico's export earnings, down from a record high of 67 percent in 1980 (Di Sbroiavacca and Girardin 2000; Halpern 2001).

The Evolution of Mexican Climate Politics

Climate change policy in Mexico has moved forward in fits and starts, and climate concerns have yet to have any significant impact on Mexican energy policy. Policy momentum created by advocates for action on climate change has been blocked by disagreement among government agencies, particularly the energy and environment ministries; by the slow progress of international efforts to regulate GHG emissions; and by changes in political leadership (see table 2.1). Nevertheless, understanding the emergence and dynamics of the constituencies pushing for action on climate change in Mexico provides a starting point for the long-term project of integrating climate considerations into Mexico's energy and development trajectories.

The Early Years, 1990–1992

Interest in climate change in Mexico dates back to the early 1990s and was motivated by events in the international arena. In December 1990, the UN General Assembly passed Resolution 45/212, initiating negotiations on an international convention on climate change (Paterson 1996). Mexico was one among the hundred countries that sent a delegation to the first round of climate negotiations in February 1991, held in Chantilly, Virginia. The Mexican delegation to the Chantilly talks, and to the subsequent ten rounds of climate negotiations held between 1991 and

Table 2.1
Key periods in the evolution of Mexican climate politics

Period	Events
Period 1: 1990–1992 The early years	**February 1991:** Mexico sends delegation to first round of international climate change negotiations **1992:** Mexico opposes binding targets in the UNFCCC negotiations
Period 2: 1992–1996 Emergence of an epistemic community	**April 1994:** First U.S. Country Studies Mexico Workshop. **1995:** Carlos Gay Garcia at UNAM establishes an ad hoc group to coordinate interministerial dialogue on climate change **May 1995:** Second U.S. Country Studies Mexico Workshop **September 1995:** INE publishes Preliminary National Inventory of Greenhouse Gases **January 1996:** Third U.S. Country Studies Mexico Workshop
Period 3: 1997–2000 Interministerial wrangling and electoral politics	**April 1997:** Ad hoc group is reorganized into a formal Interministerial Committee for Climate Change **September 1997:** SENER begins to engage in climate policy debates, challenging SEMARNAT **September 1997:** Mexico publishes First National Communication under UNFCCC **September 1997:** Mexico hosts 12th plenary session of IPCC **December 1997:** Kyoto Protocol is negotiated **December 1999:** Pemex organizes climate change workshop with SEMARNAT and UNDP **April, 2000:** Pemex publishes its first Annual Report on Health, Safety, and Environment **April 29, 2000:** Mexican Senate votes to ratify Kyoto Protocol, after battle between SEMARNAT and SENER **August 2000:** Vicente Fox elected to presidency **December 2000:** President Fox assumes office
Period 4: 2001–2005 International drivers of domestic climate politics	**March 2001:** U.S. President George W. Bush withdraws United States from Kyoto Protocol; expectations for CDM collapse **June 2001:** Pemex launches carbon emissions trading system internal to company **July 2001:** Mexico publishes Second National Communication under UNFCCC **Spring 2002:** Victor Lichtinger, Mexico's Secretary of the Environment, meets with European environmental ministers **May 2002:** EU ratifies Kyoto Protocol **2003:** National Climate Change Office established in SEMARNAT **2004:** SEMARNAT, WRI, and WBCSD launch Mexico Greenhouse Gas Program **February 2005:** Kyoto Protocol enters into force; Mexico CDM office begins issuing letters of approval for projects **April 2005:** Interministerial Commission on Climate Change (CICC) is established

Table 2.1
(continued)

Period 5: 2006–2008 National climate strategies, but little national action	**August 2006:** Felipe Calderon elected to presidency **November 2006:** CICC issues a national climate strategy **November 2006:** Mexico publishes Third National Communication under UNFCCC **December 2006:** President Calderon assumes office **May 2007:** President Calderon announces a new national strategy on climate change

1993, consisted primarily of representatives from the Secretaría de Relaciones Exteriores (SRE, the ministry for foreign affairs) and from the Secretaría de Medio Ambiente y Recursos Naturales (SEMARNAT, the ministry of environmental affairs).[1] At the time, interest in climate change was limited to a few key individuals in each ministry.

The Mexican delegation played a leadership role in the early rounds of the international climate change negotiations. Edmundo de Alba, the lead negotiator for Mexico, was elected as cochair, along with a delegate from Japan, of the negotiating group addressing the issue of binding GHG reduction commitments in the draft climate change convention. Mexico used its leadership position to promote the status quo.[2] In negotiating the 1992 UN Framework Convention on Climate Change (UNFCCC), Mexico sided with countries like the United States, Japan, Canada, Australia, and New Zealand who were opposed to binding national GHG reduction targets (Mintzer and Leonard 1994). These countries prevailed. The text of the UNFCCC did not mandate binding reductions in GHG emissions.

Emergence of an Epistemic Community, 1992–1996

Events post-1992 mark the first transition in Mexico's climate policy agenda. Between 1992 and 1995, policymaking on climate change in Mexico shifted from support of the status quo to vocal advocacy for action on climate change in both international and domestic arenas. At the international level, Mexico switched allegiances and joined the negotiating bloc that favored binding GHG emissions reductions. At the national level, there was an effort to broaden domestic interest in climate change beyond the small group of diplomats in SRE and SEMARNAT.

The climate policy shift between 1992 and 1995 was catalyzed by the emergence and consolidation in Mexico of a scientific community centered on climate research. After the adoption of the UNFCCC in June 1992 at the Rio Earth Summit, a community of scientists and environmental bureaucrats at Mexico's national university (Universidad Nacional Autónoma de México, UNAM) and at the national ecology institute (Instituto Nacional de Ecología, INE), launched a concerted climate-change-research effort. Through a collaborative effort, the UNAM Center for Atmospheric Sciences (Centro de Ciencias Atmósphera, CCA) and INE established a national scientific program on global climate change (Programa Nacional Científico sobre Cambio Climático Global) as a means to stimulate and coordinate climate change research in Mexico (Gay Garcia 1994). Their efforts received a further boost via financial support from the U.S. Country Studies Program (CSP), which provided financial and technical assistance to developing countries to support efforts to address climate change. Mexico's application for support was funded during the first round of applications in October 1993.

The efforts of the Mexican climate science research community, with funding from the CSP, produced both technical and political results. On the technical side, information was generated on three topics: (1) a GHG emissions inventory for Mexico; (2) a portfolio of GHG emissions scenarios; and (3) an update of previous studies on Mexico's vulnerability to climate change impacts (Ramos-Mane and Benioff 1995). Findings on inventories, scenarios, and vulnerability from the country study research efforts were presented at three workshops in April 1994, May 1995, and January 1996 (Benioff, Ness and Hirst 1997). The country study work also became the basis for Mexico's first official national GHG inventory, published in December 1995 (Di Sbroiavacca and Girardin 2000), its first National Communication under the UNFCCC, completed in 1996 and submitted in November 1997 (Government of Mexico 1997), and a summary report on vulnerability to climate change (Gay Garcia 2000).

Politically, the country study–process acted to centralize the group of scientists and bureaucrats working on climate change in Mexico. Individuals in UNAM and INE, the two coordinating agencies, became the central nodes in the network of Mexican scientists working on climate change. Key individuals involved in the climate research program were also able to turn their status in the scientific community into leadership roles in the policy arena. Of the thirty-five contributors to the first country study workshop, over a third remain lead experts in Mexico on climate change science and policy. In particular, UNAM Professor Carlos Gay Garcia, a lead convener of the research effort supported by the CSP, became politically prominent in both the international and domestic climate policy arenas. After the 1994 Mexican presidential election that replaced Carlos Salinas with Ernesto Zedillo, Gay Garcia became the lead technical negotiator for the Mexican delegation to the international climate negotiations and also the head of delegation at meetings of the UNFCCC scientific subsidiary bodies. Domestically, Gay Garcia organized and convened an ad hoc group to coordinate interministerial dialogue on climate change. The group included the ministries of foreign affairs (SRE), environment (SEMARNAT), energy

(Secretaría de Energía, SENER), and commerce and industrial promotion (Secretaría de Comercio y Fomento Industrial, SECOFI).[3] Under Gay Garcia's leadership, the ad hoc group prepared Mexico's policy positions in preparation for the international climate change negotiations.

Interministerial Wrangling and Electoral Politics 1997–2000

While the period from 1992 to 1996 can be described as one of increasing policy momentum, political developments from 1997 to 2000 retarded action on climate change at the domestic level in Mexico. The catalyst for the reversal was the jump in international political prominence of the climate issue in 1997, associated with the Kyoto Protocol negotiations. Increased international attention to climate change led to a widening in the scope of domestic actors and agencies that perceived themselves as having a stake in the climate policy process. In 1997, the ministries of agriculture and rural development, commerce and industrial development, communications and transport, energy, and social development all increased their participation in domestic climate politics (SEMARNAP 1998).

Among these agencies, the energy ministry in particular began to take a much more active interest in the climate issue. Observers date intensified SENER engagement on climate change to early 1997. The December 1997 Kyoto round of the international climate negotiations was the first time a representative from SENER was included in the Mexican delegation. By 1998, SENER had generated several internal documents addressing energy and climate change issues (SENER 1998). Unlike SEMARNAT, SENER was less concerned with Mexico's ecological vulnerability to climate change and more focused on the potential adverse effects of international GHG regulations on Mexico's oil economy. At the time, most oil-exporting countries were vocal opponents to action on climate change (Pershing 2000). Bureaucrats in SENER echoed this policy stance and opposed international GHG regulations.

The consequences both of the general politicization within Mexico of the climate issue and of SENER's increased involvement in Mexican climate policy debates can be seen in the fate of UNAM Professor Gay Garcia's ad hoc group for interministerial dialogue. In 1997, the informal group was converted into a formal Intersecretarial Committee on Climate Change (Comité Intersecretarial de Cambio Climático), with an expanded list of seven participating ministries, including agriculture, transport, social development, environment, energy, economy, and foreign affairs (Belausteguigoitia and Lopez-Bassols 1999; SEMARNAP 1998). At the same time, Gay Garcia was replaced by Julia Carabias Lillo from SEMARNAT as the lead coordinator of the Mexican climate policy process. Gay Garcia was ousted because of his vocal support for domestic and international action on climate change.[4]

The emergence of SENER as a prominent voice in Mexican climate politics also generated an intense struggle in 2000 between SEMARNAT and SENER over ratification of the Kyoto Protocol. As a non–Annex 1 party, Mexico's ratification of the Kyoto Protocol would place few obligations on the country, while giving it access to the Kyoto mechanisms. A cornerstone of Mexico's interest in the Kyoto Protocol was the Clean Development Mechanism (CDM), which finances clean development by enabling industrialized countries to purchase GHG emissions reductions credits generated in developing countries. CDM projects were of interest to a wide range of stakeholders in Mexico. For example, SEMARNAT officials saw potential in projects that simultaneously limited GHG emissions while furthering Mexico City's efforts to reduce local air pollution (West, Osnaya, Laguna, et al. 2004). CDM projects also had the potential to engage actors in the cement, steel, and petrochemical sectors. Some of the most vocal interest in the CDM came from Mexico's national oil company (Petroleos Mexicanos, Pemex). Pemex executives saw the CDM as a way to channel foreign investment into company operations (Pulver 2007).

The benefits of the CDM and the enhancement of Mexico's status in the international arena were arguments mobilized by SEMARNAT bureaucrats and SRE diplomats in support of Mexican ratification of the Kyoto Protocol. They were opposed by representatives from SENER and SECOFI, who emphasized the potential negative consequences of international GHG regulations on Mexico's economy. In 2000, the ratification decision came to the Mexican Senate, comprising 128 senators. The Senate was inclined to ratify the Kyoto Protocol, until an initiative led by SENER attempted to block the vote. In the evening before the scheduled vote, envoys from both SEMARNAT and SENER were invited to the presidential residence at Los Pinos. With guidance from President Zedillo, the decision was made to go ahead with the ratification vote, and on April 29, 2000, the Mexican Senate voted unanimously to ratify the Kyoto Protocol.[5] Mexico's instrument of ratification was deposited at the UN headquarters in New York in September 2000. Mexico was the twenty-ninth country to ratify the Kyoto Protocol.

Ratification of the Kyoto Protocol represented a victory for SEMARNAT and reestablished some momentum around climate policy that had been diminished by the interministerial wrangling of the previous years. In line with the renewed momentum, SEMARNAT, in collaboration with INE, developed a new national program on climate change, concurrent with their preparation of Mexico's Second National Communication under the UNFCCC. Unfortunately, when the national program on climate change was circulated for public review, the proposal generated much negative commentary, and it was recast as merely a national strategy on climate action. According to one interviewee, the whole effort has since been swept under the rug.[6] Moreover, even if the attempt to draft a national climate program had been successful, its prospects would have been uncertain at best. Zedillo's support for ratification of the Kyoto Protocol was made in the final year of his six-year term. Vicente Fox was elected to the Mexican presidency in August 2000—four months after the Kyoto ratification decision—and assumed the position of president in December of that year. The first Mexican president not from the ruling Institutional Revolutionary Party (Partido Revolucionario Institucional, PRI)) in seventy-one years, Fox declined to carry forward the proposed national climate program of the previous administration. In addition, Fox restructured the environmental leadership within Mexico. Fox named Victor Lichtinger as Minister of the Environment, replacing Carabias Lillo. In addition, under the new Fox presidency, INE, which had played a lead role in Mexican climate and environmental politics, was divested of its policy functions and retasked as a research institute. It retained only the responsibility for generating Mexico's GHG inventory data and national communications under the UNFCCC (Tudela, Gupta, and Peeva 2003).

International Drivers of Domestic Climate Politics, 2001–2005

While domestic debates dominated climate politics in Mexico from 1996 to 2000, international dynamics came to the fore in shaping the post-2000 agenda. From 2001 to 2005, interest in climate change in Mexico waxed and waned with the stop-and-go progress of the international climate negotiations. In the first year of the new Fox administration, international developments reinforced his lack of interest in the climate issue. In the longer term, international events restimulated federal interest in action on climate change.

The first key international event came in March 2001, when President George W. Bush withdrew the United States from the Kyoto Protocol. While never a leader in the international climate negotiations, U.S. participation was considered central to the effectiveness of the Kyoto Protocol. The U.S. delegation to the international climate negotiations under the Clinton Administration had agreed to adopt the Kyoto Protocol text in 1997, and President Clinton formally signed the protocol in 1998. Therefore, President Bush's announcement that the United States was withdrawing from the Kyoto treaty was a blow both to the climate change regime and to Mexico's expectations of U.S. participation. In particular, before U.S. withdrawal from the Kyoto Protocol, the size of the CDM was estimated at $2 billion to $4 billion (in U.S. dollars), translating to a price of $10 to $20 per ton of carbon (Quadri 2000). This price reflected expected demand for credits by large industrialized country emitters such as the United States. Moreover, Mexican officials expected the United States to look to its southern neighbor and partner under the North American Free Trade Agreement for CDM opportunities (CCA/CEC 2001). When the United States pulled out from the Kyoto Protocol in 2001, the projected price of carbon credits under the CDM declined substantially.

President Bush's decision also undermined prospects for a North American GHG emissions trading scheme (Betsill, this volume). In addition to the CDM, Mexican policymakers were interested in the Kyoto Protocol's emissions trading mechanism. As a non–Annex 1 country, Mexico was not eligible to participate in emissions trading. However, some individuals in the SEMARNAT were thinking beyond the first commitment period under the Kyoto Protocol, which ends in 2012, to a second commitment period, when Mexico might take on a binding GHG emissions reduction target. One future scenario envisioned by regulators included a burden-sharing agreement across North America, similar to the EU's burden-sharing arrangement, linked to a North American emissions trading program (CCA/CEC 2001). Although the prospects of such a trading program were always tenuous, they vanished when the United States withdrew from the Kyoto Protocol.

Mexican authorities did not regain confidence in the CDM until late 2002. This renewed interest was once again linked to international events. Lichtinger, Fox's appointed Secretary of the Environment, had been a member of the Mexican delegation to five rounds of the international climate negotiations in 1991 and 1992. Yet, when he first assumed his position a Secretary of the Environment, he chose not to make climate change a high profile issue. A presidential visit to Europe in the spring of 2002 changed that decision. During the visit, Lichtinger met with European environment ministers who at the time were pushing for ratification of the Kyoto Protocol within Europe. The European Union's decision to ratify was announced in May 2002, followed by Japan and Canada. With Russian ratification, the Kyoto Protocol entered into force in February 2005. International endorsement of the Kyoto regime reenergized domestic action within Mexico. A Climate Change Office, which serves as Mexico's CDM project-approval authority, was created within SEMARNAT in 2003, after contentious negotiations with SENER about which ministry should house the unit (SEMARNAT 2006). The office ramped up activities in late 2005 and began issuing letters of approval for CDM projects in Mexico. Also, a formalized interministerial commission on climate change (Comisión Intersecretarial de Cambio Climático, CICC) was established. In 2008, Mexico was host to approximately 10 percent of the 1,190 CDM projects officially approved and registered by the CDM Executive Board worldwide, with another 86 CDM projects in the approval pipeline. Mexico's CDM activities are projected to yield almost 80 million tons of CO_2 equivalent reductions by 2012 (Fenhann 2008).

National Climate Strategies, but Little National Action, 2006–Present

The most recent period in the evolution of Mexican climate politics was heralded by Mexico's 2006 presidential election. While neither the winning candidate, Felipe Calderón, nor his main rival, Andrés Manuel López Obrador, made climate change a central plank of his campaign, President Calderón has broken with the past by articulating a national climate strategy early in his administration. Unlike his predecessors, who postponed action on climate change to the final years of their presidencies,

Calderón announced his new national strategy on climate change in May 2007, six months into his new administration (CICC 2007). The strategy builds on a November 2006 planning document produced by the CICC and on Mexico's Third National Communication under the UNFCCC (CICC 2006; Government of Mexico 2006).

The 2007 national strategy, which projects forward to 2014—two years after the end of Calderón's term—is wide-ranging in its goals. It outlines specific mitigation options and GHG reduction targets for several energy generation, energy use, and vegetation and land-use activities, yielding total emissions reductions of 124 million tons of CO_2 equivalent by 2014. Particular attention is focused on Pemex and the state-owned electric utility (Comision Federal de Electricidad, CFE). Furthermore it specifies a climate change adaptation strategy and identifies future research priorities (CICC 2007). Finally, the strategy also contains plans for the central government to assist Mexico's thirty-one states in developing state-specific climate programs (González-Dávila 2007). The extent to which this strategy's proposed mandates and initiatives will be realized has yet to be seen. Critics are skeptical of the strategy's voluntary status and doubt the likelihood that it will meet the 2014 target (CCC Newsdesk 2007).

Nonstate Actors (I): Climate Change Is Not a Civil Society Priority

Given the stop-and-go nature of federal government action on climate change, there are two other possible sites for climate leadership in Mexico: civil society and the private sector. Surprisingly, the latter has been the more prominent. Two striking features of Mexican climate politics are the active engagement of certain key industries in climate policy debates and the absence of a civil society-led climate campaign.

The NGO community in Mexico is vibrant, yet still in the early stages of development. Delgado (2001) identifies the 1980s and particularly the battle against the Laguna Verde nuclear power plant as the beginning of a self-identified environmental NGO community in Mexico. This community continued to thrive and expand in preparation for the 1992 UN Conference on Environment and Development in Rio de Janeiro and for the 2002 World Summit on Sustainable Development in Johannesburg. A Mexican NGO (Grupos de Estudios Ambientales, GEA) sent a representative to multiple rounds of the international climate change negotiations. However, international activities did not lead to an active domestic climate campaign. A combination of factors accounts for the absence of a climate focus among Mexican environmental groups.

First and most importantly, Mexican NGOs assign responsibility for climate change to the world's major GHG emitters and rely on their NGO colleagues in industrialized countries to push forward on climate change advocacy. Second, unlike in the United States, the Mexican government has generally been forward-thinking on the climate issue at the domestic and international levels, and the environmental community has not needed to mobilize against a recalcitrant federal government. Third, there is little public pressure for action on climate change. As a result, Mexican NGOs focus their efforts on local environmental concerns that are perceived as more pressing and deserving of attention than climate change. Conservation issues have a high profile in Mexico, and many domestic NGOs work on conservation projects, collaborating with both local community groups and large international NGOs, such as Conservation International and the World Wide Fund for Nature. As one of twelve "mega-diverse" countries, Mexico is a biodiversity "hotspot" (Ramamoorthy, Bye, Lot, et al. 1993). Other Mexican NGOs focus on local pollution issues and rarely link with international campaigns.[7] For example, in Mexico City, the primary issue of concern is local air pollution, and the focus of activities is on redesigning the city's transportation infrastructure. These activities are recognized as generating climate benefits, but the drivers for action are local pollution concerns (West, Osnaya, Laguna, et al. 2004). In the oil-producing state of Tabasco, there is a long history of activism focused on the adverse environmental effects of oil extraction and refining activities. The target of activism is Pemex, Mexico's national oil company (Town and Hanson 2001).

Within the Mexican environmental NGO community, there are three groups that are well positioned to potentially organize a climate campaign. They are Greenpeace Mexico, the Centro Mexicano de Derecho Ambiental (CEMDA), and the Unión de Grupos Ambientalistas (UGA). These three groups employ staff with advanced degrees, can access the international environmental advocacy community, and have relevant experience in Mexican politics; in other words, they have the necessary resources and expertise to campaign on climate change. However, policy directors and campaigners from Greenpeace Mexico, CEMDA, and UGA all reiterated that climate change is simply not a priority issue for their organizations.[8]

Nonstate Actors (II): Industry Follows Pemex's Lead

In contrast to civil society groups, the private sector in Mexico has been surprisingly engaged in climate policy debates. Most recently, thirty-five Mexican companies, primarily from energy-intensive sectors, were lauded publicly for participating in a GHG inventory initiative (Programa GEI Mexico 2007). The business pioneer in Mexican climate change politics is Petróleos Mexicanos (Pemex), Mexico's national oil company. As a national oil company, Pemex is not per se a private-sector actor. However, the company models itself on publicly traded oil corporations, and its climate policy approach has become path-setting for other Mexican companies (Pulver 2007).

Pemex began seriously engaging with climate change in 1999. That December, Pemex organized a workshop on climate change with SEMARNAT and the United Nations Development Program (UNDP). Four months later in April 2000, Pemex launched its 1999 Annual Report on Safety, Health, and Environment—a first for the company. The report laid out Pemex's proactive stance on climate change (Pemex 2000). The next major step in Pemex's climate program came in 2001. In June of that year, Pemex initiated an internal emissions trading system, with the pledge to reduce CO_2 emissions from its facilities by 1 percent by the end of the year (Pemex 2002). While 1 percent may appear to be a small reduction, Pemex's GHG emissions are quite significant. Estimated carbon dioxide emissions for 1999 amounted to almost 40 million tons of CO_2—equivalent to the annual GHG emissions of Ireland (Pemex 2000; WRI 2003).

Pemex took on the emissions reductions target and launched the trading system with the support of Environmental Defense (ED), a Washington, D.C.–based environmental NGO. Pemex also became a member of the ED Partnership for Climate Action (PCA), a collaboration among large firms like British Petroleum, Shell International, DuPont, and Alcan, who are committed to reducing their GHG emissions. Pemex's decision to take on a target and engage with the PCA was profiled in an article in the *Washington Post* (Pianin 2001). In developing its climate policy, Pemex acted as a "close follower" of British Petroleum, the oil major that Pemex managers identified as the industry leader in the climate arena (Pulver 2007).[9]

A variety of factors contributed to Pemex's proactive engagement with the climate issue. First, the particular evolution of the climate issue in Mexico caused Pemex to rely on input from SEMARNAT rather than SENER when initially formulating its climate policy. Structurally, SENER should have had the determining voice. Pemex, the national electricity company (CFE), and Luz y Fuerza del Centro (LFC), the electricity and power company serving Mexico City, are very large and semi-independent suborganizations within SENER. However, despite their organizational links, Pemex and SENER developed their climate policies relatively independently. When SENER began voicing opposition to action on climate change from 1997 onward, Pemex had already been drawn into the climate issue via the process of assembling information for Mexico's first national GHG emissions inventory in 1995. As described above, this process was coordinated by Mexico's emerging epistemic community, who presented climate change as an issue that merited domestic and international action. Based on that initial contact, Pemex and SEMARNAT maintained a cooperative relationship.

A second driver of the company's proactive approach to climate change was that the international climate regime offered Pemex managers tools by which to advance preexisting business objectives. Pemex has long been recognized as one of the most inefficient companies in the oil industry. Moreover, foreign investment in the petroleum sector is constitutionally restricted (Halpern 2001; Smith 2004). In the late 1990s, Pemex was under strong pressure to reform it operations, in order to increase operational efficiency and to access foreign investment (Shields 2001). Pemex managers saw the Kyoto mechanisms, particularly the CDM and emissions trading, as means to meet both goals. By 1999, Pemex had already been approached by both Canadian and Norwegian companies interested in investing in Pemex to create CDM emissions reduction projects. CDM project investments in operational efficiency improvements and in the development of natural gas facilities were defined as the "sale of environmental services" and framed as bypassing the constitutional restriction on foreign investment in the oil sector.[10] Likewise, Pemex's internal emissions trading system was designed to identify opportunities for energy efficiency projects.

Third, Pemex leadership was receptive to environmental initiatives. Rafael Fernandez de la Garza, Pemex's Director of Environmental Health and Safety, came to the company from a regulatory position in the nuclear industry. During his tenure as a nuclear regulator, he was the target of ongoing environmental protests against the Laguna Verde nuclear power plant, Mexico's only nuclear energy facility. Interviewees reported that his experience in the nuclear industry made him very environmentally aware.[11]

Pemex's forward-looking approach to climate change has made it a leader within government, within the Mexican private sector, and within the region. Within the governmental context, Pemex has been a frequent advocate for action on climate change. For example, the timing of Pemex's 2000 formal public announcement of its climate policy was critical (Gomez Avila, Martinez, Guzman, et al. 2001). Observers report that Pemex's support for international GHG regulations factored into President Zedillo's decision in April 2000 to support SEMARNAT over SENER and approve Mexico's ratification of the Kyoto Protocol.[12] Likewise, Pemex was closely involved in discussions with Mexico's CDM office to ensure that Pemex projects are eligible under Mexico's CDM rules. Finally, Pemex's action on the CDM and emissions trading has become a site for positive interaction between SEMARNAT and SENER. In September 2002, preliminary discussions were held within the two ministries regarding the expansion of the Pemex emissions trading system to include CFE, the national electricity company. To date, these discussions have not materialized in a concrete program.

Pemex has also played a pioneering role in the Mexican private sector. Following Pemex's lead, several private-sector associations including a sustainable business NGO (Centro de Estudios del Sector Privado para el Desarrollo Sustenable, CESPEDES) and the Mexican Employer's Association (COPARMEX) have promoted the business opportunities presented by climate change among their membership. In 2004, SEMARNAT, the World Resources Institute (WRI), and the World Business Council for Sustainable Development (WBCSD) jointly launched the Mexico Greenhouse Gas Program, under which participating companies compile corporate GHG inventories—the necessary precursor to emissions reductions (WRI 2006; Ozawa-Meida, Fransen, and Jimemez Ambriz 2008). Pemex was one of the twelve initial participants in the program. By 2008, over thirty-five leading Mexican companies participated in the program, accounting for approximately 120 million tons of CO_2 equivalent, or 35 percent of all GHG emissions from industrial processes and stationary sources in Mexico (Programa GEI Mexico 2007).

In addition to efforts within Mexico, Pemex has also been a regional industry leader on climate change. The company has hosted several international workshops on the CDM and emissions trading for the oil and gas industry association of Latin America and the Caribbean (ARPEL). Finally, Pemex may turn out to be a leader among state-owned oil companies. Although no other state-owned oil companies in developing countries have publicly announced support for climate change, they are beginning to educate themselves. For example, in 2006, Saudi Aramco, Saudi Arabia's national oil company, convened a conference of experts to brief corporate executives on the climate change issue.[13]

The history of Pemex as an industry pioneer has a cautionary epilogue. While Pemex has continued its active engagement in tracking GHG emissions and in sponsoring CDM projects, the company did not follow up its 1 percent CO_2 reduction target with a more stringent 10 percent target, as was being discussed in 2002. Interest in emissions trading within Pemex is likely to be revived if the international emissions trading system proves to be robust.

Future Prospects for Climate Change Action in Mexico and Beyond

Understanding the history and evolution of Mexican climate politics in the scientific, political, market, and civil society arenas sheds light on possibilities for future climate change action in Mexico, North America, and globally. First, both in the public and private sectors, climate action in Mexico has been based on win-win programs and projects that have both GHG and other economic, environmental, or social cobenefits. The CDM has been a mechanism to finance such projects, and it has worked to build common ground between SEMARNAT and SENER. Looking to the future, there is interest in expanding the scope of the CDM from the current project-based structure to a broader sectoral and programmatic structure (Figueres 2006). For example, a programmatic CDM structure could help the Mexican government secure the foreign financing it needs to develop its domestic production of natural gas, a stated goal of the energy ministry. Increased use of natural gas would both help to reduce Mexico's GHG emissions and economically compensate for declines in oil production.

Financial support from industrialized countries for such large-scale initiatives is crucial. More generally, Mexican interest in the climate issue has been driven by the actions of Annex 1 countries, including but not limited to the United States. For example, the U.S. Country Studies Program was central to organizing Mexico's climate research community, a constituency that played a galvanizing role in Mexico's initial response to climate change. Likewise, federal level government interest in climate change in Mexico has followed a stop-and-go pattern that reflected the negotiation of the UNFCCC in 1992, the U.S. withdrawal from the Kyoto regime in 2001, the EU's decision to ratify the Kyoto Protocol in 2002, and the protocol's entry-into-force in 2005. Looking to the future, this pattern suggests the general importance of Annex 1 country leadership on climate change as a stimulus to action on climate change not only in Mexico but also in other developing countries.

A third widely applicable lesson emerging from the Mexico case is the prominent role of scientists in the early agenda-setting phase in Mexican climate politics. Scientists had a long-term impact in establishing climate change as an issue of concern that merited action. While climate policy developments around the globe follow particular trajectories specific to national contexts, the trend of early involvement by climate scientists is observed across a range of developing countries. Developing-country delegations to the UN climate negotiations in the early to mid-1990s generally included only representatives from the national meteorological office or the environmental ministry. Tapping into a domestic scientific community can be a viable strategy to mobilize developing country support for action on climate change.

Finally, the future of climate action in Mexico has to be considered within a North American context. Canada, the United States, and Mexico have highly integrated economies and energy systems. Moreover, Mexico and Canada have more in common on the climate issue than first appears. Both are oil-producing countries, whose economies are relatively energy inefficient and for whom the United States is the primary export market. Both countries are potentially very vulnerable to the biophysical effects of climate change; Mexico due to widespread poverty and high levels of biological diversity and Canada because of the vulnerability of social and ecological communities in the Arctic (Nilsson, this volume; Stoett, this volume). Finally, in both countries, climate politics is influenced by the climate policy choices of the United States. Preexisting economic integration and common interests could serve as a basis for cooperation between Mexico and its northern neighbors, yet so far these efforts have failed (Betsill, this volume). International action among the three countries has been limited to bilateral statements of cooperation. For example, in 2003, the United States and Mexico pledged to strengthen bilateral cooperation on climate change, creating a Bilateral Working Group on Climate Change (U.S. Department of State 2003). Likewise, Canada and Mexico signed a joint statement on climate change cooperation at the December 2005 round of the international climate negotiations (Government of Canada 2005).

International cooperation has been hindered by at least two factors. First, in all three countries, support for climate change has been subject to the vagaries of electoral politics. Mexico's ratification of the Kyoto Protocol came at the end of Zedillo's presidential term. Likewise, Canada's contentious decision to ratify the Kyoto Protocol was an end-of-term effort by Prime Minister Jean Chrétien (Harrison 2007). Only recently has the climate change issue become a focus of policy action at the beginning of presidential administrations. Consistent interest in the climate issue needs to be separated from the ebb and flow of electoral politics. Second, both the United States and Canada are, at best, weak participants in the global climate regime. The United States is party only to the UNFCCC and not the Kyoto Protocol, despite varied GHG regulatory initiatives at the subnational level. Canada, while a Kyoto Protocol signatory, is likely to default on its target and has expressed an unwillingness to meet its target via the Kyoto mechanisms (Stoett, this volume). A decision by the United States to reengage in the Kyoto process or by Canada to meet its Kyoto target via significant CDM investment in Mexico might serve as a basis for international cooperation with Mexico. However, absent such dramatic changes, prospects of an integrated North American climate regime are limited.

In place of international cooperation, transnational action across public-private and market-civil society divides offers more fruitful prospects. In particular, the private sector remains the most promising arena in which to promote bottom-up action on climate change in Mexico in the short term. Activity in this area has always been based on transnational linkages. International and U.S. NGOs were central to both Pemex's climate policy program and the private-sector GHG inventory project. With the upsurge in subnational climate change activities in the United States, there are several prospects for partnerships within the transportation and energy sectors between U.S. states, Canadian provinces, and Mexican states.

Notes

1. Over the period discussed in this chapter, Mexico's ministry of environmental affairs has undergone several name changes. The current federal environment agency, SEMARNAT, was established in 1994 as the Secretaría de Medio Ambiente, Recursos Naturales, y Pesca (SEMARNAP). Prior to 1994, environmental issues were under the purview of the Subsecretaría de Desarollo Urbano y Ecología (SEDUE), i.e., the ministry of urban development

and ecology, established in 1982. In 1992, SEDUE was transformed into the Secretaría de Desarollo Social (SEDESOL), i.e., the ministry of social development. At the same time, two independent technical bodies were created to support SEDESOL: the Instituto Nacional de Ecología (INE), an environmental research institute, and the Procuraduría Federal de Proteccion al Ambiente (PROFEPA), an environmental enforcement agency. To date, INE and PROFEPA remain operational and support the activities of SEMARNAT. For clarity, this chapter always refers to Mexico's ministry of environmental affairs as SEMARNAT.

2. Author's interview with government representative (GOVT10), July 30, 2002.

3. Author's interview with government representative (GOVT2), June 5, 2002.

4. Author's interview with government representative (GOVT2), August 12, 2002.

5. Author's interview with academic expert (UNIV4), August 22, 2002.

6. Author's interview with government representative (GOVT10), July 30, 2002.

7. Greenpeace is the one exception to this rule. The Greenpeace Mexico office campaigns on conservation and pollution issues, mobilizing local groups as well as resources from Greenpeace International.

8. Author's interviews with environmental nongovernmental organization representatives (ENGO1), July 24, 2002; (ENGO2), September 17, 2002; (ENGO 3), July 23, 2002.

9. Author's interview with Pemex representative (OIL2), August 14, 2002.

10. Author's interview with Pemex representative (OIL2), August 14, 2002.

11. Author's interview with environmental nongovernmental organization representative (ENGO3), July 23, 2002.

12. Author's interview with Mexican academic (UNIV4), August 22, 2002.

13. Personal communication with Kelly Sims Gallagher, director of Energy Technology Innovation Project, Belfer Center, Harvard University, March 23, 2006.

References

Belausteguigoitia, Juan Carlos, and Indira Lopez-Bassols. 1999. Mexico's Policies and Programs That Affect Climate Change. In *Promoting Development While Limiting Greenhouse Gas Emissions*, edited by J. Goldemberg and W. Reid. New York: UN Development Programme and World Resources Institute.

Benioff, Ron, Erik Ness, and Jessica Hirst. 1997. *National Climate Change Action Plans: Interim Report for Developing and Transition Countries*. Washington, DC: U.S. Country Studies Program.

CCA/CEC. 2001. *México y el Incipiente Mercado de Emisiones de Carbono*. Mexico City: Comisión para la Cooperación Ambiental.

CCC Newsdesk. 2007. Latin America Special Report: Mexico's Flimsy Raft of Climate Change Measures. August 16, 2007. http://www.climatechangecorp.com/content.asp?ContentID=4897.

CCS. 2008. *UNFCCC Data Interface*. Climate Change Secretariat. http://unfccc.int/ghg_data/ghg_data_unfccc/items/4146.php.

CICC. 2006. *Towards a National Climate Change Strategy*. Mexico City: Interministerial Commission on Climate Change, SEMARNAT.

CICC. 2007. *National Strategy on Climate Change: Mexico*. Mexico City: Interministerial Commission on Climate Change, SEMARNAT.

Climate Group. 2004. Mexico City Case Study. http://www.theclimategroup.org/reducing_emissions/case_study/mexico_city/.

Delgado, Martha. 2001. El Papel de las Organizaciones de la Sociedad Civil ante el Cambio Climático Global. Mexico City: Unión de Grupos Ambientalistas.

Di Sbroiavacca, Nicolas, and Leonidas Osvaldo Girardin. 2000. Mexico. In *Confronting Climate Change: Economic Priorities and Climate Protection in Developing Nations*, edited by B. Biagini. Washington, DC: National Environmental Trust.

EIA. 2007. Country Analysis Briefs: Mexico. Energy Information Administration. http://www.eia.doe.gov/emeu/cabs/Mexico/Background.html.

Fenhann, Jørgen. 2008. The CDM Pipeline. UNEP Risø Centre. http://www.cdmpipeline.org/.

Figueres, Christiana. 2006. Sectoral CDM: Opening the CDM to the Yet Unrealized Goal of Sustainable Development. *Journal of Sustainable Development Law and Policy* 2(1): 5–27.

González-Dávila, Germán. 2007. México Ante el Cambio Climático: Opciones de Desarrollo. Workshop presentation at SEMARNAT, Mexico City, December 5.

Gay Garcia, Carlos. 1994. Propuesta de Programa Nacional sobre Cambio Climático Global. Paper presented at Primer Taller de Estudio de País: México Ante el Cambio Climático, April 18–22, Cuernavaca.

Gay Garcia, Carlos, ed. 2000. *México: Una Visión Hacia el Siglo XXI—El Cambio Climático en México*. Mexico City: UNAM Programa Universitario de Medio Ambiente.

Gomez Avila, Salvador, Nicolas Rodriguez Martinez, Francisco Guzman, and Mariano Bauer. 2001. Petróleos Mexicanos: A National Oil Company Committed to Improve Its Environmental Performance. In *Values Added: Ethical Experience in the Energy Sector*. London: World Energy Council.

Government of Canada. 2005. Canada and Mexico Sign Joint Statement on Climate Change Cooperation during the United Nations Climate Change Conference in Canada. December 8. http://www.canadianenvironmental.com/bin/cf_external_frameset.cfm?new_url=http://www.ec.gc.ca/press/2005/051208-4_n_e.htm.

Government of Mexico. 1997. *First National Communication under Framework Convention on Climate Change*. Mexico City: Secretaría de Medio Ambiente, Recursos Naturales, y Pescas.

Government of Mexico. 2006. *Third National Communication under Framework Convention on Climate Change*. Mexico City: Secretaría de Medio Ambiente y Recursos Naturales.

Halpern, John. 2001. Energy. In *Mexico: A Comprehensive Development Agenda for the New Era*, edited by M. Giugale, O. Lafourcade, and V. Nguyen. Washington, DC: World Bank.

Harrison, Kathryn. 2007. The Road Not Taken: Climate Change Policy in Canada and the United States. *Global Environmental Politics* 7(4): 92–117.

Martinez, Julia, and Adrian Fernandez Bremauntz, eds. 2004. *Cambio Climático: Una Vision desde México*. Mexico City: Instituto Nacional de Ecología.

Mintzer, Irving M., and J. A. Leonard, eds. 1994. *Negotiating Climate Change: The Inside Story of the Rio Convention*. Cambridge: Cambridge University Press and Stockholm Environment Institute.

Ozawa-Meida, Leticia, Taryn Fransen, and Rosa Maria Jimenez-Ambriz. 2008. The Mexico Greenhouse Gas Program: Corporate responses to Climate Change Initiatives in a 'Non-Annex I' Country. In *Corporate Responses to Climate Change: Achieving Emissions Reductions Through Regulation, Self-Regulation, and Economic Incentives*, edited by Rory Sullivan. Sheffield: Greenleaf Publishing.

Paterson, Matthew. 1996. *Global Warming and Global Politics*. London: Routledge.

Pemex. 2000. *Safety, Health, and Environment Report 1999*. Mexico City: Petróleos Mexicanos.

Pemex. 2002. *Safety, Health, and Environment Report 2001*. Mexico City: Petróleos Mexicanos.

Pershing, Jonathan. 2000. Fossil Fuel Implications of Climate Change Mitigation Responses. *Sectoral Economic Costs and Benefits of GHG Mitigation, Proceedings of an IPCC Expert Meeting*, edited by L. Bernstein and J. Pan. Bilthoven, Netherlands: Intergovernmental Panel on Climate Change, WGIII, RIVM.

Pianin, Eric. 2001. Mexican Company Agrees to Reduce Emissions. *The Washington Post*, June 5.

PIW. 2006. PIW's Top 50: How the Firms Stack Up. Energy Intelligence Group. http://www.energyintel.com/documentdetail.asp?document_id=137158.

Programa GEI Mexico. 2008. Tercera Entrega de Reconocimientos del Programa GEI México. November 6, Programa GEI Mexico. http://www.geimexico.org/.

Pulver, Simone. 2007. Importing Environmentalism: Explaining Petroleos Mexicanos' Proactive Climate Policy. *Studies in Comparative International Development* 42 (3/4): 233–255.

Quadri, Gabriel. 2000. Climate Change: Mexico and the Kyoto Flexibility Mechanisms. Comisión de Estudios del Sector Privado para el Desarrollo Sustentable. http://www.cce.org.mx/cespedeshttp://www.cce.org.mx/cespedes.

Ramamoorthy, T. P., Robert Bye, Antonio Lot, and John E. Fa, eds. 1993. *Biological Diversity of Mexico: Origins and Distribution*. New York: Oxford University Press.

Ramos-Mane, Cecilia, and Ron Benioff. 1995. Interim Report on Climate Change Country Studies. Washington, DC: U.S. Country Studies Program.

Romero Lankao, Patricia. 2007. How Do Local Governments in Mexico City Manage Global Warming? *Local Environment* 12(5): 519–535.

SEMARNAP. 1998. *México ante el Cambio Climático*. Mexico City: Secretaría de Medio Ambiente, Recursos Naturales, y Pescas.

SEMARNAT. 2006. Dirección General Adjunta para Proyectos de Cambio Climático. Secretaría de Medio Ambiente y Recursos Naturales. http://www.semarnat.gob.mx/queessemarnat/politica_ambiental/cambioclimatico/Pages/cicc.aspx.

SENER. 1998. Climate Change and the Energy Sector. Mexico City: Secretaría de Energía, Subsecretaría de Políticas y Desarrollo de Energéticos, Dirección General de Política y Desarrollo Energético.

SENER. 1998. Principales Políticas Energéticas Enfocadas al la Disminución de las Emisiones de Gas Efecto Invernadero (1993–1998). Mexico City: Secretaría de Energía, Subsecretaría de Políticas y Desarrollo de Energéticos, Dirección General de Política y Desarrollo Energético.

Shields, David. 2001. Mexican Pipeline: The Future of Oil under Vicente Fox. *NACLA: Report on the Americas* (January/February): 31–37.

Smith, Geri. 2004. Pemex May Be Turning from Gusher to Black Hole. *Businessweek Online*, December 13. http://www.businessweek.com/magazine/content/04_50/b3912084_mz058.htm.

Town, Sarah, and Heather Hanson. 2001. Oil at the Grassroots: Report from Tabasco. *NACLA: Report on the Americas* (January/February): 34–35.

Tudela, Fernando, Shreekant Gupta, and Valya Peeva. 2003. Institutional Capacity and Climate Actions: Case Studies on Mexico, India, and Bulgaria. Paris: Organization for Economic Cooperation and Development.

U.S. Department of State. 2003. Joint Statement of Enhanced Bilateral Climate Change Cooperation between the United States and Mexico. March 18. http://www.state.gov/r/pa/prs/ps/2003/18801.htm.

West, Jason J., Patricia Osnaya, Israel Laguna, Julia Martinez, and Adrian Fernandez Bremauntz. 2004. Co-control of Urban Air Pollutants and Greenhouse Gases in Mexico City. *Environmental Science and Technology* 38(13): 3474–3481.

WRI. 2003. *World Resources 2002–2004. Decisions for the Earth: Balance, Voice, and Power*. Washington, DC: World Resources Institute.

WRI. 2006. Mexican Industry Takes Voluntary Action against Climate Change; Government Gives Public Recognition. February 22. http://www.wri.org/press/2007/10/mexican-industry-takes-voluntary-action-against-climate-change-government-gives-public.

WRI. 2007. Earth Trends: Climate and Atmosphere Searchable Database: World Resources Institute. http://earthtrends.wri.org/.

3

Looking for Leadership: Canada and Climate Change Policy

Peter J. Stoett

Introduction

Climate change is emerging as one of the most potent and divisive political issues in Canada, reflecting both concerns over the potential costs of climate change and the difficulties inherent in developing national policy in a federal political system. The stakeholder gallery is diverse, including each level of government (municipal, provincial/territorial, and federal), small and large businesses, First Nations peoples, farmers, environmental activists, and other groups that are animated by their support or opposition to more aggressive climate change policy. A situation in 2007 and 2008, when the minority federal government and the political leadership of the Albertan government were in ideological agreement, did not mask the most challenging and divisive issue: distinct policy preferences among the provinces. The federal government also acknowledges this problem. Former Environment Minister John Baird, reacting to the development of carbon taxation policies in Québec and British Columbia, in February 2008 stated, "What works in British Columbia may not necessarily work in Nova Scotia. What works in Ontario may not work in Alberta. Different provinces are coming forward with different approaches that suit their needs" (Hunter 2008).

As varied as Canada itself, the predicted impacts of climate change include extreme weather events, excessive hail, drought in the prairies, increased smog over large cities, lower water levels in the Great Lakes, severe impacts on aboriginal peoples and northern regions, rising coastland, and increases in invasive alien species (Bruce and Cohen 2004; Lonergan 2004; McBean 2006). Efforts to reduce Canadian greenhouse gas (GHG) emissions, however, have had dismal results: in 2005, total emissions of carbon dioxide equivalent were estimated at 747 megatons, up by 25 percent from 1990 and 33 percent above Canada's target under the Kyoto Protocol of 563 megatons (CESI 2007, 3). A combination of factors, including the federal structure, governmental procrastination, citizen complacency, and a misleading reliance on Kyoto Protocol ratification, has led to a lack of firm national guidance. Though climate change is undoubtedly attracting growing political attention in Canada, it will certainly remain the subject of partisan debate and political divisions for some time to come. As such, it is likely that sustained federal leadership will remain elusive.

At the heart of this national political contention is the reality that much of Canada remains wedded to resource extraction as a principal means of income generation. A large proportion of Canada's export revenues come from natural resources, including oil (Howlett and Brownsey 2008). An obvious indication of this is the rapid development of oil extraction from the Albertan tar sands, which are projected to contribute almost half of Canada's total increase in GHG emissions between 2003 and 2010 (Woynillowicz 2005). This contribution will furthermore increase beyond 2010 as both expanded surface mining and new deep mining commence in Alberta. The vast majority of that oil is intended for export to the United States and other foreign oil markets. Efforts to implement the Kyoto Protocol may have steered several European countries toward a future that is less dependent on fossil fuels, but this achievement has not been matched by a series of federal Canadian governments.

This chapter suggests that local initiatives—many of which are in fact anticipatory of proactive adaptation strategies (Cohen et al. 2004)—will serve as the predominant climate change policy in Canada for some time to come. In 1994, Emery Roe (1994, 20) wrote that climate change signaled "a wider 'analytic tip' taking place ... a tip toward the notion that issues that can still be effectively dealt with locally, regionally, or nationally must now first be addressed globally." This was not because it was necessarily more prudent to begin globally, but because it is often easier to start there as individual stakeholders are less likely to be held directly accountable at the global level. Many local initiatives are furthermore greatly dependent on federal funding (Sancton 2006). Barring a radical shift in the policy orientation at the federal level, however, the short-term prognosis is bleak for those demanding serious national Canadian mitigation efforts. Some positive change can, however, be noted, including the development of more ambitious provincial, territorial, and municipal policy initiatives that may help to shape a Canadian political context more conducive to serious climate change action in the future.

Federalism and Environmental Policymaking in Canada

Canada's political structure is a source of great frustration for those who want to see a strong national policy on climate change. The relationship between the federal government and the provinces and territories has been dynamic and evolving, described variously as colonial, classical, cooperative, competitive, and collaborative (Simeon and Robinson 2004). The British North America Act of 1867 (now the Constitution Act of 1982) gives the provinces jurisdiction over natural resources, local works and undertakings, and property and civil rights, though the federal government has jurisdiction over sea coasts and fisheries as well as foreign affairs, including the right to conclude treaties. Clearly, "since before the Second World War, many environmental functions, including resource management, pollution regulation, wilderness and species regulation, and parks services, have been legislated and administered largely at the provincial level" (McKenzie 2002, 107).

In addition, provinces and territories face unique challenges regarding climate change impacts and GHG reductions. Provincial disharmony has been a constant factor in Canadian climate change policy as provinces differ in terms of their resource and energy production bases, natural habitats, links to continental energy markets, and other pertinent factors. In general, federal efforts to harmonize provincial policies have been made by either a multilateral process of dealing with all the provinces at once or with individual bilateral federal-provincial agreements. The style has fluctuated over the years, but the Conservative government under Stephen Harper elected in 2006 and again in 2008 often prefers the multilateral approach. There are some bilateral agreements between Ottawa and the provinces and territories, but these are largely just memorandums of understanding and do not come with much federal funding (Winfield and Macdonald 2008, 278).

Importantly, there are major differences among the Canadian provinces regarding their reliance on carbon-intensive industry, the impact of various stakeholder groups, and political and economic history. Though GHG emissions increased in all provinces and territories between 1990 and 2005, the most noticeable increases are found in Alberta, Ontario, and Saskatchewan (Environment Canada 2007). Historically, heavy industries such as the automobile and steel industries have been centered in Ontario, which still relies on coal-fired power plants. Alberta has emerged as the carbon champion, largely because of the tar sands oil projects. Québec has developed its own variation of energy nationalism based on hydropower; and it has been the strongest proponent (alongside Manitoba) of the Kyoto Protocol (members of the Bloc Québécois also hold seats in the national parliament). While some provinces have pushed climate change policy to the forefront of policy planning (for example, Québec and British Columbia have implemented contentious carbon taxes), others (such as Alberta) have openly campaigned against the Kyoto Protocol, and Newfoundland and Labrador are becoming increasingly dependent on offshore oil drilling proceeds.

The reality of Canadian federalism means that there is no easy path to policy consensus on such an economically sensitive issue as climate change. However, though federalism is the policy context that overshadows any effort from Ottawa to push toward a national plan on climate change, leadership from the federal level has been confused, confrontational, and lukewarm. This should not be surprising, given that the federal Conservative government has made it clear that it intends Canada to be known as the next "emerging energy superpower" (Brownsey 2008, 246–251). Federal-provincial relations on climate change take place within a broad and complex set of stakeholder interests and institutions, which are described below.

The Stakeholder Complex

Canadian environmental policy reflects the federal nature of the political system described above, the resource dependency and size of the state, and continued tension in the citizenry due to diverse worldviews and human-nature conceptions. Many analysts have examined the evolution of environmental policy in Canada (Boardman 1992; Doern and Conway 1995; Dwivedi et al. 2001; Harrison 1996; McKenzie 2002; Parson 2001). In short, the dominant normative context of Canadian environmental politics has been a liberal/utilitarian approach to the environment as a storehouse of resources that can be transformed to commodities and sold domestically and internationally for economic development. Growing awareness of the ecological repercussions of extractive activity in various sectors, including fisheries, forestry, and oil and gas exploration, has challenged the hegemony of the utilitarian paradigm, but it arguably remains the dominant perspective. Concern over climate change may be encouraging a more holistic trend, but it would be premature and optimistic to assume this to be the case. First Nations peoples also have diverse views on how the environment should be regarded, though they hold a general consensus that long-term thinking and the precautionary principle are integral elements of resource management.

There are also widely divergent views in Canada on the proper level of governmental intrusion in economic affairs. In general, there is a marked distinction between traditional left- and right-wing viewpoints. One federal party (the Bloc Québécois) is also committed to the separation of one of the largest provinces, Québec. There is furthermore an undeniable regional dimension to party politics in Canada; the Harper government generated much of its initial support from the province of Alberta, the home of Canada's crude oil and tar sands oil industries. With the Liberals, New Democrats, and Bloc Québécois aligned against the Conservatives, climate change perhaps more than any other issue put the Conservative minority government status in peril. Since all economic sectors and governing bodies are affected by the environment in which they operate, this "ultimate horizontal issue" (Winfield 1994) has been an awkward fit with Canada's federal governmental structure. On the issue of climate change, the sheer complexity of not only Canadian federalism but also the vast shareholder pool proves daunting for analyst and politician alike. Reduction targets for GHG emissions affect a wide range of people and corporate bodies, from individual homeowners to large automobile manufacturers to educational institutions, and as a result policy implementation is multilayered, slower than desired by most, and often contentious.

Business lobby organizations, including those of natural resource extracting industries, have always played a large role in federal and provincial legislative processes. Indeed, much of rural Canada remains highly dependent on resource extraction for local employment and economic growth. Lobby groups, including the Canadian Association of Petroleum Producers, the Canadian Petroleum Products Institute, and the Canadian Renewable Fuels Association, are highly active on issues related to climate change. Environmental groups such as the Sierra Club and Greenpeace and think tanks like the Pembina Institute and the David Suzuki Foundation are also vocal participants in domestic climate change debates. However, it is clear that, as the world's third-largest producer of natural gas and ninth-largest producer of crude oil, the petroleum industry in Canada often has a considerable impact on decision-making processes and policy outcomes at all political levels.

In addition, Canada is heavily trade-dependent and operates in effect as a member of a continental economy dominated by the U.S. marketplace under the North American Free Trade Agreement. This fact penetrates each area of Canada's political and cultural existence, including natural resource and environmental policy (LePrestre and Stoett 2006). Thus, many of the stakeholders involved in climate change issues are not Canadian; they are foreign firms and managers with considerable investments in various natural resource sectors, consumers of imported Canadian products, and transnational environmental activists. There are also stark contrasts among pro- and anti-Kyoto voices regarding Canada's multilateral political, economic, and environmental obligations.

The Evolution of National Policy on Climate Change

In 1987, Canadian Prime Minister Brian Mulroney, during a multilateral conference in Toronto, characterized the threat from climate change as "second only to a global nuclear war" and called for a 20 percent cut in GHG emissions by 2005. This started a fairly consistent pattern of Canadian rhetoric outweighing policy formulation and implementation on climate change. Canada promised to stabilize GHG emissions at 1990 levels immediately prior to the 1992 United Nations Conference on Environment and Development, at which the United Nations Framework Convention on Climate Change (UNFCCC) was signed. The subsequent Liberal government went further, promising a 20 percent reduction from 1990 levels and introducing the first National Action Program in Climate Change. As it became apparent that this goal was rather unrealistic, the Kyoto negotiation target slipped in 1997 to 6 percent below 1990 levels by 2012. Though Canada signed the Kyoto Protocol on April 29, 1998, it did not ratify the treaty until December 17, 2002.

Actual policy initiatives on emissions reduction have been slow in coming, but many environmentalists were somewhat heartened by the later steps taken by the Liberal governments under Chrétien and Martin. These included the Action Plan 2000 on Climate Change, which committed $500 million; and, in November 2000, the Climate Change Plan for Canada, which promised (but did not deliver) annual cuts of 240 megatons of emissions. Total spending on Kyoto-related programs neared $4 billion by 2003. In 2004 the One Tonne Challenge, which sought voluntary efforts by citizens to reduce their own emissions but came with energy conservation initiatives and other incentives, was released. Later the same year, Environment Canada released its 2002 greenhouse gas inventory, indicating that Canada emitted 731 megatons of greenhouse gases that year, up 2.1 percent over 2001, and 28 percent above the Kyoto target of 572 megatons that Canada committed to reach by 2012.

In March 2005, the federal government reached an agreement with Canadian automakers that contained voluntary commitments to GHG reductions from automobiles. Later a plan was released that increased government spending and decreased the obligation of large emitters to reduce emissions. The government of Alberta, however, made it clear it had no intention of enforcing Kyoto provisions in its jurisdiction, and proceeded to intensify the highly pollutive and energy-consumptive process of extracting oil from tar sands. With Canadian GHG emissions well over 750 megatons, reaching the Kyoto Protocol target of 571 megatons per year would be a colossal achievement within the time span permitted.

In January of 2006, Stephen Harper's Conservative party won a minority government, and with it came a shift from an ostensibly Kyoto-oriented policy platform to an essentially anti-Kyoto platform. The release of yet another action plan in 2006, part of a broader Clean Air and Climate Change Act, which relies on intensity targets instead of caps, and very long-term hopeful thinking (50 percent GHG reductions from 2003 levels by 2050), did little to assuage concerns that the Harper government is not committed to serious legislative remedies to the climate change problem. When it became clear the Clean Air and Climate Change Act would not pass in parliament without significant modification, it is was more or less shelved. The Canadian delegation in Bali in 2007, there ostensibly to help draw up a road map for the development of a post-Kyoto agreement, acted deliberately to obstruct any hard targets in post-Kyoto discussions.

Abandoning Kyoto

It is premature to declare the Canadian implementation of Kyoto dead, in part because the federal government has not officially withdrawn from the treaty, but the Harper government has made audible funeral arrangements. Indeed it moved with rather remarkable speed toward dismantling whatever scaffolding previous Liberal governments had managed, in their procrastinate manner, to erect. Though many of the discontinued programs have been resurrected under new titles, the government has balked at keeping hefty fiscal promises, such as $538 million to Ontario (to help phase out coal-fired power plants) and $328 million to Québec (to help implement its climate change plan). Public comments by former Environment Minister Rona Ambrose and Natural Resources Minister Gary Lunn made it clear that "less Kyoto, more Washington" is the preferred approach. A "made in Canada solution" emerged as the mantra for the development of a new set of policies, which include an overhaul of the Canadian Environmental Protection Act (CEPA) and a focus on air and Great Lakes pollution; some critics are already labeling it a "made in Washington" approach (the review of CEPA is not a Conservative initiative—CEPA is subject to mandated, scheduled reviews).

The approach of the Harper government differs significantly from the seemingly false promises made by the Chrétien and Martin governments, and the Harper government is even less willing to direct onerous responsibilities toward the large final emitters that contribute just under half of all Canadian emissions. Large final emitters are found in the primary energy production, electricity production, mining, and manufacturing sectors. This covers about 700 companies operating in Canada; approximately 90 of these companies account for approximately 85 percent of the large final emitters' GHG emissions. Cuts have included the much-publicized One Tonne Challenge, forty public information offices across the country, several scientific and research programs on climate change, and a home conservation rebate plan. Government websites devoted to climate change policy remained "under construction" ten months after the Harper government assumed office. The Clean Air and Climate Change Act that was tabled in October 2006, which promised to achieve an absolute reduction in GHG emissions between 45 and 65 percent from 2003 levels by 2050, was loudly rejected by environmentalists, opposition parties, and even many industry representatives as far too little, far too late for effective action.

Of course, the death of Kyoto has long been predicted by many observers, especially once the administration of George W. Bush rejected the treaty (Soroos 2001). Critics argue that the Kyoto Protocol suffers from several embedded problems, including a lack of mandatory GHG reduction requirements by key developing countries with rapidly expanding economies and a reliance on market mechanisms to control GHG emissions with insufficient infrastructure to avoid corruption. Without a Herculean effort and the complete participation of every provincial and municipal government, Canada's commitment of 6 percent below 1990 levels will be an embarrassing failure; even Liberal officials said as much prior to ratification. Herb Dhaliwal, who later served as Minister of Natural Resources in the Martin government, noted in September 2002: "Canada has no intention of meeting the conditions of the Kyoto Protocol on greenhouse gases even though the government hopes to ratify it this fall" (Cheadle 2002).

The Harper government has argued that it faces the stark choice of admitting defeat in terms of the specific goals, or of pretending Canada can meet the targets and facing certain embarrassment at a later date. In this context they can at least be commended for an honest assessment and statement of their capabilities. Environmentalists argue this is no excuse for not taking more robust short-term action. Cutbacks to Liberal Kyoto-inspired programs have proceeded at breakneck speed under the Harper government, ostensibly to make budgetary room for a tax subsidy for citizens willing to take public transit on a regular basis. Nearly every statement from Baird, Ambrose, and Lunn about Kyoto has at least mentioned the sheer futility, and implied folly, of trying to meet the original goals. Environmental groups were not even permitted to join the Canadian delegation to the conference of the parties for the UNFCCC in Nairobi, Kenya, in 2006. When Ambrose arrived she was jeered by international NGOs in attendance as Canada was awarded the infamous "fossil of the day" award distributed by the Climate Action Network; similar jeering from the environmental NGO community was heard in Bali a year later.

Leadership Questions and Questionable Leadership

Many casual observers have been hampered by the erroneous belief that federal or national leadership is just a matter of time on climate change. The leap of faith here is that visible extreme weather events will force politicians to lead. Yet political logic suggests that politicians will not cut off the branches on which they sit. In the Conservatives' case, this includes the oil wealth and tar sands development in Alberta, and an ideological platform encouraging deregulation. Short-term and relatively minor infusions of cash into research and development aside, we will not see major leadership initiatives by Ottawa on climate change. The jurisdictional limitations imposed by federalism described above further hamper any serious effort to develop a comprehensive national approach. In short, a lack of Canadian national leadership on climate change issues is a safe assumption to make looking forward.

Given the lack of federal leadership, a decentralized vision of a climate change action strategy has begun to emerge. This remains one of the most sensitive political issues in a federalist state. For example, if much of the action on climate change will take place at the city level, the federal government needs to find innovative ways to support local initiatives without soliciting provincial territorialism. This is, however, easier said than done. Nonetheless, mayoral leadership in the United States is impressive: more than 800 American mayors have signed the U.S. Mayors Climate Protection Agreement, and more than 100 Canadian municipalities belong to the Climate Change Program run by the International Council for Local Environmental Initiatives (Gore and Robinson, this volume).

Further, transgovernmental initiatives between the Canadian provinces and U.S. states may hold greater promise than multilateral ones at this stage, since the Kyoto commitment demands a national response that is simply not forthcoming. For example, in 2001 the New England Governors and the Eastern Canadian Premiers adopted a joint Climate Change Action Plan, committing to reduce GHG emissions to 1990 levels by 2010, and 10 percent below 1990 levels by 2020 (Selin and VanDeveer, this volume). Also, California is leading an initiative among Western U.S. states and Canadian provinces such as British Columbia and Québec that include discussions about establishing a regional GHG emissions trading scheme. It is time to recognize that most leadership on this issue is not national but provincial and municipal—Canadians cannot rely upon the inevitability of national leadership. Nonetheless, such action must involve extensive consultative relations with the federal government.

Federal-Provincial Coordination

In Canada, the lead-in to the negotiations of the Kyoto Protocol is often referred to as a textbook example of how *not* to conduct the complex interplay between foreign affairs commitments and federal-provincial relations. Arguably, it was a squandered opportunity for serious cooperation. There are of course several interpretations of this, with some portraying the provinces' recalcitrance as the main culprit and others insisting Ottawa made all the wrong moves in its lackluster effort to achieve provincial harmony.

From an outsiders' viewpoint, it is clear that both the federal government and several provincial governments are to blame for the essential disconnect. The 1995 National Action Program in Climate Change, resulting from federal-provincial ministerial dialogue, did nothing to decrease emissions, which were almost 10 percent above 1990 levels in 1999. A 1997 agreement, sans Québec, to stabilize emissions by 2010 had some promise, but the federal government unilaterally declared its intention to agree to a 3 percent reduction instead of stabilization at Kyoto. Once there, it went a step further, effectively doubling that commitment to 6 percent. No doubt this description misses much of the nuance behind the process, but it remains an event that most provincial historians note as a federal betrayal (Macdonald and Smith 1999–2000).

As always, relations between Canada and the United States and their public optics are interesting facets of the story. Much of the Canadian federal oscillation during the Kyoto negotiations seemed to be predicated on shifts within the Clinton administration. Likewise, Harper is sensitive to the popular suspicion that his approach is based largely on the Bush administration's approach to energy policy and changing U.S. policies are likely to engender Canadian federal reactions. Also constant are concerns that a majority government will be impossible to achieve if the rest of Canada perceives the administration as excessively Albertan. But open consultation and, on some key issues, negotiation with the provinces will be both essential and strained. A rapprochement between Ottawa and Québec since 2007 will be tested as Québec's moral high ground on climate change (afforded by its immense hydropower development) is particularly irksome to westerners.

There is more than mere political territoriality involved in differing federal and provincial perspectives. When it comes to GHG reductions, common terminology may be found, but common understandings will be a much more painful and localized process. For example, Manitoba claims it is on a path to exceed Canada's Kyoto commitments in large part because of its use of hydropower, but there remains ample controversy over the exact level of GHG contributions made by hydropower. This is a fact that Québec also routinely ignores in its often self-congratulatory assessment of its leading role on renewable energy and climate change issues. Indeed, it is regularly assumed that there are commonly agreed-upon methodologies for measuring emissions, and even this is false.

Similarly, domestic debates over carbon sequestration and sinks leave room for innovation and compromise. Canada could push its expertise through international organizations and educational development. A cost-sharing agreement reached between Alberta and Ottawa in 2007 to implement carbon sequestration technology in the tar sands region is simultaneously promising to some and viewed as a stalling tactic or federal favoritism toward Alberta by others. The United Nations International Energy Agency estimates that Canadian sequestration deposit sites could hold 1,300 billion tons of carbon dioxide; even offshore injection is a "live option," but startup costs are very high (Sheppard 2007). An agreement with the Obama administration on these technological developments is possible by 2010.

One possibility is that a national cap-and-trade scheme can be established that unites provincial jurisdictions (though under the Harper government this would likely entail intensity-based caps only). Also, a Montreal Exchange futures market for carbon dioxide emissions (a joint venture with the Chicago Climate Exchange) was established in the spring of 2008. This should be received cautiously, however, as the growing number of emissions trading systems in North America and elsewhere has generated an entirely new field of economics, based almost entirely on derivatives and futures. The idea is borrowed from U.S. efforts to reduce air pollution, but its Kyoto variations have spawned a frenzy of potential investors, chartered accountants, financial advisers, and lawyers. It is speedily becoming a major industry in itself.

However, it would be imprudent to put too much stock in emissions trading as either a profitable activity or a sincere effort to reduce global warming. This is a market-based incentive compromise that is both lambasted by the right (who feel it gives undue credit to overpopulated developing states and deindustrialized Cold War losers) and the left (who view this as yet another way to escape the demands of emissions reductions at home and carry on business as usual). Québec's historically contentious relationship with Ottawa will only be exacerbated if a GHG trading system does not recognize its lower emission base and investment in hydropower. Even the Clean Development Mechanism (CDM), an international mechanism under the Kyoto Protocol generating tradable carbon allowances, has raised serious concerns among environmentalists that it could be used to "avoid Kyoto action" while contributing to projects in the southern hemisphere with dubious ecological and human rights implications (Suzuki Foundation 2003).

Ultimately, the debate over the most appropriate political level of environmental governance in Canada may be fruitless. It is clear that all levels are heavily involved and none has sufficient leadership capacity to firmly take the helm. Local initiatives, which are flourishing, offer the best hope for an effective GHG reduction program. Ronnie Lipshutz (1994) offered five essential arguments in favor of local approaches that focus on the bioregional level of implementation. They allow for the scale and practices of ecosystems; more effectively assign property rights to local users of resources; locate local and indigenous knowledge; increase participation of stakeholders; and display greater sensitivity to feedback. But there is no doubt that some form of national or federal level leadership is instrumental, since "pollution lies substantially within federal jurisdiction. Pollution and the protection of habitat are very much a part of providing peace, order, and good government" (Paehlke 2001, 118).

Edward Parson (2001, 355) concluded a major research project on environmental governance with the thought that "a promising direction for resolving competing claims of environmental authority at multiple scales would be to construct cross-scale networks of shared authority and negotiated joint decisions that mirror the complex cross-scale structure of environmental issues. Canada's loose federal structure may facilitate such an approach, or indeed compel it if redrawing the lines of formal environmental authority is out of the question." He adds that the Canadian Council of Ministers of the Environment held such promise in the 1980s and early 1990s, as it "helped build technical capacity in smaller jurisdictions; it invested provincial and territorial officials with a national perspective when they held the rotating chair; and it provided key research and analysis to address technically challenging problems shared by multiple jurisdictions" (Parson 2001, 355).

The Canadian Council of Ministers of the Environment could be rejuvenated if there was political will to do so. This body includes federal, provincial, municipal, and aboriginal participation, and in the case of an issue so obviously global in scope, the participation of the foreign policy community is essential as well; in total this has been referred to as the "microfederalism of environmental policy" (Gillroy 1999, 360). Given the lack of national leadership, this is not necessarily a bad thing. Some combination of unwieldiness and pragmatic cooperation is a hallmark of democracy, and few of us are convinced of the need for radical centralization at this stage. Even with regional variations, public opinion is fairly strong on the issue of climate change, and NGOs can keep genuine pressure on politicians at the federal, provincial, and municipal levels to engage in serious discussions.

Looking Ahead: Action and Adaptation

This chapter is written in a time of great flux in Canadian climate change politics, as yet another minority Conservative government has been elected at the national level in October 2008, and policy developments at the provincial level are ongoing. However, some longer-term factors can be discussed. In general, climate change may produce some nasty surprises, but uncertainty is also an opportunity to promote climate change policy by directly appealing to threats-to-livelihood issues.

There is a need to start talking openly about adaptation to climate change (Bell 2006), a topic the Inuit will no doubt become very familiar with as their way of life is further altered by climatic shifts. Some scholars have been doing this for some time (Pielke 1998), but generally it has been taboo among environmentalists to seriously discuss adaptation, since it implies resignation to the fate of global warming and might discourage more active prevention programs. The norm of stopping global warming is pitted against the relatively mild, even acquiescent need to limit human damage, and naturally the former appears more robust. This is often referred to as "norm entrapment" (Risse 2000). However, given the lack of political leadership, the immensity of the problems associated with climate change mitigation, and the continued drive for industrialization, adaptation will become one of the more pressing policy concerns Canadians will face in coming decades.

More importantly, however, openly discussing adaptation will frame the issue as a mainstream concern, and provoke more reasonable demands on successive federal governments to begin assessing and thinking aloud about these issues. The Arctic Climate Impact Assessment concluded that air temperatures in Alaska and western Canada have increased as much as three to four degrees Celsius in the past fifty years, leading to an estimated 8 percent increase of precipitation across the Arctic; as this falls as rain it increases snow melting and dangers of flash flooding. Melting glaciers, reductions in the thickness of sea ice, and thawing of permafrost are all occurring, and they will in turn exacerbate the warming trend (ACIA 2005). Rather quickly, the Arctic has become what is perhaps the most visible climate-related issue-area for Canadians, illustrated by a *TIME* magazine cover depicting a lonesome and, perhaps, doomed polar bear (Nilsson, this volume): "Should the Arctic Ocean become ice-free in summer, it is likely that polar bears and other northern species would be driven toward extinction" (Canada 2005).

Arctic disturbance also raises various national and international security issues for Canada. Both opponents and proponents of more aggressive political action on climate change and the Kyoto Protocol seek to publicize this fact. Oil and gas companies will strive to demonstrate their ecological consciousness with television commercials; environmental advocacy groups will use the Artic as a platform to raise broader awareness of their concerns; the military will request additional funding for increased surveillance of ice-free Arctic waters. What might get lost in all of these competing perspectives and issues, however, are the actual effects of climate change upon northern indigenous peoples and communities. Here we have both a local constituency, albeit a small one, and a global human rights concern that could prove to be a great embarrassment for an ostensibly progressive country such as Canada.

While improving, Canadian commitment to solar power, wind power, geothermal activities, and hydrogen fuel cell development has also been limited. Although wind technologies are beginning to penetrate utility markets and catch the eye of domestic policymakers, and companies such as Ballard have emerged as world leaders in the development and employment of fuel cell technologies, the Canadian International Development Agency remains actively engaged in developing the oil and gas sector abroad, from Bolivia to Kazakhstan. Given the immense potential for solar power and biomass development in Africa and elsewhere, it might be wise at this juncture to investigate more seriously the option of redirecting resources into these emerging fields.

Considering Canada's potential contribution on international technology transfers, and that any global agreement based on emissions reductions may prove futile in the face of expanding industrialization in Asia and Latin America, it is in

Canada's long-term interest to encourage states to either limit or rapidly bypass the oil-based technological culture that characterized North American and European development. The current federal government's disdain for the Clean Development Mechanism need not preclude more creative, non-Kyoto-based efforts to promote clean energy technologies abroad. For example, many Canadian firms are moving toward carbon-neutrality in their operations by planting trees and developing other efforts to offset their GHG emissions (Hoag 2006).

Meanwhile, a combination of economic globalization and climate change increases the likelihood of bio-invasions at both microbial and species levels (Price-Smith 2002). There is evidence that warming trends will induce North American species migration northward, raising concerns about disease and threats to native species (Hughes 2000). However, such "unassisted migration" will prove difficult for rare species of plants and trees, and adaptation or extinction is likely (Iverson, Schwartz, and Prasad 2004). Not so for many troublesome species. For example, a warming climate may cause migration of zebra mussels, which have already clogged entire swaths of the Great Lakes. Increased flooding could also expand zebra mussel territory even further (Kolar and Lodge 2000; Sutherst 2000).

Furthermore, warming patterns have vastly extended the range of the mountain pine beetle, ravaging Yoho National Park in British Columbia and threatening forests in Washington; officials in Alberta are "setting fires and traps and felling thousands of trees in an attempt to keep the beetle at bay" (Struck 2006).[1] A government official involved in the negotiations with the United States over softwood lumber tariffs even mentioned the possibility that the 2006 agreement was provoked at least partly by the pine beetle—or, rather, the urgent need to clear potentially infected forests, paving the way for resuming large-scale exports.

Conclusion

The nature of Canadian federalism and extreme stakeholder complexity will continue to characterize climate change policy discussions and actions. Indeed, this chapter has only touched on the range of actors necessary for a serious, post-Kyoto Canadian effort to take shape for the purpose of combating global warming. It is certainly necessary to involve small businesses, educational institutions, the NGO community, heavy industry, foreign investors, aboriginal groups, tourist industries, building and renovation industries, the Canadian military, labor unions, all levels of government, and many more organized and unorganized actors. Indeed most of these sectors will involve themselves, with or without invitation, through courting public opinion.

The big question may well be whether the Conservatives—or indeed any of the major national political parties—have the legitimacy or ability to pursue such a broad agenda when it is related to a topic they do not find particularly galvanizing in the first place. Many citizens have adapted a Kyoto-based litmus test for environmental concern, and await national leadership to get there. Though the Liberal Party adopted a "Green Shift," including a proposal for a carbon tax, as its central election plank in 2008, its defeat at the polls has returned climate change to its usual secondary status at the federal level. Given the evident leadership vortex on this issue, it is time to look elsewhere for both leadership and cooperative possibilities; indeed this is happening with unprecedented frequency within civil society and the private sector.

Thankfully, Canadians have more than Kyoto with which to approach issues of climate change mitigation and adaptation. Transgovernmental and community-level programs can both set regulatory examples and reduce GHG emissions. There are also many other international agreements related to technology, trade, human rights, the law of the sea, and biodiversity conservation that have a direct or indirect impact on shaping climate change–related policies. Canadians would be remiss to mourn the failure of Kyoto without some optimistic referral to the opportunities other multilateral arrangements offer (Doelle 2006). In some cases, there is a blatant advocacy role; the Coalition of Small Island States has thrust global warming onto the human rights agenda, and Canadian Inuit and other northern dwellers have begun a similar process. In other cases there are incidental benefits. For example, efforts to curtail the loss of biodiversity must be explicitly tied to habitat preservation, which protects carbon sinks.

But none of this subverts the fact that the "Kyoto gap" between pledged promises and reality will increase as long as Canada pursues increased fossil fuel production (McKenzie 2002, 230). The Albertan tar sands have become nothing short of an oil rush, drawing capital and further entrenching Canada's dependence on the U.S. economy. The goal of producing a million barrels of oil per day was set in 1995 and surpassed in 2004; 5 million barrels a day is projected for 2030 (Woynillowicz 2005). According to the Canadian Association of Petroleum Producers, by 2008 the oil sands accounted for nearly 45 percent of Canada's oil production.[2] Meanwhile, deep oil sands development will even outstrip the surface mining for conventional tar sands, scarring the Boreal Forest and contaminating groundwater in the process. The U.S. oil market will remain voracious for decades to come, and the Canadian economy under the North American Free Trade Agreement remains highly dependent on this source of income.

At the same time, a strong federal climate change mitigation or adaptation program has yet to be developed, and a growing number of provincial initiatives have assumed primacy. If Canada is reluctant to further embrace the Kyoto Protocol and associated political efforts, it can nevertheless improve the odds of climate change mitigation and adaptation by pursuing a robust and bold sustainable-development agenda that is multilaterally oriented. Most Canadians, still convinced Canada is or could be a world leader in environmental policy, would support this.

Notes

1. The pine beetle has swept across British Columbia and scientists fear it will "cross the Rocky Mountains and sweep across the northern continent into areas where it used to be killed by severe cold.... U.S. Forest Service officials say they are watching warily as the outbreak has spread." The United States is less vulnerable because it "lacks the seamless forest of lodgepole pines that are a highway for the beetle in Canada" (Struck 2006).

2. Industry facts and information from the Web site of the Canadian Association of Petroleum Producers, http://www.capp.ca.

References

ACIA. 2005. *Arctic Climate Impact Assessment*. Cambridge: Cambridge University Press.

Bell, Ruth. 2006. What To Do about Climate Change. *Foreign Affairs* 85(3): 105–113.

Boardman, Robert, ed. 1992. *Canadian Environmental Policy: Ecosystems, Politics and Process*. Oxford: Oxford University Press.

Brownsey, Keith. 2008. The New Oil Order: The Staples Paradigm and the Canadian Upstream Oil and Gas Industry. In *Canada's Resource Economy in Transition: The Past, Present, and Future of Canadian Staples Industries*, edited by Michael Howlett and Keith Brownsey. Toronto: Emond Montgomery Publications.

Bruce, James, and Stewart Cohen. 2004. Impacts of Climate Change in Canada. In *Hard Choices: Climate Change in Canada*, edited by Harold Coward and Andrew J. Weaver. Waterloo: Wilfred Laurier Press.

Canada, Government of. 2005. *Action on Climate Change: Considerations for an Effective International Approach*. Discussion paper for the preparatory meeting of ministers for Montreal 2005: UN Climate Change Conference. Ottawa: Environment Canada and Foreign Affairs Canada.

CESI. 2007. *Canadian Environmental Sustainability Indicators, 2007*. Ottawa: Environment Canada.

Cheadle, Bruce. 2002. Canada to Sign Kyoto, but Won't Abide by It. *Toronto Star*, September 5.

Cohen, Stewart, Brad Bass, David Etkin, Brenda Jones, Jacinthe Lacroix, Brian Mills, Daniel Scott, and G. Cornelis van Kooten. 2004. Regional Adaptation Strategies. In *Hard Choices: Climate Change in Canada*, edited by Harold Coward and Andrew J. Weaver. Waterloo: Wilfred Laurier Press.

Doelle, Meinhard. 2006. *From Hot Air to Action? Climate Change, Compliance and the Future of International Environmental Law*. New York: Carswell.

Doern, Bruce, and Thomas Conway. 1995. *The Greening of Canada*. Toronto: University of Toronto Press.

Dwivedi, O. P., Pat Kyba, Peter Stoett, and Rebecca Tiessen. 2001. *Sustainable Development and Canada: National and International Perspectives*. Toronto: Broadview Press.

Environment Canada. 2007. *National Inventory Report 1990–2005: Greenhouse Gas Sources and Sinks in Canada*. Gatineau, Québec: Environment Canada.

Gillroy, John. 1999. American and Canadian Environmental Federalism: A Game-Theoretic Analysis. *Policy Studies Journal* 27(2): 360–388.

Harrison, Kathryn. 1996. *Passing the Buck: Federalism and Environmental Policy*. Vancouver: University of British Columbia Press.

Hoag, Hannah. 2006. Helping Companies Clean Up Their Acts. *Montreal Gazette*, November 7.

Howlett, Michael, and Keith Brownsey. 2008. *Canada's Resource Economy in Transition: The Past, Present, and Future of Canadian Staples Industries*. Toronto: Emond Montgomery.

Hughes, Lesley. 2000. Biological Consequences of Global Warming: Is the Signal Apparent Already? *Trends in Ecology and Evolution* 15(1): 56–61.

Hunter, Justine. 2008. Provinces Free to Tackle Climate, Ottawa Says. *Globe and Mail*, February 21.

Iverson, Louis, M. W. Schwartz, and Anantha M. Prasad. 2004. How Fast and Far Might Tree Species Migrate in the Eastern U.S. Duc to Climate Change? *Global Ecology and Biogeography* 13(3): 209–219.

Kolar, Christopher, and David Lodge. 2000. Freshwater Nonindigenous Species: Interactions with Other Global Changes. In *Invasive Species in a Changing World*, edited by Harold Mooney and Richard J. Hobbs. Washington, DC: Island Press.

LePrestre, Philippe, and Peter Stoett, eds. 2006. *Bilateral Ecopolitics: Continuity and Change in Canadian-American Environmental Relations*. Aldershot, UK: Ashgate.

Lipschutz, Ronnie. 1994. Bioregional Politics and Local Organization in Policy Responses to Global Climate Change. In *Global Climate Change and Public Policy*, edited by David L. Feldman. Chicago: Nelson-Hall.

Lonergan, Steve. 2004. The Human Challenges of Climate Change. In *Hard Choices: Climate Change in Canada*, edited by Harold Coward and Andrew J. Weaver. Waterloo: Wilfred Laurier Press.

Macdonald, Douglas, and Heather A. Smith. 1999–2000. Promises Made, Promises Broken: Questioning Canada's Commitments to Climate Change. *International Journal* 55(1): 107–124.

McBean, Gordon. 2006. An Integrated Approach to Air Pollution, Climate and Weather Hazards. *Policy Options Politiques* 27(8): 18–24.

McKenzie, Judith. 2002. *Environmental Politics in Canada: Managing the Commons into the Twenty-First Century*. Oxford: Oxford University Press.

Paehlke, Robert. 2001. Spatial Proportionality: Right-Sizing Environmental Decision-Making. In *Governing the Environment: Persistent Challenges, Uncertain Innovations*, edited by Edward A. Parson. Toronto: University of Toronto Press.

Parson, Edward A. 2001. Persistent Challenges, Uncertain Innovations: A Synthesis. In *Governing the Environment: Persistent Challenges, Uncertain Innovations*, edited by Edward A. Parson. Toronto: University of Toronto Press.

Paterson, Matthew. 2001. Climate Change as Accumulation Strategy: The Failure of COP6 and Emerging Trends in Climate Politics. *Global Environmental Politics* 1(2): 10–17.

Pielke, Roger. 1998. Rethinking the Role of Adaptation in Climate Policy. *Global Environmental Change* 8(2): 159–170.

Price-Smith, Andrew. 2002. *The Health of Nations: Infectious Disease, Environmental Change, and Their Effects on National Security and Development*. Cambridge, MA: MIT Press.

Risse, Thomas. 2000. Let's Argue! Communicative Action in World Politics. *International Organization* 54(1): 1–39.

Roe, Emery. 1994. Global Warming as Analytic Tip. In *Global Climate Change and Public Policy*, edited by David L. Feldman. Chicago: Nelson-Hall.

Sancton, Andrew. 2006. Cities and Climate Change: Policy-Takers, not Policy-Makers. *Policy Options* 27(8): 32–34.

Sheppard, Robert. 2007. Piping Carbon Back into the Ground. *CBC News*, March 9.

Simeon, Richard, and Ian Robinson. 2004. The Dynamics of Canadian Federalism. In *Canadian Politics*, edited by James Bickerton and Alain-G. Gagnon. Toronto: Broadview Press.

Soroos, Marv. 2001. Global Climate Change and the Futility of the Kyoto Process. *Global Environmental Politics* 1(2): 1–9.

Struck, Doug. 2006. "Rapid Warming" Spreads Havoc in Canada's Forests: Tiny Beetles Destroying Pines. *Washington Post*, March 1.

Sutherst, Robert. 2000. Climate Change and Invasive Species: A Conceptual Framework. In *Invasive Species in a Changing World*, edited by Harold Mooney and Richard J. Hobbs. Washington, DC: Island Press.

Suzuki Foundation. 2003. *Risky Business: How Canada Is Avoiding Kyoto Action with Controversial Projects in Developing Countries*. Vancouver: Suzuki Foundation.

Winfield, Mark. 1994. The Ultimate Horizontal Issue: The Environmental Policy Experiences of Alberta and Ontario, 1971–1993. *Canadian Journal of Political Science* 27(1): 129–152.

Winfield, Mark, and Douglas Macdonald. 2008. The Harmonization Accord and Climate Change Policy: Two Case Studies in Federal-Provincial Environmental Policy. In *Canadian Federalism: Performance, Effectiveness, and Legitimacy*, edited by Herman Bakvis and Grace Skogstad. Oxford: Oxford University Press.

Woynillowicz, Dan. 2005. *Oil Sand Fever: The Environmental Implications of Canada's Oil Sand Rush*. Calgary: Pembina Institute.

II

States and Cities Out Front

4

Second-Generation Climate Policies in the States: Proliferation, Diffusion, and Regionalization

Barry G. Rabe

Introduction

The conventional depiction of the American governmental role in climate policy reflects disengagement, based on the federal government's withdrawal from the Kyoto Protocol and a prolonged inability to consider serious policy initiatives. The United States is, however, a federal system. Thus an examination of the American government's engagement on climate policy shifts significantly when one expands the perspective to include state governments. This chapter examines trends in state climate policy formation and implementation in the United States, which include the following developments: a continuing proliferation of a diverse array of greenhouse gas (GHG) reduction policy measures; the diffusion of particular tools, such as renewable energy mandates, in a large number of states; and multistate collaboration, which brings a regional dimension to these state efforts. The chapter also examines alternative venues for state climate policy development, including direct democracy through ballot propositions and litigation through elected attorneys general. It concludes with a comparison of the evolving U.S. system with other multilevel governance systems such as the Canadian system, along with a discussion of the potential stumbling blocks that may arise from a bottom-up approach to policy development as the federal government becomes more engaged.

More than fifteen years after the adoption of the Rio Declaration on Environment and Development and over a decade after the signing of the Kyoto Protocol, the U.S. federal government has maintained an inability to enact climate change policy. The U.S. Congress held 175 hearings on climate change between 1975 and 2006 but repeatedly deflected legislative proposals that would have established modest targets for containing the growth of GHG emissions from major sources. Congressional passage of any one of these bills would likely have been blocked by a presidential veto. Congressional deliberations increased markedly in the 110th Congress (2007–2009), but they did not result in any clarity on expanding the federal role, and there was only preliminary consideration of how any future federal engagement might influence existing state policies. Indeed, the climate policy stance of the federal government as of 2008 may be characterized as a blend of international disengagement and a continuing reluctance to enact any domestic policies beyond research on the issue and voluntary efforts to reduce GHGs, although the November election of Barack Obama to the presidency raised anew the possible expansion of the federal role.

This summary, though, does not provide a complete picture of the evolving American governmental role in climate policy development and implementation. At the very time that federal institutions continued to stall, major initiatives were launched, often with bipartisan support, in a number of state capitals, including Sacramento, California; Carson City, Nevada; Austin, Texas; Springfield, Illinois; Harrisburg, Pennsylvania; and Albany, New York. By late-2008, nearly half of all U.S. states were actively involved in climate change, implementing two or more from a set of eight major policies that promised to significantly reduce their levels of GHG emissions (Rabe 2008). Virtually all the states were beginning to at least study the issue of climate change and explore modest remedies, while some—such as California, Connecticut, New Jersey, and New York—were equally engaged on multiple policy fronts as many European countries.

As these state programs begin to move from enactment into early implementation, they are clearly having some effect on stabilizing emissions in their jurisdictions. Indeed, many states are major sources of GHG emissions, with considerable potential for reduction. If the fifty states were to secede and become sovereign nations, thirteen of them would rank among the top forty nations of the world in GHG emissions, led by Texas in seventh place between Germany and the United Kingdom (Rabe 2004). This policy reality emphasizing the GHG emissions of states has an increasingly international aspect. There have been active negotiations among seven states in the American West and four Canadian provinces to establish a unified region for a carbon cap-and-trade program, known as the Western Climate Initiative. State governments in the United States have also begun to explore avenues of possible collaboration with other nations in addition to Canada.

There are, of course, limitations on what U.S. states, acting individually or collectively, can do to reverse the steady growth of GHG emissions. For example, states face constitutional constraints on restricting commercial transactions that cross state boundaries. Moreover, states cannot encroach on regulatory areas in which the federal government has established a primary role, such as setting standards on fuel economy for motor vehicles or sustaining extensive subsidies and incentives for the continued use of fossil fuels. States may also face disincentives for taking unilateral early action and reducing emissions, without guarantees that their steps will receive credit under subsequent federal regulatory programs. Given these limitations and the indifference to climate change at the federal level, why would so many states attempt to take action, redefining American climate policy in the process?

This chapter responds to that question and considers both the historic role of U.S. states in national policy development and the particular drivers that appear pivotal in the case of climate policy. The chapter also chronicles the evolving role of states in developing climate policy, with particular attention to new trends that appear to be emerging in the latter years of the first decade of the twenty-first century. In addition, our discussion in this chapter looks ahead to consider possible limitations facing state-driven policy as well as the opportunities for state-level developments to continue to expand and ultimately define a unique American response to the enormous challenge of devising climate policy.

The continuing expansion of the states' role in U.S. climate change policy underscores the importance of examining both national and subnational engagement in evaluating policy responses from multilevel systems of government. Indeed, formal commitment to international agreements, such as ratification of the Kyoto Protocol, thus far means little in terms of translation into actual policy and attendant reduction of GHG emissions. Among neighboring North American countries, Canada continues to occupy the moral high ground on climate policy through its Kyoto ratification. But Canada's rate of GHG emissions growth continues to outpace the United States; its decentralized system of environmental governance has resulted in far less provincial policy development compared with that of the U.S. states (Selin and VanDeveer, this volume; Stoett, this volume). Comparisons between U.S. and EU governance systems suggest far more parallels in terms of climate policy development, and variations in commitment to implementation, than is commonly assumed (Rabe 2008).

Policy from the Bottom Up

Many accounts of American public policy are written as if the United States operated as a unitary system, with all innovations and major departures emanating from the federal government. However, a more nuanced view of American federalism demonstrates that states have often served a far more expansive and, at times, visionary role. Many major developments of the late nineteenth and early twentieth centuries, from child labor reform to women's voting rights, were launched through movements in which policy was first formed in various state capitals (Peterson 2003; Walker 2000). In environmental policy, many states took early action on particular concerns years or even decades in advance of the first Earth Day and the advent of federal command-and-control regulatory programs. Indeed, some current federal policies that are widely heralded for their effectiveness are based in large part on earlier state experiments, such as the Toxics Release Inventory that mandates toxic pollution disclosure and market-based emissions trading programs for air contaminants.

The potential for early state engagement on policy issues has only intensified in recent decades. Many states have either drafted entirely new constitutions or dramatically revised existing ones to expand areas for state policy involvement (Rosenthal 2004; Teske 2004). This has led to dramatic increases in state revenues and the expansion of state agencies with oversight in areas relevant to GHGs, including environmental protection, energy, transportation, and natural resources. Even in areas with significant federal policy oversight, states have become increasingly active and, in some cases, fairly autonomous in interpretation and implementation. States collectively issue more than 90 percent of all environmental permits in the United States, complete more than 75 percent of environmental enforcement actions, and rely on their own fiscal sources for more than 75 percent of funding (Rabe 2004).

Extending such resources and powers into the realm of climate change can be a fairly incremental step. For example, in regulation of the electricity sector, the state government role has been dominant for decades (Gormley 1983). The burgeoning state role is not merely an extension of existing authority, but rather a new kind of movement driven by factors distinct to the issue of climate change. These factors have proven increasingly effective in building broad coalitions for action that frequently cross partisan divides. In some jurisdictions, this dynamic has advanced so far that one of the greatest conflicts in climate policy innovation is determining which political leaders get to "claim credit" for taking early steps. In California, for example, Republican Governor Arnold Schwarzenegger has grappled with Democratic legislators, such as Assemblywoman Fran Pavley and Democratic Assembly Speaker Fabian Nunez, to determine whose name will become synonymous with far-reaching climate change initiatives that are almost unthinkable at the federal level but have proven quite popular in Sacramento (Farrell and Hanemann, this volume). The following factors appear to be pivotal in driving much state action on climate policy.

Immediate Signs of Climate Impact

While climate science has been the subject of acrimonious debate in Washington, D.C., policy circles, individual states have begun to feel the impact of climate change in more immediate ways. These various impacts are often brought to the states' attention by state-based researchers, who work cooperatively with state regulatory agencies to discern early and localized indicators of climate impact. In the coastal states, for example, concern is often concentrated on the impact of rising sea levels, particularly given the substantial amount of economic development concentrated along many shores located at relatively low sea levels in the United States. This dynamic has influenced state governments from Honolulu, Hawaii, to Trenton, New Jersey.

In other states, the greatest concern is the decline and loss of key species, due to the migration of plants, birds, insects, and animals as climatic conditions change. A number of states have focused their concerns on dramatic shifts in weather patterns that put the agricultural sector at considerable risk, particularly the Great Plains states, including Nebraska, North Dakota, and South Dakota. No two states have faced identical experiences, but a common theme has emerged as individual states and regions have faced direct impacts from climate change, thereby taking the climate policy debate from the realm of graphs and charts toward real-life experience that merits a serious policy response.

Economic Development Opportunities

Virtually all states that have responded to the challenge of climate change have done so through methods that they deem likely to reduce GHG emissions but simultaneously foster alternative forms of economic development. States have actively promoted renewable energy through a combination of mandates and financial incentive programs, which have focused upon the development of "home grown" sources of electricity that promise to simultaneously stabilize local energy supplies and promote significant new job opportunities for state residents. Many states with active economic development programs have concluded that investment in the technologies and skills that will be necessary in a less carbonized society in coming decades is a sound bet. In response, they have advanced many policy initiatives in large part due to anticipated economic benefits. Even some states with substantial sectors that generate massive amounts of GHG emissions, such as coal mining and usage in Pennsylvania, have begun to shift their thinking toward the opportunities for longer-term economic development presented by investment in renewable energy (Rabe and Mundo 2007).

State Agency Support and Advocacy

Many states have worked intensively in recent decades to build an in-house capacity to support the environment, energy, and other related areas that are now relevant to climate change. Consequently, state agencies are increasingly fertile areas for advocates of policy change to develop ideas that are tailored to their state's needs and opportunities. These ideas can then be translated into legislation, executive orders, and pilot programs. Such advocates also have proven effective in forming coalitions, often working across partisan lines in the legislature and engaging supportive interest groups where feasible (Mintrom 2000; Rabe 2004). No two states have assembled identical constituencies for climate change, just as no two states have devised identical climate policies.

State agencies have been significant drivers behind innovations in climate policy change, whether in the stages of developing policy ideas, seeing those ideas through to policy formation, or moving into policy implementation. State-based environmental advocacy groups as well as businesses and industries that might benefit financially from climate policy have become increasingly visible and active in bringing about far-reaching initiatives. This has created broader, more supportive coalitions for new policy development, although schisms can emerge among likely allies in finalizing the details of policy design, requiring deft political maneuvering by state officials (Rabe and Mundo 2007).

Entering the Second Generation of State Climate Policies

The sheer volume and variety of state climate initiatives is staggering, difficult to measure with precision, and subject to expansion. Much policy analysis has been so heavily focused on federal- or international-level actions that state or other subnational policies have received markedly less attention. This chapter relies heavily upon the ongoing refinement of climate policy profiles established for all fifty states since early 2000 at the University of Michigan. These profiles have been drawn from elite interviews, government documents and reports, and legislative histories, as well as data gathered by nationally based organizations that represent the state agencies in particular sectors relevant to climate policy, such as the Environmental Council of the States and the National Association of State Energy Officials. The study also utilizes ongoing state policy inventories maintained by the Pew Center on Global Climate Change. The developments and trends seen in this study point toward a "second generation" of state climate policy development.

The Beat Goes On: Continuing Proliferation

Perhaps the most evident trend in state policies on climate change is the sheer number of states involved as well as the aggregate number and range of policies, which continue to grow on a monthly basis. As of late 2008, this trend showed no signs of slowing and appeared to be accelerating in reaction to the federal government's continued inability to engage the issue. Well over half of the states have enacted at least one piece of climate legislation or issued at least one executive order that set formal requirements for reducing GHG emissions; twenty-two states have passed two or more laws designed to achieve such reductions; forty-seven states have completed GHG inventories; and twenty-two states have devised action plans to guide future policy. Six states have formally established statewide reduction commitments that span future years and decades, linked to policies designed to attain these reduction pledges.

Renewable energy, discussed further below, has been a particularly intensive area of engagement. Twenty-eight states have enacted renewable portfolio standards (RPS), which mandate a formal increase in the amount of electricity distributed in a state that must be generated from renewable sources. Furthermore, fifteen states have established their own version of carbon taxes, by applying social benefit charges on electricity and allocating the revenues to renewable energy development or energy efficiency projects. In transportation, fourteen states have agreed to follow the lead of California in establishing the world's first carbon dioxide emissions standards for vehicles if a federal waiver can be secured, and twenty-three states are engaged in some form of a cap-and-trade system for carbon emissions from major industrial sources.

Alongside the sheer magnitude of state policies, these efforts are generally becoming more rigorous. There has been a gradual shift in state policy over the past decade, with voluntary initiatives increasingly supplanted by more ambitious regulatory efforts. Most of these policies retain considerable flexibility in terms of compliance, consistent with the credit-trading mechanisms dominant in most Western governments that have ratified Kyoto. But their regulatory rigor is steadily increasing, along with the likely impact on GHG emissions if they are faithfully implemented. The states may have multiple motivations for pursuing these policies, but they are becoming increasingly explicit and forceful in specifying climate change as a central concern.

This runs somewhat contrary to earlier practice, whereby many states were aware of the potential climate impact of a proposed policy but said little if anything about this element. In these "stealth" cases, such as the RPS that Texas established in 1999, initial emphasis was placed almost exclusively on the nonclimate benefits of enactment (Rabe 2004). That approach has begun to give way as policy proponents see more advantage in being explicit about the GHG ramifications, among other climate impacts, of various policy tools. This is particularly evident among current and recent state governors with prominent national profiles, such as Schwarzenegger, Bill Richardson of New Mexico, and National Governors' Association cochairs Tim Pawlenty of Minnesota and Kathleen Sebelius of Kansas.

The number of states with multiple (and rigorous) policies continues to expand. These states increasingly function like major organizational players in climate policy development and implementation. Many tend to resemble the more active members of multilevel federations that have ratified Kyoto, such as the more engaged member-states of the EU, with sizable climate policy teams of administrators and analysts. These clusters of policy professionals not only develop policy within their own boundaries but also increasingly look to other states (and even some foreign governments) as potential partners. In some instances, this reflects economies of scale among neighboring states in an established region, but it may also represent mutual perceived gains among states in noncontiguous jurisdictions, including those lodged on opposite coasts and in the Great Lakes Basin. Consequently, a growing number of states have begun to operate more like semi-independent entities on an international stage.

Diffusion: Spreading across the States

Much of the existing infrastructure of state climate programs has been individually tailored to the circumstances of a particular state. However, there is increasing evidence that some policies enacted in one state are later replicated in one or more additional states. There is, in fact, abundant precedent in other policy arenas for such policy diffusion to spread across the nation and become, in effect, a de facto national policy (Mossberger 2000). Under such circumstances, it may be possible for the states to simply negotiate interstate differences and implement these interrelated programs. There may also be some tipping point at which diffusion reaches enough states for the federal government to conclude that it should respond by drawing from these state models and establishing a national version of climate policy. In the late 1980s, for example, the Reagan administration actively opposed a federal role in increasing energy efficiency standards for a wide range of household appliances. After more than two dozen states responded with some form of state-specific regulation, the Congress and President Reagan negotiated a federal bill that drew heavily on state experience but preempted all existing state laws in the process.

There are several areas in which enactment of a climate policy in one jurisdiction has already been duplicated elsewhere. In 2000, Nebraska enacted carbon sequestration legislation, designed to promote changes in agricultural practice that could result in less use of fossil fuels in farming and an increased capacity of state-grown crops to sequester carbon through growing plant material. Shortly thereafter, three other states adopted essentially identical legislation, although there was virtually no contact between officials in the respective states during this period. However, the policy tool that appears to be diffusing most rapidly is the renewable portfolio standard, which was established in twenty-eight states and the District of Columbia by late-2008. The first RPS was enacted in 1991 in Iowa, with little if any attention to GHG impacts. Subsequently, the pace of adoption has intensified, with many new programs added and many existing ones significantly expanded since 2002 (Rabe and Mundo 2007). Collectively, these state policies are projected to increase the national level of electricity provided by renewable sources from less than 1 percent in 2000 to approximately 5 percent by 2020, even when including assumptions of increased electricity demand, no additional state or federal policies, and no significant changes in the cost-effectiveness of renewable technology.

Particular RPS features vary by state but all such programs mandate a certain increase over time in the level of renewable energy that must be provided by all electricity providers in a state. For example, Nevada passed legislation in 2005 that requires its two primary utilities, Nevada Power Corporation and Sierra Pacific Power Corporation, to gradually increase their supply of renewable energy over the following decade, ultimately reaching a level of 20 percent by 2015. This legislation passed with unanimous support in both legislative chambers in Nevada and was signed into law by Republican Governor Kenny Guinn. It built on earlier laws enacted in 1997, 2001, and 2003, each expanding the state's commitment to formally promote renewable energy. Nevada, like virtually every other state that has enacted an RPS, provides regulatory utilities with considerable flexibility in finding ways to meet renewable mandates through renewable energy credit programs. These programs function much like other market-based programs and promise to lower compliance costs significantly.

Renewable portfolio standard programs appear likely to continue to diffuse in coming years, reflecting recent legislative enactments and the continuing exploration of this approach as a policy option in a number of other state legislatures. In turn, numerous states with established RPS programs, such as Texas, have found them so successful in terms of their ability to add renewable energy at reasonable costs that they are looking actively to "increase the bar," building on the exponential rate of renewable energy growth of recent years with a substantial increase in future mandate levels (Texas Public Utility Commission 2005). Interestingly, this U.S. state pattern coincides closely with the experience of the EU, where a growing number of nations—including Denmark, Sweden, and the United Kingdom—have adopted RPS programs as central components of their plans for meeting Kyoto Protocol obligations.

This growing trend toward state diffusion may explain why recent U.S. Congresses have begun to consider seriously proposals to enact a federal version of this policy. One growing challenge that has arisen with the proliferation of state RPS is the existence of differential state requirements, ranging from varied definitions of what constitutes renewable energy to state efforts to maximize generation of instate renewable sources for economic development reasons. The former issue poses challenges for renewable energy market development in areas where generators serve multiple states, whereas the latter raises questions of state adherence to the Commerce Clause of the U.S. Constitution (Rabe and Mundo 2007). But any future federal policy will have to address whether to allow for a two-track system that would allow states with RPS to maintain their own policies and, where applicable, exceed federal standards.

Regionalism: Between Nation and State

There is also ample precedent in American federalism for states to work cooperatively on common concerns and, in some instances, to formalize regional approaches involving two or more states (Derthick 1975; Zimmerman 2002). Some regional strategies take a formal structure, such as interstate compacts, which involve an agreement ratified by participating states and ultimately by Congress. These have been used extensively among states that share responsibility for an ecosystem, such as the Great Lakes Commission, which was established in 1955 to promote the environmental well-being of the Great Lakes Basin. Other strategies may entail establishing a multistate organization or commission to facilitate ongoing negotiation over particular issues or memoranda-of-understanding concerning reciprocal policy commitments. An obvious rationale for regional action involves those instances in which participating jurisdictions see a common advantage to working cooperatively rather than independently on a particular policy issue.

As state climate policies proliferate and diffuse, it is entirely possible that certain clusters of states may become, in practice, regions even in the absence of formal agreements. All Southwestern states between California and Texas, for example, have an RPS program. It is increasingly possible to envision interstate trading of renewable energy credits and other forms of cooperation that link these state boundaries and programs. But more formal regional arrangements are also under consideration. Perhaps most notable is an arrangement among the Northeastern states, where relatively small physical size and heavy population densities foster considerable economic and environmental interdependence. States in this region have a strong tradition of working together on issues, whether campaigning for federal air emission standards to deter acid rain or negotiating common regional standards with the U.S. Environmental Protection Agency's office for the region (Scheberle 2004).

For over three decades, the New England governors have further formalized this partnership through an organization, the Conference of New England Governors, that links them in cooperative ventures with the five eastern provinces of Canada. In fact, the respective provincial premiers and governors meet annually, with environmental and energy concerns often paramount. In 2001, the leaders of these jurisdictions, representing five different political parties, agreed to common GHG reduction goals, beginning with a pledge to stabilize at 1990 levels by 2010, reach at least 10 percent below 1990 levels by 2020, and achieve more significant reductions thereafter (Selin and VanDeveer, this volume). These goals are not formally binding, even in Canada where provinces are in fact obligated by Kyoto after federal ratification of the Protocol in 2002. But they have triggered exploration of common strategies and prodded all of the participating states to take more aggressive steps on climate policy than ever before, even though the participating provinces have proven much less capable of backing their pledges with policy initiatives.

Perhaps the most vibrant regional initiative is the Regional Greenhouse Gas Initiative (RGGI) (Selin and VanDeveer, this volume). This effort was launched in April 2003, when George Pataki, at the time governor of New York, invited his counterparts from ten states and the mayor of the District of Columbia to explore the possibilities of establishing a regional cap-and-trade program for reducing carbon dioxide emissions from all fossil-fuel-burning power plants located within the region. Some states like Massachusetts and New Hampshire had already taken action to cap GHG emissions from their own coal-burning plants, while other states in the region, including Maine and New York, explored a number of renewable energy initiatives. But state policy analysts concluded that a regional approach might be most cost-effective. In response, New York reached agreement in December 2005 with six other states (Connecticut, Delaware, Maine, New Hampshire, New Jersey, and Vermont) on a regional cap-and-trade program. Massachusetts, Rhode Island, and Maryland later joined. Development of a model rule and regulatory provisions continued thereafter. The RGGI is designed to cap emissions at 2009 levels through 2014, and then to reduce these 10 percent below that level by 2018.

This process would follow some of the framework for interstate coordination in reducing nitrogen oxides emissions in the northeastern Ozone Transport Region (Burtraw and Evans 2004), although RGGI entails almost exclusively a negotiation among states with no significant input from federal officials. One formal RGGI goal is to establish and implement a regional cap on carbon emissions, while accommodating, to the extent feasible, the diversity in policies and programs in individual states. In that regard, it bears a rather significant resemblance to the EU Emissions Trading Scheme that was launched in 2005 and has triggered "informal contacts between state officials and representatives of the European Commission and European member states" (Kruger and Pizer 2005, 6). California and six other Western states have followed a similar pattern in launching a Western Climate Initiative, including neighboring Canadian provinces and Mexican states in deliberations, as have a series of Midwestern state governors.

Yet another variant of a multistate approach involves an extension of regionalism to include states that are not necessarily contiguous with one another. Under federal air pollution legislation, for example, California enjoys unique status among the fifty states that it can parlay to establish a network of states with regulatory standards more stringent than those of the federal government. This exemption stems from congressional recognition in the 1970s that California had acted in an early and assertive way on confronting air emissions and so was entitled to take any emerging federal air standard as a minimum from which it could establish its own regulations. The remaining states would then be free to adhere to federal standards or to join forces with California, often setting up a dynamic of upward bidding in air regulatory standards.

California chose in July 2002 to revisit those powers, becoming the first government in the world to establish carbon dioxide emission caps for motor vehicles (Farrell and Hanemann, this volume). This took the form of legislation, signed by former Democratic Governor Gray Davis, that emphasized that the goal of the legislation was to control air emissions. This legislation went to considerable lengths to characterize carbon dioxide as a natural extension of earlier regulatory efforts. The state has continued to assert that it is in no way encroaching on fuel economy standards, which clearly remain under federal control. Since the legislation was enacted, the California Air Resources Board has moved toward implementation, which is scheduled to go into effect toward the end of the current decade and could achieve reductions of up to 25 percent in vehicle emissions in future fleets.

At least fourteen states have signaled their support for the California legislation, which has remained subject to prolonged litigious and intergovernmental bargaining as the state attempts to secure a waiver from the U.S. Environmental Protection Agency to pursue this policy. This process creates the very real possibility of two separate regional standards for vehicular emissions, further illustrating the possibilities of multiple states working collaboratively. Of course, the 2002 California legislation was only the prelude for an incredibly ambitious set of subsequent policies, most notably the state's 2006 Global Warming Solution Act, which sets major statewide reduction targets for 2020 and 2050 and introduces a wide range of policies designed to attain these targets (Farrell and Hanemann, this volume).

Direct Democracy: Taking It to the People

The overwhelming majority of state climate policies have been enacted through traditional mechanisms of representative democracy, involving either legislation or executive orders issued by governors. But many state and local governments possess an alternative route for making policy not constitutionally available to Washington, D.C., which is direct democracy through ballot proposition. State constitutions define provisions such as a referendum or an initiative in differing ways, but a common theme is to set before the electorate of a state a question or policy proposal. In the event that a majority of participating voters support the proposition, it becomes law, no different than legislation fashioned through representative institutions. Forty-nine states also have provision for some form of direct democracy for approval of constitutional amendments.

Direct democracy has been an alternative route for policymaking in more than thirty states for nearly a century, reflecting its origins in the Populist and Progressive movements (Gerber 1999; LeDuc 2003). But its use in the American state context has grown at an exponential rate over the past two decades, particularly in controversial arenas such as environmental and energy policy (Guber 2003). In recent decades, states have used the tools of direct democracy to take decisions on such issues as nuclear plant closure and waste management, state land use policy and public land acquisition, and the disclosure and regulation of toxic substance releases, among numerous others. Indeed, state constitutions impose few if any restrictions on the kinds of policy questions that can be addressed through direct democracy. A number of states, such as California and Oregon, make extensive and regular use of this feature.

In November 2004, state climate policy moved from the exclusive realm of representative institutions to the expressed will of the people. Colorado voters, by a 54 to 46 margin, approved Proposition 37, which established an RPS for that state. The ballot proposition set forth an ambitious target for steadily increasing the level of electricity in the state derived from renewable sources—from a current level of approximately 2 percent to 10 percent by 2015. Many other provisions in this legislation are comparable to RPS programs in Nevada and other states. What makes Colorado unique is that proponents turned to direct democracy after repeated efforts to enact such a statute were blocked in the Colorado legislature. Opposition by the state's dominant electric utility company, an affiliate of Xcel Energy, and coal-mining interests, was instrumental in blocking the legislative proposals, in a manner similar to efforts to block climate policy in Congress in the face of ferocious interest group opposition.

Colorado had generally proved to be among the least supportive states in climate policy development more generally during the past decade, due in large part to its major coal mining sector (Rabe 2004). In the RPS case, however, a bipartisan group led by the Republican speaker of the Colorado House of Representatives and a Democratic member of the U.S. House assembled a very broad coalition, attracting a mixture of environmental and agricultural interests, citizen groups, and public health organizations, as well as manufacturers of renewable energy systems that stood to gain from the legislation. Most major media outlets in the state offered strong endorsement. Despite a massive media opposition campaign led by the state's largest utility,which clearly reduced the margin of support as measured through late shifts in polling, the proposition passed and has since passed through an extensive rule-making process.

In numerous other areas of environmental policy, once one state turns to the ballot on a particular issue stalled in representative institutions, others often follow suit. The RPS approach continues to move apace in many jurisdictions, with several states following the Colorado example—through conventional methods—shortly thereafter. But the Colorado case establishes an important precedent and further underscores the possibilities for expanding the state role in climate policy development. Indeed, climate policy proponents in other states, most notably Washington and Missouri, have already replicated the Colorado experience by passing similar ballot propositions of their own to create an RPS after legislative deadlock.

Litigation from State Attorneys General: Taking It to the Courts

Alongside citizen-driven policy, states also have turned increasingly to litigation against their neighbors or the federal government for actions—or inactions—seen to cause environmental harm to their states and citizenries. The vast majority of state attorneys general are elected officials and many of them are very prominent figures in state governance (Provost 2003). They often possess considerable independence from their respective governors and have proven increasingly bold in expanding the definition of what warrants a state litigation strategy. Huge shifts in policy have followed attorney general–led interventions in such areas as regulation of the tobacco and financial services industries (Derthick 2005; Greenblatt 2003a). There are strong indicators that climate policy is emerging as the next target for this type of policy engagement.

In recent years, a loose coalition of attorneys general has formed and begun to explore ways in which they might use litigation to force the federal government to act. For example, in 2003, attorneys general from California, Connecticut, Illinois, Maine, Massachusetts, New Jersey, New Mexico, New York, Oregon, Rhode Island, Vermont, and Washington filed suit in federal court challenging a Bush administration decision to exclude carbon dioxide as a pollutant regulated under the 1990 Clean Air Act Amendments. In 2007, the Supreme Court largely ruled in favor of the states, and this decision remains central to the challenge of California and other states to secure a federal waiver for their proposed carbon emission standards for vehicles. Other initiatives have followed, whereby the lead legal officers of various states contend that climate change is posing a direct environmental and human threat to state residents and seek a judicial remedy that would force some degree of active federal engagement or, in some instances, press localities for expanded activity.

Such steps have often been endorsed and supported by coalitions of environmental groups and state regulatory agencies, which often supply detail and expertise in fashioning the litigation strategy. It remains much too soon to discern what impact, if any, these respective approaches might have, since they move the federal courts into new policy terrain and are likely to receive very different hearings in respective federal judicial districts. Nonetheless, they represent yet another strategy that states appear increasingly willing to employ in assuming a lead role in American climate policy formation. This approach, of course, appears particularly unique in that it is designed not to result in intrastate action or interstate cooperation. Instead, the focus is finding state-based policy levers that might compel a recalcitrant federal government to take action on the climate issue.

Looking Ahead: The Second Generation and Beyond

There is at present no sign of a slowing pace in state engagement on climate change. Long-active states are expanding their efforts and elevating their reduction commitments. Long-dormant states are, in some instances, showing signs of engagement. Consequently, one could increasingly envision an American climate policy system emerging on a bottom-up basis, with a permanent role for states to play in continued policy development and implementation. In certain respects, this appears to parallel the experience in other federal or federated systems. For example, in Australia, which generates a level of GHGs per capita similar to the United States, the six states developed an increasingly diverse array of policy initiatives, all before a 2008 election led to a shift of government and a federal decision to ratify Kyoto.

At the same time, this phenomenon is not universal in federal systems, reflected in the remarkably slow pace of climate policy development in many Canadian provinces despite federal government ratification of Kyoto. This relative lag in Canadian policy development provides a partial explanation as to why Canadian GHG emissions growth has significantly surpassed that of the United States since 1990. Ironically, Canadian provinces retain vast constitutional authority over many policy areas relevant to GHG emissions, with more extensive powers in this arena than their American state counterparts. But they have proven very reluctant to use them, in part because of widespread provincial opposition to the federal government decision to ratify Kyoto. Many provinces have been adamant about leveraging maximum federal incentives and subsidies in exchange for any actions that they might take to contribute to emission reductions, and they have generally not developed the ministerial competence in climate that has proven such a stimulus for action among the states (Rabe 2007).

As a result, provinces have generally worked to protect traditional energy and industrial entities, rather than to seize climate change as a possibility for economic development and diversification, as so many states have done (Stoett, this volume). Moreover, whereas U.S. states most vocal on climate change, such as California and New York, have launched aggressive policies, the most active province on this issue has been Alberta, which has been outwardly hostile to Kyoto ratification and any policy steps beyond voluntary reductions. As a result, there has been little subfederal momentum in Canada at the provincial level, although municipalities have indeed been more engaged (Gore and Robinson, this volume). There were indications in

2008 that this pattern might be beginning to change, most notably in a number of new climate policies emanating from British Columbia.

Even in Europe, striking parallels exist with the American case. Despite the shaky constitutional standing of European-wide environmental decision-making, the EU has remained formally bound to meeting Kyoto reduction targets between 2008 and 2012. This led to the launch of the EU Emissions Trading Scheme (ETS) in early 2005 and the first volley of cross-national emission trading in carbon. However, each EU member state has a different reduction target and is free to establish its own internal policies to achieve it. This has resulted in a tapestry of different strategies and varying degrees of success in individual nations in approaching their pledged reductions. In renewable energy, for example, EU member states have adopted variations of RPS programs, direct subsidies for development, and various taxes on consumption of conventional energy sources. Just as some states lead while others lag in U.S. climate policy development, a similar dynamic operates among EU member states. Even in the second and expanded round of ETS development, the majority of EU emission reductions are expected to come from other policies solely at the discretion of individual member states.

At the same time, there may be two distinct challenges facing continued or expanding state involvement on climate policy, some unique to the American context. These have yet to have any demonstrable effect on state policy engagement but could potentially have a chilling impact. First, a consortium of well-funded organizations hostile to any action by any American government to reduce GHG emissions has become increasingly vocal and visible in the state policymaking process. Organizations such as the Heartland Institute and the Competitive Enterprise Institute have released a series of studies and reports that not only challenge prevailing scientific views on climate change but also portray state-based initiatives as posing dire economic and social consequences. Such groups have roundly condemned most existing state policies as "mini-Kyoto regimes," with extraordinarily high estimates of their projected economic impacts.

Perhaps most importantly, the American Legislative Exchange Council (ALEC) has launched an aggressive campaign to reverse or rescind existing state climate laws. The council contends that it draws roughly one-third of all American state legislators into its orbit, which includes hosting of conferences and study trips, as well as draft legislation that can easily be modified for an individual state. Such model legislation includes a "Resolution in Opposition to Carbon Dioxide Emissions Standards," which, if enacted, would essentially eliminate existing state GHG reduction programs of either a mandatory or voluntary nature and prohibit any future policy of this sort (ALEC 2003). ALEC contends that it introduces approximately 1,500 bills into state legislatures each year, although it has had little demonstrable effect on state climate policy to date (Greenblatt 2003b). Thus far, however, these well-heeled efforts have had little measurable impact and, in a few instances, appear to have further propelled states to enact policies in the face of opposition deemed lacking in credibility.

Second, and most significantly, it appears increasingly likely that various interest groups and the federal government may join forces in challenging many state climate policy initiatives on constitutional grounds. A major confrontation involved the California vehicle emissions program, whereby the Bush administration and a consortium of major vehicle manufacturers contended that California had usurped federal jurisdiction by enacting what was portrayed as a fuel economy standard rather than an air pollution bill. But other challenges are also possible. For example, the Commerce Clause allows for few restrictions on the movement of interstate goods. A number of state climate policies potentially cross the line of a narrowly interpreted Commerce Clause, such as RPS programs that arguably discriminate against electricity generated by fossil fuels in other states in favor of homegrown renewables.

In both sets of cases there are compelling constitutional arguments to sustain state policy (Engel 1999; Engel and Saleska 2005), but federal efforts could attempt to quash certain state initiatives. Indeed, the possibility of new federal climate legislation could feature various provisions designed to preempt existing state policies to secure national uniformity, raising many questions about credit for those states that took early policy steps and secured emission reductions when others stood by silently (Rabe 2008). This issue received remarkably little attention in the 110th Congress, with both House and Senate proponents of new federal legislation being remarkably vague on the ramifications for existing state policy of any future federal efforts, but is certain to resurface in the 111th Congress and beyond.

Concluding Remarks

This chapter has explored some of the underlying rationale behind this robust and rather unexpected set of developments, as well as highlighting possible future trends. Indeed, one can further envision other forms of policy development, including indicators that clusters of states are beginning to work formally with other foreign governments at various levels. All of this suggests that the political context for climate policy is far more complex—and far less fruitless—than many conventional depictions would suggest. Moreover, there remain abundant precedents in other areas of public policy for states to take the lead and often remain active in continuing policy development and implementation. Consequently, there is considerable reason to suspect that states will remain central players in the evolution of U.S. and North American climate policy, with substantial potential for achieving emission reductions as well as offering a host of lessons and models worthy of consideration in Washington, D.C., and around the world.

References

American Legislative Exchange Council. 2003. *Energy, Environment, and Economics: A Guide for State Legislators*. Washington, DC: American Legislative Exchange Council.

Burtraw, Dallas, and David A. Evans. 2004. NOX Emissions in the United States: A Potpourri of Policies. In *Choosing Environmental Policy: Comparing Instruments and Outcomes in the United States and Europe*, edited by Winston Harrington, Richard D. Morgenstern, and Thomas Sterner. Washington, DC: Resources for the Future Press.

Derthick, Martha A. 1975. *Between State and Nation*. Washington, DC: Brookings Institution Press.

Derthick, Martha A. 2005. *Up in Smoke: From Legislation to Litigation in Tobacco Politics*. Rev. ed. Washington, DC: Congressional Quarterly Press.

Engel, Kirsten H. 1999. The Dormant Commerce Clause Threat to Market-Based Environmental Regulation: The Case of Electricity Deregulation. *Ecology Law Quarterly* 26(2): 243–349.

Engel, Kirsten H., and Scott R. Saleska. 2005. Subglobal Regulation of the Global Commons: The Case of Climate Change. *Ecology Law Quarterly* 32(2): 183–233.

Gerber, Elizabeth R. 1999. *The Populist Paradox*. Princeton, NJ: Princeton University Press.

Gormley, William T. 1983. *The Politics of Public Utility Regulation*. Pittsburgh: University of Pittsburgh Press.

Greenblatt, Alan. 2003a. The Avengers General. *Governing* (May): 52–56.

Greenblatt, Alan. 2003b. What Makes ALEC Smart? *Governing* (October): 30–34.

Guber, Deborah. 2003. *The Grassroots of a Green Revolution: Polling America on the Environment*. Cambridge, MA: MIT Press.

Kruger, Joseph, and William A. Pizer. 2005. Regional Greenhouse Gas Initiative: Prelude to a National Program? *Resources* (Winter): 4–6.

LeDuc, Lawrence. 2003. *The Politics of Direct Democracy*. Toronto: Broadview Press.

Mintrom, Michael. 2000. *Policy Entrepreneurs and School Choice*. Washington, DC: Georgetown University Press.

Mossberger, Karen. 2000. *The Politics of Ideas and the Spread of Enterprise Zones*. Washington, DC: Georgetown University Press.

Peterson, Paul E. 2003. The Changing Politics of Federalism. Paper presented at the International Seminar Series, University College London, September 26.

Provost, Colin. 2003. State Attorneys General, Entrepreneurship, and Consumer Protection in the New Federalism. *Publius: The Journal of Federalism* 33(2): 37–53.

Rabe, Barry G. 2004. *Statehouse and Greenhouse: The Emerging Politics of American Climate Change Policy*. Washington, DC: Brookings Institution Press.

Rabe, Barry G. 2007. Beyond Kyoto: Implementing Greenhouse Gas Reduction Pledges in Multi-Level Governance Systems. *Governance: An International Journal of Policy, Administration and Institutions* 20(3): 423–444.

Rabe, Barry G. 2008. States on Steroids: The Intergovernmental Odyssey of American Climate Policy. *Review of Policy Research* 25(2): 105–128.

Rabe, Barry G., and Philip A. Mundo. 2007. Business Influence in State-Level Environmental Policy. In *Business and Environmental Policy: Corporate Interests in the American Political System*, edited by Michael E. Kraft and Sheldon Kamieniecki. Cambridge, MA: MIT Press.

Rosenthal, Alan. 2004. *Heavy Lifting: The Job of the American Legislature*. Washington, DC: Congressional Quarterly Press.

Scheberle, Denise. 2004. *American Federalism and Environmental Policy*. Washington, DC: Georgetown University Press.

Teske, Paul. 2004. *Regulation in the States*. Washington, DC: Brookings Institution Press.

Texas Public Utility Commission. 2005. *Scope of Electricity Market Competition in Texas*. Austin: Texas Public Utility Commission.

Walker, David B. 2000. *The Rebirth of Federalism*. New York: Chatham House.

Zimmerman, Joseph. 2002. *Interstate Cooperation: Compacts and Administrative Agreements*. Westport, CT: Praeger.

5

Field Notes on the Political Economy of California Climate Policy

Alexander E. Farrell and W. Michael Hanemann

"I say the debate is over. We know the science. We see the threat. And we know the time for action is now."
—Governor Arnold Schwarzenegger, World Environment Day, San Francisco, June 1, 2005

Introduction

Driven by this call to action from Governor Schwarzenegger, the California legislature passed AB32, the California Global Warming Solutions Act, on August 31, 2006. Together with related regulatory actions by California's state government (see tables 5.1 and 5.2), these laws constitute the most ambitious and comprehensive effort to control greenhouse gas (GHG) emissions in North America. California often has an unusual influence because of its size: it has a population of about 35 million people (or 12 percent of the U.S. population) and an economy about as large as that of France ($1.6 trillion). Although California's size makes it a substantial source of GHGs, it has a low level of per capita emissions compared to most of Canada and the rest of the United States because of energy efficiency and clean air policies combined with a mild climate and relative lack of heavy industry. Transportation is the largest source of GHG emissions, accounting for 41 percent, followed by electricity generation, which accounts for 20 percent (including emissions from imported electricity). Other sources of GHG emissions include agriculture, landfills, wastewater treatment, and a range of industrial processes (CEC 2005).

The provisions of AB32 place a cap on all GHG emissions in California and require that they be reduced to their 1990 level by 2020. This is expected to be a reduction of about 25 percent compared to emissions projected for 2020 under business as usual without AB 32. Other state laws prohibit any electrical-load-serving entity based in California from entering into a long-term contract for base-load generation unless the GHG emissions are as low as or lower than those from new, combined-cycle natural gas power plants. This performance standard applies to

Table 5.1
Key events in the development of California's climate change policy

1988: AB4420 is passed, leading to publication of the report *The Impacts of Global Warming on California* in 1989.
2000: California electricity crisis.
2001: California Climate Action Registry is established (SB1771/SB527).
2002: AB1493 (Pavely) is signed by Governor Davis to control vehicle GHG emissions.
2003: Recall gubernatorial election is won by Arnold Schwarzenegger.
2003: The Public Interest Energy Research program (created in 1996) expands its activities on climate change research.
2002: California Energy Commission and California Public Utilities Commission publish the first Energy Action Plan.
2002: California Public Utilities Commission starts proceedings on GHG performance standard.
2004: West Coast Governors' Global Warming Initiative is launched.
2004: California Public Utilities Commission "carbon adder" requires investor-owned utilities to account for financial risk of future policies.
2004: "Turning the Corner on Global Warming Emissions" is published by the Tellus Institute.
2004: California climate change impacts study is published in the *Proceedings of the National Academy of Sciences.*
2004: Nine auto manufacturers sue California to block the implementation of AB1493.
2005: Executive Order S-3-05 establishes GHG emission targets and the Climate Action Team.
2005: Four Schwarzenegger-backed ballot initiatives are rejected by voters.
2006: California Public Utilities Commission announces "threshold decision" to cap GHG emissions from investor-owned utilities.
2006: Climate Action Team report is published.
2006: California Solar Initiative establishes goal of 3GW of solar electricity capacity for investor-owned utilities and commits $3.2 billion in incentive funds; SB 1 extends the initiative to municipal-owned utilities and sets goal of photovoltaic installations in 50 percent of new homes by 2018.
2006: AB32 (Núñez/Pavley) California Global Warming Solutions Act sets 2020 GHG reduction target.
2006: SB1368 prohibits new long-term contracts for baseload electricity generation if GHG emissions exceed those from combined-cycle natural gas.
2007: Executive Order S-1-07 proposes a low carbon fuel standard.
2007: California attorney general sues San Bernardino County for failure under the California Environmental Quality Act to analyze GHG emissions resulting from its general plan amendment.
2007: Auto manufacturers' suits rejected in Vermont and California.
2007: California Air Resources Board identifies early action measures for reducing GHG emissions.

Table 5.1
(continued)

2007: US Environmental Protection Agency denies California's request for a waiver of an exemption for implementation of AB1493.
2008: California files suit challenging the U.S. Environmental Protection Agency denial of waiver request.
2008: California Air Resources Board issues draft Scoping Plan for implementing AB32.

both municipal-owned and investor-owned utilities and also affects imported electricity. By contrast, the Regional Greenhouse Gas Initiative (RGGI) in northeastern North America only caps emissions from electric power generation in the region, which account for about 27 percent of those states' total GHG emissions. The RGGI cap is also less stringent than that in AB32: under RGGI, power sector emissions will have to be reduced to 10 percent below their 2006 level by 2019 (Selin and VanDeveer, this volume).

To achieve these comprehensive GHG emission reduction goals, California has adopted a broad set of sectoral policies in addition to AB32 (see table 5.2). Most major sectors are covered by these policies, except for agriculture. Energy consumption is addressed largely by setting product and building efficiency standards. Transportation emissions are addressed by vehicle and fuel performance standards, while electricity and large industrial point sources will be addressed through a cap-and-trade system.[1] Dealing with these sectors individually is advantageous, because each offers extremely different options for reducing emissions in the short and long term. For this reason, it is unwise to rely on a single instrument to achieve GHG emissions reductions and technological innovation in all relevant sectors at the same time.

California's history is key to understanding how and why California came to take these steps to control its GHG emissions. The state has played an important role in advancing environmental and energy policy in advance of the U.S. federal government, including efforts to improve air quality starting in the 1950s and energy efficiency since the 1970s (Krier and Ursin 1977; Rosenfeld 1999). The California Global Warming Solutions Act (AB32) and related state actions in California are inspired by this history and can be viewed as a natural extension of it. This history, in light of federal inaction, motivated California lawmakers and provided them with confidence that they could succeed in their efforts to reduce GHG emissions across the state and also to achieve reductions in collaboration with other U.S. states and Canadian provinces.

This chapter provides a brief sketch of the scientific and political background of AB32 and other elements of California's climate policy, arguing that it emerged in

Table 5.2
Key pieces of California's sectoral climate change policy

Research portfolio: Various California agencies support relevant research, including the California Energy Commission and California Air Resources Board; a key example is the Public Interest Energy Research program.

Efficiency standards for buildings and appliances: The California Energy Commission has supported efficiency-related research for decades and together with the California Public Utilities Commission has set efficiency standards and overseen energy efficiency programs by electric utility companies. These programs pay the additional costs for more efficient end-use technologies.

Energy Action Plan: The California Energy Commission and the California Public Utilities Commission cooperate to adopt an action plan to eliminate future energy outages, decrease per capita energy use, ensure reliable energy supply, and reduce emissions. Establishes a "loading order" for electricity of preferred options: efficiency, renewables, and natural gas. Advisory only.

Executive Order S-20-06 Establishes the California Environmental Protection Agency as lead agency, transfers GHG inventory duty to California Air Resources Board, and orders that there should be a market-based emission trading program.

AB32 Lowers statewide GHG emissions to 1990 levels by 2020 (about a 25 percent cut); enforceable in 2012. Covers all GHGs, most stationary sources, and all electricity; California Air Resources Board is lead agency. Environmental justice concerns must be addressed. Governor can delay attainment date.

Electricity

Renewable portfolio standard (SB1078 in 2002 and SB107 in 2006). Achieve 20 percent renewable energy by 2010 using tradable credits. Applies to investor-owned utilities, but not municipal-owned utilities.
California Solar Initiative (California Public Utilities Commission decision). Provides for ~$3.2 billion in subsidies for solar electricity, especially photovoltaic. Must be 1/3 residential, with low-income set asides. SB1 extends program to municipal-owned utilities and requires developers of >50 new single-family homes to offer a solar option.
Carbon Risk Adder (California Public Utilities Commission procurement order D 04-12-048). To protect consumers investor-owned utilities must explicitly account for the risk of carbon regulation in the future, assuming $8/ton CO_2. No cash payments.
GHG performance standard (California Public Utilities Commission decision, SB1368). Limits CO_2 emissions for baseload electricity to equal to or less than emissions from a combined cycle gas turbine plant. Applies to both municipal-owned and investor-owned utilities.
GHG emission cap (California Public Utilities Commission decision, AB32). California Public Utilities Commission's "threshold decision" to adopt a load-based GHG cap for all electricity. AB32 is statewide and gives California Air Resources Board authority.

Transportation

Vehicle GHG performance standard (AB1493 Pavley). Automakers must reduce vehicle GHG emissions of new cars beginning with model year 2009, leading to ~30 percent reduction by model year 2016. Alternative fuel compliance path.
Low Carbon Fuel Standard (Executive Order S-1-07) Refiners, blenders, importers must lower GHG content of fuel by at least 10 percent by 2020. Emission reduction credits allowed.

---

large part due to the character of California's polity and institutions. It also argues that climate policy is likely to affect large parts of California's public policy agenda and that these effects are largely unavoidable. The next section sets the stage for California climate policy to emerge, and the third section documents the initial steps to be taken. Sections four and five describe Governor Arnold Schwarzenegger's appearance on the scene and how he came to make the dramatic statement that opens this chapter. The sixth section describes some of the complex interactions between energy policy and climate policy in California. The final sections explain how AB32 came to be passed and provide some closing thoughts on the lessons from this process and what might be expected in the future.

Setting the Stage

Several factors set the stage for climate policy action in California. First, Californians see themselves as innovators and cultural leaders, and they expect their political leaders to act accordingly, especially in terms of environmental policy. California has often been the starting place for new political trends, such as tax reform (1978), term limits (1990), and electricity restructuring (1996). The state is a well-established leader on environmental policy. Because of the strong public interest in environmental protection in the state, both major political parties compete for environmental votes, especially in statewide races.

Second, California has a number of institutions that enhance the state's capacity to make climate policy. These include strong research institutions, such as the University of California, Stanford University, and the Lawrence Berkeley National Laboratory, as well as government agencies such as the California Air Resources Board (CARB) and the California Energy Commission (CEC). These two agencies are somewhat unique to California in their large size and wide-ranging scope of activities and authority (although most states have some sort of air quality agency). The CARB was created in 1967 and has developed much scientific and technical competence. It sponsors peer-reviewed research, and its regulatory actions are preceded by a carefully organized program of scientific and engineering research. The CARB is widely regarded as a model of an aggressive, independent, science-based regulatory agency.

California is in an unusual position to influence policy on air pollution, under U.S. federal law, due to its "compelling and extraordinary conditions." Under the U.S. Clean Air Act, it has the authority to request a waiver from the automobile emission standards set by the U.S. Environmental Protection Agency (EPA) in favor of more stringent state standards. Once the waiver is granted, other states can, if they choose, copy the California standard. Since 1967, the U.S. EPA has granted California's waiver request, either in whole or in part, on fifty-three occasions, and many of its waivers have been adopted by other states or incorporated into federal law (McCarthy 2007), including nitrogen oxide (NO_x) standards for cars and light trucks (1971), catalytic converters (1975), unleaded gasoline (1976), and low-emission vehicles (1994).

In addition to statewide energy forecasting, planning, and policy formulation, the CEC is empowered to establish building- and appliance-efficiency performance standards. Its authority covers not just investor-owned utilities but also municipally owned utilities. No other state has an agency like the CEC, and no other state has been as active and effective in regulating energy efficiency. The CEC often partners with the California Public Utilities Commission (CPUC), which also has a national reputation for leadership in energy conservation. One of the most important CPUC innovations was rate decoupling (separating utilities' profits from volumes of sales), introduced in 1978 for natural gas and in 1982 for electricity. Together, the two agencies have had a significant impact on electricity use: per capita electricity use in California has stayed constant since 1975, while nationally it has risen by about 50 percent.

Third, relatively high energy prices and complex energy regulation contributed to the complex path to climate change policy in California. In the mid-1990s, the combination of high electricity prices, a general trend toward deregulation of industry, and California's tendency toward policy innovation led to a restructuring (some say deregulation) of the electricity industry. One of the concerns about California's restructuring was that it would end the strong energy efficiency programs that the three investor-owned utilities had developed over the previous twenty years under pressure from environmental groups and government policy. To resolve this problem, the electricity restructuring law (AB1890) included provisions for "public benefit charges" to be collected from customers of investor-owned utilities for the specific purposes of supporting energy efficiency programs (~$230 million/year), the Public Interest Energy Research program (~$80 million/year), and renewable electricity generation (~$110 million/year). However, electricity restructuring in California was poorly implemented and led to energy and fiscal crises for the state (Sweeney 2002). These crises helped create the conditions for California's climate policy leadership.

From the Bottom Up

As Rabe (2004) has noted, many of the earliest state policies that affect climate change were not explicitly or solely devised to do so, such as renewable portfolio standards for electric power. California followed this pattern, especially with energy efficiency policies. The state developed its own efficiency standards for appliances and buildings. Beginning in the mid-1970s, in response to the energy crises (Rosenfeld 1999), the state forced the investor-owned utilities to support the installation of efficient energy-using devices (such as compact fluorescent light bulbs and building efficiency standards). In the 1980s and 1990s, these policies helped reduce energy consumption in the state to well below predictions, and energy efficiency programs became an important part of California's proposed approach to meeting the Governor's GHG targets for 2020.

One explicit climate policy was the creation of the California Climate Action Registry, a nonprofit organization that works with private companies and other organizations to design GHG emissions reporting procedures. Establishing the registry, however, did not create a mandatory reporting requirement for GHGs. Similar to other U.S. policies at the time, organizations that chose to adopt the registry's protocols and report their GHG emissions did so voluntarily. Moreover, President Bush's reversal of his campaign promise to support climate policy in March 2001 signaled to many observers that climate policy would have to be pursued at state and municipal levels. By this time, several states and cities had already begun to enact various types of climate policy (Gore and Robinson, this volume; Rabe, this volume; Selin and VanDeveer, this volume). In California, the CEC updated the state emission inventory and developed a five-year research plan on climate change under the Public Interest Energy Research program (Franco 2002; Franco, Wilkinson, Stanstad, et al. 2003).

Another major step in climate policy was taken by the enactment of a renewable portfolio standard in September 2002 through SB1078. Although the text of this law does not mention climate change, it requires the investor-owned utilities to increase the amount of renewable energy they use each year (excluding large hydroelectric dams), from about 11 percent to 20 percent by 2017.[2] Meanwhile, term limits in California were forcing many longtime politicians out of office and opening up the electoral process to new candidates. One such new candidate was Fran Pavley, who won an open state assembly seat in 2000. Soon after, Assemblywoman Pavley began working on a bill to reduce GHG emissions from automobiles. Pavley argued that GHGs should be considered air pollutants under the U.S. Clean Air Act, which provided the state of California with an exception to federal preemption of automobile emission regulations.

The Pavley Bill (AB1493) directed the CARB to "develop and adopt regulations that achieve the maximum feasible and cost-effective reduction in GHGs emitted by passenger vehicles" (2002). It was supported by Democratic legislative leaders, passed, and was signed by Governor Gray Davis. In this context, "cost-effective" means that if the regulation imposes capital costs on consumers (higher-priced vehicles), then the savings in fuel costs must at least compensate. The CARB proposed

GHG emission reductions that would reach about 30 percent for new cars in 2016, and it calculated that the fuel savings would be over four times the additional costs, assuming a gasoline price of $1.74 per gallon. Thus, the Pavley Bill fit into the model that climate policy had to be free or to save money.

This law marked a major turning point in climate policy in the United States. It was popular with the public, garnering a great deal of positive media attention. It may also have marked the start of a deliberate strategy by environmental groups (who strongly supported the Pavley Bill) to create domestic pressure for national climate policy. The Clean Air Act requires California to obtain an exemption from the U.S. Environmental Protection Agency (EPA). If this exemption is granted, other states are permitted to adopt California's vehicle emission standards. By 2007, fourteen states had applied to do so, representing about 40 percent of new vehicle sales in the United States. In December 2007, the U.S. EPA denied California's waiver request outright, the first time it has ever done so. California subsequently filed a suit challenging the EPA's denial. Some Canadian provincial leaders have suggested that the California standards may also be applied in Canada.

State agencies were also active in the aftermath of the 2000–2001 California electricity crisis. The state's principal energy regulatory agencies came together to prepare a plan with specific goals and actions that would eliminate future energy outages and excessive price spikes in electricity or natural gas (California Energy Commission and California Public Utilities Commission 2003). The plan identified increased energy efficiency and price-based demand response as the state's preferred future energy resource. It called for reduced per capita electricity use both to save energy and to minimize emission of pollutants including GHGs, and it recommended accelerating the 20 percent renewable resource goal from 2017 to 2010. Similar themes were echoed in two other reports that year. The CEC and the CARB produced a joint report *Reducing California's Petroleum Dependence*, as required by AB2076, which stressed the need to reduce the growth in demand for petroleum by raising new vehicle fuel economy standards and, also, increasing the use of alternative fuels and advanced vehicle technologies (CEC/CARB 2003). The CEC also issued its *2003 Integrated Energy Policy Report*, which stressed the seriousness of climate change as a risk to California (CEC 2003).

Thus, by the beginning of 2003, policies that would tend to reduce GHG emissions significantly in most of the energy sector—electricity, private vehicles, buildings, and industry—were in place in California. Not all of these were explicitly identified as climate policies—this would come later—fitting the pattern that Barry Rabe discusses in more detail in this volume. In addition, the Public Interest Energy Research program had begun to develop much more detailed, California-specific information about the effects of climate change (Roos 2003; Shaw 2002). This "bottom-up" information was far more salient than previous, larger-scale studies. Because it was developed in a publicly open and transparent process, it also tended to be viewed as legitimate. A peer review process, associated journal publications, well-respected researchers, and the fact that the California climate research corroborated other climate change research provided credibility. These three factors—saliency, legitimacy, and credibility—have been identified as three necessary properties of effective environmental assessments (Farrell, Jaeger, and VanDeveer 2006).

Crisis and Opportunity

A political crisis in 2003 led to an unprecedented gubernatorial recall election in California, giving Arnold Schwarzenegger—bodybuilder, businessman, and movie star—an opportunity to exercise leadership on climate policy (Gerston and Christensen 2004). In its first few years, the restructured electricity industry seemed to operate successfully, but by late 2000 major problems erupted (Sweeney 2002). Electricity prices skyrocketed, rolling blackouts were imposed, and numerous companies, including the largest investor-owned utility in California, Pacific Gas and Electric (PG&E), started down the road to bankruptcy. Governor Gray Davis came under increasing criticism for failing to act effectively. By the spring of 2001, the state stepped in on an emergency basis to purchase electricity contracts, and the restructuring process was put on hold. The electricity crisis contributed to the poor public opinion of Governor Davis and his later recall from office, as did the economic recession the state experienced at the same time.

The 2003 recall election provided several office-seekers a chance to run for governor (Gerston and Christensen 2004). In particular, it offered an opportunity for moderate Republicans to compete for the governor's seat without having to win a primary election, which more conservative candidates generally win. Schwarzenegger quickly became the front-runner, winning by a substantial plurality. He promised bold action in reforming government and achieving the policy goals that Californians wanted. Candidates were asked whether they supported the Pavley Bill. Schwarzenegger's main environmental advisor, Terry Tamminen, recommended that Schwarzenegger not only support the Pavley Bill but also push further, making climate change an important policy initiative in the campaign. Schwarzenegger did this, and he appointed Tamminen as secretary of the California Environmental Protection Agency (CalEPA), the umbrella organization under which the CARB operates, after taking office in November 2003.

An influential development involved the Union of Concerned Scientists, an organization that publicly promotes scientific research on climate change. In 2003, the Union of Concerned Scientists initiated a study in California, partly to generate scientific information that could be informative when the implementation of AB1493 came up for consideration in the summer and fall of 2004. This study was conducted by eighteen scientists and developed specific, quantitative estimates of how California might be affected by climate change, using the projections of global circulation models and downscaling them to a relatively fine spatial scale in California (Hayhoe et al. 2004). It showed that under usual conditions, summertime temperatures in California might rise by as much as 18 °F toward the end of the twenty-first century, with serious implications for heat waves and fires and consequently for water supply and demand. Also, California might lose about 89 percent of the snowpack in the Sierras, which would significantly deplete California's water supply.

The study's results were published in the *Proceedings of the National Academy of Sciences* in August 2004, receiving much scientific and media attention, and the results were cited in rulemaking by the CARB to implement AB1493. There were also presentations by the researchers to high-level officials in state agencies and the governor's office. Governor Schwarzenegger was briefed on climate science in 2004. Staff scheduled a thirty-minute discussion, but Schwarzenegger interrupted an oral briefing with senior staff to read supporting documents and ask detailed questions, turning a short meeting into a lengthy one. In 2004, Tamminen and his staff began considering specific GHG emission reduction targets, using information developed from a study by the Tellus Institute in support of the West Coast Governor's Initiative and subsequently refined for use by the Schwarzenegger administration (Bailie, Dougherty, Heaps, et al. 2004). These studies and many other proclimate policy activities were funded by the Energy Foundation, which exerted considerable influence in this way.

A landmark event was the publication of the CARB staff recommendations for the implementation of AB1493 in June 2004 and their adoption by the CARB Board in September (California Air Resources Board 2004). The CARB evaluated vehicular GHG emissions and the technologies available to reduce them, focusing only on technologies currently in use on some vehicle models or demonstrated by auto companies and/or vehicle component suppliers in at least prototype form. The CARB found that GHG emission reductions by 2016 of 30 percent would be cost effective—the added cost of the vehicles would be more than paid back to the consumer in fuel savings. The automobile industry strongly criticized both the technology analysis and the cost estimates, but at a 2004 CARB hearing, the board voted unanimously to adopt the staff recommendations. In December 2004, the automobile industry filed suit against the CARB claiming that forcing them to reduce GHG emissions was akin to setting fuel economy standards, which was preempted by the federal government. The Schwarzenegger administration responded by vigorously defending the law.

During 2004, the CPUC was also expanding its activities relating to climate change. In June 2004, it requested that the regulated investor-owned utilities—which accounted for approximately 75 percent of electricity load—address key issues pertaining to climate change as part of their long-term energy procurement planning. Building on information developed by the utilities and other stakeholders, in December 2004 the CPUC issued Procurement Order D 04-12-048, which requires investor-owned utilities to employ a "greenhouse gas adder" when evaluating competitive bids to supply energy. The adder, intended to reflect the financial risk to ratepayers of emitting GHGs given the likelihood that these emissions will be limited by regulation in the future, was to be determined subsequently by the CPUC (California Public Utilities Commission 2004). Secondly, the CPUC announced its intention to investigate, as part of a general framework of incentives to promote the selection of environmentally sensitive resources, the creation of a carbon cap to be applied to each regulated utility's resource portfolio.

The governor and the CPUC worked together to go around the legislature to implement climate policy. Initially announced by the governor in August 2004, the "Million Solar Roofs" effort was designed to subsidize rooftop photovoltaic installations with about $3 billion as a way to rapidly increase demand and encourage private research and investment to bring down costs. However, the Democratic speaker of the House, Fabian Núñez, inserted a provision widely seen as prounion into the bill, which the Governor found unacceptable. The governor encouraged the CPUC to the take the opportunity to create a very similar California Solar Initiative without the labor provisions disliked by the business community; this was approved by the CPUC in January 2006. In August 2006, the legislature enacted SB1, which extends the California Solar Initiative to municipal-owned utilities and requires a developer of more than fifty new single-family homes to offer buyers a solar option.

In sum, climate change politics are complicated, offering competing incentives to policymakers. On the one hand, if the governor set a GHG reduction target for California, he wanted it to be bolder than what other states were doing. On the other hand, California already had lower GHG emissions per capita than almost any other state because much of the low-hanging fruit of energy efficiency had already been plucked. In addition, conservative elements in the governor's office were not keen to embrace something that the California Chamber of Commerce would oppose.

We Know the Time for Action Is Now

Governor Schwarzenegger unveiled his climate change policy on June 1, 2005, at the United Nations World Environment Day conference in San Francisco, a celebration of the fiftieth anniversary of the founding of the United Nations. At the ceremony, Schwarzenegger made the statement given at the beginning of this chapter and then signed Executive Order S-3-05, establishing the following GHG reductions targets for California: by 2010, reduce emissions to the level in 2000; by 2020, reduce emissions to the level in 1990; and by 2050, reduce emissions to 80 percent below the level in 1990 (Schwarzenegger 2005). By adopting these goals, Schwarzenegger put California out in front of all other U.S. states.[3] In addition, CalEPA was designated the lead agency for climate affairs and the Climate Action Team was created. Schwarzenegger's announcement received much media attention, in part because it directly confronted federal U.S. policy. Among the most famous politicians in the country, Schwarzenegger made it more difficult for other Republicans to continue to deny the scientific case for climate policy or to argue that action to mitigate climate change was infeasible.

Thus, the twin goals of California's climate policy are (1) emission reductions to meet the 2020 target and (2) innovation to prepare to meet the 2050 target. Because of their scope, the governor's targets require a portfolio of different strategies. The strategy to meet the 2020 targets is to draw on technologies that are presently in use or on the verge of deployment and to fashion a set of incentives that shift the normal pattern of economic activity and growth toward less carbon-intensive or less energy-intensive technologies. To meet the 2050 target, a different energy system is required, plus many changes across the economy to reduce GHG emissions directly and consume less of what will likely be more expensive energy. New technologies, new institutions, and new industries will also be required to meet this target.

While the governor's Executive Order set targets, it did not specify how those targets were to be met. The Secretary of the CalEPA was directed to report to the governor and the legislature by January 2006, and biennially thereafter, to provide updates on the impacts to California of global warming. However, no studies of these goals had ever been undertaken, including the ambitious 2050 goal. The literature focused on the effects of the Kyoto Protocol on national economies, which offered only limited insights into California's unique situation. In the technology-forcing style of environmental laws of the 1970s, Schwarzenegger had decided on the policy goal first and then asked for analysis of the means to achieve those goals. Thus, a key practical effect of the Executive Order was that it started a scramble for relevant scientific and policy information.

At about this time, several studies more definitively establishing (and quantifying) California's serious vulnerability to climate change began to emerge (Fried, Torn, and Mills 2004; Hayhoe et al. 2004). The Public Interest Energy Research program funded much of this research. The biennial impacts assessments have the effect of regularly updating and reinforcing to California's elected leaders the importance of the problem. The first one was completed between June and December 2005 by a broad research effort of seventy researchers from multiple institutions and scientific disciplines, leading to the publication of twenty technical reports (CEC 2006). However, for many participants the key question was, "How much will climate policy cost?"

Much of the analysis of the Kyoto Protocol splits along the line of whether "no regrets" options—which cost zero or have negative costs—exist or not. Some analysts argue that climate policy must by definition impose economic costs because GHG-free energy supplies are not seen as the least-cost approach in today's energy sector, and that these costs may be large. This view is supported by some analyses using computable general equilibrium models (Energy Information Administration 1998; Nordhaus and Yang 1996). Others focus on energy efficiency and argue that large-scale market failures and barriers prevent consumers in the United States from choosing energy-efficient alternatives (Brown 2001; Howarth, Haddad, and Paton 2000).[4]

When Schwarzenegger's GHG goals were announced, there was one study of the costs of reducing GHGs in California, performed by the Tellus Institute for the West Coast Governors' Global Warming Initiative (Bailie, Dougherty, Heaps, et al. 2004). An updated and more focused study was released shortly after the announcement, to which the administration likely had access beforehand. Like much of the climate policy analysis and advocacy in California, the Energy Foundation and the Hewlett Foundation sponsored these reports. The Tellus study investigated a range of strategies, stressing energy efficiency and renewable energy. It found that significant GHG emission reductions were possible in California (about 25 percent below business as usual) at a net savings in energy costs (about $7 billion per year). However, the Tellus report used an economic methodology that did not allow for the impact of program costs on the rest of the California economy. Hence, it did not address impacts on statewide economic growth, employment, or overall impacts on consumers.

Avoiding Another Energy Crisis

A key issue in California and elsewhere is how tightly climate policy and energy policy are linked and how challenging it will be to work out conflicting objectives. California faces two problems—(1) high prices for electricity and (2) a perceived need for more generation capacity in the future. These problems would be exacerbated if some of the older existing power plants in California were retired. This issue became apparent in two actions that were hard to reconcile with the governor's climate policy actions—as illustrated by the Frontier Line proposal and short career of Joe Desmond as chair of the CEC.

The Frontier Line is a proposal for new transmission lines to bring electricity to California from Colorado, Montana, Wyoming, and other Western states, although it is not clear if this would be mostly wind power, coal power, or another source. Schwarzenegger supported this proposal and signed a Memorandum of Understanding with three other Western Governors to find ways to develop the Frontier Line. This presents a puzzle. Coal-fired electricity will undermine the governor's climate policy goals unless it employs expensive new technologies, but these expensive technologies would undermine one of the main reasons for seeking coal-fired electricity —to lower costs.

Shortly after signing this agreement, Schwarzenegger appointed Joe Desmond to the open chairman's position on the CEC. The former CEO of an energy consulting and software development firm, Desmond supported the Frontier Line and the possibility that this power might be from advanced "clean coal" plants. The CEC is one of two commissions (the other being the CPUC) that must approve new transmission lines in California. However, any CEC action on the Frontier Line will be taken without Desmond. California law requires energy commissioners to be confirmed by the state Senate. Desmond's support of the Frontier Line and clean coal displeased the Democrats who controlled the Senate, and a confirmation hearing was never scheduled (Martin 2006).

Political Winds Shift

Schwarzenegger called a special election for November 8, 2005, to allow voters to decide on eight ballot propositions, including a "reform package" of four propositions pushed by the governor.[5] The landslide defeat for the governor's propositions had two important consequences: (1) the governor moved quickly to reposition himself in the political center; and (2) the state's Democrats were emboldened to confront the governor head-on. Climate change was one area for this confrontation.

Through the fall 2005 elections and into spring 2006, the CPUC was moving energetically to regulate GHG emissions from new power plants used by investor-owned electric utilities. In October 2005, it issued a Policy Statement on Greenhouse Gas Performance Standards directing staff to investigate the adoption of a GHG emissions performance standard for investor-owned utility procurement that would be "no higher than the GHG emissions levels of a combined-cycle natural gas turbine" for all procurement contracts that exceeded three years in length and for all new investor-owned generation. In February 2006, the CPUC announced that it would develop a cap on all GHG emissions from investor-owned utilities and other load-serving entities, including emissions associated with energy imported from out of state (California Public Utilities Commission 2006). Later, SB1368 gave legislative force to the CPUC's performance standard and extended it to municipal utilities in California.

Also in 2005, Climate Action Team staff prepared a report on implementation of the Governor's GHG emission targets, identifying specific regulatory policies within the purview of state agencies that, in aggregate, could achieve the 2020 target for GHG emission reductions—in effect, demonstrating the feasibility of this target. The Climate Action Team came up with a laundry list of thirty-eight possible regulatory actions by the CEC, the CARB, the CPUC, and other state agencies (Climate Action Team 2006). Some were large, including implementation of AB1493 plus additional measures that could be implemented after 2016, the reduction of hydrofluorocarbon emissions from vehicular and commercial refrigeration systems, and forest management and afforestation; others were small, such as regulations to control the handling of wet manure at animal facilities in California.

Implicit in this approach was an emphasis on "cap" rather than a market-based approach such as cap-and-trade. There were two reasons for this emphasis. First, it is the logical extension of the regulatory approach traditionally used with air pollution and energy efficiency in California. Second, strong opposition to emission trading was expressed by certain groups with which the Climate Action Team consulted, especially the environmental justice community. They argued that market-based approaches had failed both in controlling emissions, citing Southern California's experience with market-based emissions reductions incentives (Chinn 1999), and as energy policy, citing the electricity crisis (Sweeney 2002); thus, they advised, these approaches should be viewed with skepticism now. In particular, they were concerned with the possibility that strict caps on GHG emissions could lead to reductions of conventional pollutants as a cobenefit and did not want to lose this opportunity.

Initially, no economic analysis was planned, but concerns from the business community led to a rushed effort to look at these issues. Eventually, two computable general equilibrium analyses were performed. One by the CARB staff analyzed all thirty-eight regulatory actions. The other, by David Roland-Holst at University of California, Berkeley, focused on eight regulations for which more substantial documentation was available, accounting for about half the targeted emission reduction (Climate Action Team 2006; Roland-Holst 2006). Both studies reached a similar conclusion: the regulatory actions to reduce GHG emissions in California would lead to small increases in statewide employment and gross state product by 2020.

This finding attracted widespread attention, including skepticism from industry sources. However, there are three specific economic reasons why this outcome occurs. First, a significant portion of the programs by which the emission reduction would be effectuated involve regulations to promote fuel and energy efficiency; these can save money for fuel and energy users by lowering their cost of doing business, and this stimulates economic growth. Second, it turns out that, of the goods and services consumed in California that are relatively GHG-intensive, a significant fraction are produced out of state, whereas a significant fraction of the goods and services produced instate tend to be less GHG-intensive. Hence, a limit that raises the relative price of GHG emissions redirects consumption away from imported production and toward domestic production, boosting the domestic economy. Thirdly, if the emission limit induces technological innovation, this also promotes economic growth.

From Targets to Law

By 2006, the governor and the Democrats in the state assembly were competing to make climate change policy. The Democrats' vehicle was AB32, a bill originally introduced in December 2004 by Fran Pavley to revise some of the functions and duties of the California Climate Action Registry. Now, the bill was being amended to direct the CARB to adopt regulations by January 2008 for reducing statewide GHG emissions to their level in 1990 by 2020. Moreover, Pavley obtained the support of Speaker Núñez, the governor's opponent in the partisan wars of 2004–2005, who became a cosponsor of the bill.

The governor's office stopped short of endorsing AB32 but said that the governor looked forward to working with the legislature to curb GHG emissions in California. The governor's continuing caution, or ambivalence, was visible at a "global warming summit" held a week after his statement regarding AB32. At this event, he called for California to become a national leader in combating global warming; "Let's work together to create the world's best market-based system to limit and slash emissions." He went on to say that the state should move slowly in imposing controls on industries that emit GHGs because "we could really scare the business community ... we don't [want to] have businesses leaving the state." He concluded: "I think we should start out without the caps and ... shoot for certain goals." Speaker Núñez commented that the governor "needs to walk the walk, not just talk the talk" on climate change.[6]

As the legislature considered AB32 during the summer of 2006, there were two major sets of disagreement. One was between supporters and business opponents of the bill. Industry critics argued that global warming was not a local problem and that it should be regulated by the federal government. By itself, they maintained, California could accomplish little, since it accounted for a small fraction of global GHG emissions. They also said that emission regulation would destroy many jobs in California. Supporters of AB32 responded that California was the world's twelfth-largest source of GHG emissions and had been highly influential in setting standards for the regulation of air pollution and energy efficiency nationally and internationally. Present action would help California become a leader in the emerging global market for GHG control technologies, they said, and a well-designed strategy to limit emissions in California could yield net economic growth in California.

The second skirmish was between the administration and the legislature. The governor's office indicated that the governor wanted several provisions inserted in AB32 and that he would not sign the bill if it lacked them. Núñez rejected the provisions, leading to a stalemate. The dispute centered on three specific issues: which agency would be in charge of implementing the emissions cap set by AB32; the role of emissions trading in this implementation; and the question of a "safety valve." The structure of AB32 was similar to that of AB1493: it set a goal, in this case reducing statewide GHG emissions to their 1990 level by 2020, but it left the details of how the goal was to be met for a state agency to determine within a set period of time and subject to some specific restrictions. The Democrats wanted the agency to be the CARB, the agency that implemented AB1493. In contrast, the governor wanted it to be the Climate Action Team led by CalEPA. The Democrats had more confidence in the CARB, considered it more independent, and liked that its decisions were made in a public process. As a cabinet officer, the secretary of CalEPA was closer to the governor.

On the issue of market-based regulation (such as cap-and-trade), the Democrats were willing to state that the implementing agency "may include" the use of market-based mechanisms, provided it took certain prior steps including examining the possibility that this could exacerbate localized "hot spots" and designing a mechanism to rule out any increase in the emissions of toxic air contaminants or criteria air pollutants. The governor wanted the bill to state that the CARB "shall include" market mechanisms. The governor's office was sensitive to the drumbeat of criticism from the Chamber of Commerce and other business groups that a cap on GHG emissions would cause a calamitous increase in energy prices in California, causing businesses to leave the state and creating job losses. As such, the governor's office wanted a "safety valve" written into AB32 whereby the emission cap could be relaxed if there were going to be harmful economic consequences.

The stalemate continued until the last minute. A week before the 2005–2006 legislative session ended on August 31, the *Sacramento Bee* and the *Los Angeles Times* ran editorials urging the governor and the legislature to resolve their differences and pass landmark GHG legislation. Agreement was reached on August 30. The Democrats got their way on the CARB as the implementing agency and on the use of the wording "may" rather than "shall" with regard to considering emission trading. The governor got his way on the safety valve. This compromise has potential pitfalls. It could hinder investments in GHG controls if the safety valve is viewed by industry as an escape hatch. On the other hand, making the governor decide (rather than an interagency board, which was the administration's preference) greatly increases the political costs of weakening the program. The bill passed and was signed it into law on September 27, 2006.

The following month there was an unexpected addendum to this flurry of activity. The governor issued Executive Order S-17-06 on October 10, 2006, reaffirming the primacy of the secretary of CalEPA as "statewide leader for California's GHG emission reduction programs." It directs the CARB to "work with" the secretary of CalEPA in developing measures to implement AB32 and, in particular, to collaborate with the secretary "to develop a comprehensive market-based compliance program with the goal of creating a program that permits trading with the European Union, the Regional Greenhouse Gas Initiative and other jurisdictions." The Executive Order also states that the CalEPA secretary should "facilitate and coordinate" the CARB and the CPUC as they develop regulations that affect electricity and natural gas providers in order to avoid duplicative or inconsistent requirements. As such, the governor was having the last word in the confrontation with the Democrats regarding the terms of AB32, although the force of the AB32 statute ultimately supersedes that of the Executive Order.

Looking Forward

The scope of climate change policies in place and in development in California is broad and their number is bewildering and growing. Implementing them all will be challenging, stimulating demand for new technologies for low-GHG energy supply, energy efficiency, and to reduce GHG emissions from agriculture and manufacturing. This should spur innovation and lower costs, which could have major implications for climate policy nationally and perhaps globally. Looking over the events, three issues with broader relevance beyond the vagaries of political maneuvering in California stand out: scope, consumption, and preemption.

First, mitigating GHG emissions is an enormous task that will affect many parts of the economy, much more so than any other environmental policy. This scope underlies the bureaucratic tussles over turf, which is indicative of the larger problem: no single state agency is fully adequate for the task of managing GHG emissions across the breadth of the California economy. Because GHG emissions are emitted from many industries—including agriculture, electricity, manufacturing, and transportation—devising an appropriate regulatory structure will be difficult. There is a genuine policy dilemma regarding the weight to be placed on regulatory policies versus emission trading in reducing GHGs. A key challenge for GHG reduction policy is likely to be designing the right blend of regulatory and market-incentive programs. This, however, this does not mean specifying an exhaustive set of regulatory policies such as the thirty-eight programs listed in the Climate Action

Team Report, nor does it mean relying on a single policy instrument such as an economywide carbon tax or cap-and-trade scheme.

To simultaneously reduce GHGs to meet the 2020 reduction target and to stimulate innovation broadly across the economy to prepare for longer-term emission cuts, a combination of cost-effective regulatory policies and a well-designed cap-and-trade system (or tax) is needed. The sectoral approach taken in California follows this model: energy consumption is addressed largely by product and building efficiency standards, transportation emissions will be addressed by separate vehicle and fuel performance standards, and electricity and large point sources are likely to be addressed with a cap-and-trade system. Dealing with these sectors individually is advantageous because they offer quite different options for reducing emissions in the short and long term, and they have a very different industrial organization. It appears unwise to rely on a single instrument to achieve emissions reductions and technological innovation in all relevant sectors. A single policy instrument applied to the entire economy (whether a tax or an upstream emissions cap) would likely result in significant changes in some parts of the economy (such as electricity generation) but little change in others (such as transportation fuels). As such, sectoral approaches are likely to become the norm for climate policy. Sectoral approaches also appear to be emerging in the European Union, where a combination of regulatory policies plus the Emission Trading System is being implemented.

Ideally, a single economywide price signal approach would be efficient at achieving the goal of reducing GHG emissions by 2020 and beyond. However, because the real world entails imperfect information, agency problems within large corporations, transaction costs, differential taxes, different regulatory structures, and other less-than-ideal conditions, a single economywide price signal would be relatively ineffective at stimulating the targeted reduction in GHG emissions. That said, sectoral policies should still be designed to be as economically efficient as possible. A sectoral approach is also likely to be significantly better than a single economywide price signal at achieving the second policy goal, technological innovation, because social discount rates are much lower than private discount rates, research into environmental technologies is a public good, and the potential costs of reducing GHG emissions vary greatly among different sectors.

Second, with regard to consumption, some observers argue that high levels of consumption are incompatible with low GHGs and that smaller cars, smaller houses, and simpler lifestyles are necessary. Others argue that solving the global climate change problem is largely one of technological and business innovation to develop and deploy low-GHG energy supplies. So far, much of the public debate about the costs of climate change policies has centered on the question of whether they will enhance economic growth or merely slow the rate of growth. California political leaders have been unwilling to ask the public for any sort of sacrifice to solve the global warming problem, and we expect this state of affairs to continue for some time.

Third, much of California's (and other jurisdictions') climate change policy is vulnerable to federal preemption (Rabe, this volume). Climate change policy in the United States is developing quickly, as illustrated by many of this volume's chapters. In part this is due to the 2006 and 2008 federal elections, which replaced several Republicans who were climate-science skeptics with Democratic legislators who have placed climate policy higher on their agenda.[7] Increased interest in climate policy may also be partly a result of the most recent IPCC report, which found that climate change was "very likely" to be anthropogenic and to have significant negative consequences. Whatever the cause, more climate policy bills circulated during the 2007–2008 session of Congress than ever before and the Obama administration appears likely to continue accelerated federal activity.

California and other states will probably remain in the lead in U.S. climate policy development for several more years, as lawmakers and regulators get to work. During this time, state climate policies will begin to come into effect. How much these serve mostly as policy experiments and as lessons, and how much they become integrated into national (or international) policy, is left to be determined. Perhaps the most contentious cases will be where states are willing to take on more ambitious policy goals than the federal government. While California and other U.S. states have important roles to play, they cannot substitute for global leadership by the United States government. President Obama will face a difficult task, but he will at least be able to point to the development of strong climate policies in the states in returning the United States to the international effort to fight global warming.

Notes

The authors were participants in some of the activities described in this chapter. They conducted research supported by the Public Interest Energy Research program and were codirectors of the 2006 study *Managing Greenhouse Gas Emissions in California*. The authors thank Jason Mark and Alan Sanstad for helpful suggestions on earlier versions. Support was provided by the Climate Decision Making Center created through a cooperative agreement between the National Science Foundation (SES-034578) and Carnegie Mellon University. All opinions and any remaining errors are ours alone. An earlier version of this chapter was written in March 2007; it was revised and updated in June 2008. Alex Farrell passed away tragically and unexpectedly in April 2008. It is an immeasurable loss to the climate and energy policy communities. He is sorely missed, and this chapter is dedicated to his memory.

1. The Proposed Scoping Plan, issued on October 15, 2008, calls for transportation and natural gas to be covered by the emission trading system in 2015.

2. In 2006, SB107 accelerated this target to 20 percent by 2010. Governor Schwarzenegger has set a longer-term state goal of 33 percent by 2020 and the CPUC and the CEC are working to achieve that goal.

3. In December 2004, the Governor of Oregon had set a target of reducing the state's GHG emissions 10 percent below the 1990 level by 2020 and 75 percent below the 1990 level by 2050. Governor Schwarzenegger wanted his action to exceed that of any other state, which he accomplished with the 2050 target. Oregon's targets were enacted into law by HB3543 in August 2007.

4. The market failure/barrier argument raises two questions. First, is it empirically true that these barriers exist—that consumers, for example, apply a high discount rate when comparing operating cost savings versus an increased up-front purchase cost? Here, the evidence seems unambiguous. Second, would there be a large and lasting welfare loss if, say, appliance standards forced cheap but low-energy-efficiency items off the market, or would consumers adjust their preferences and soon get used to the forced change in their choice set? Conventional economic models, including computable general equilibrium models, rule out preference change, but this seems unreasonable as an empirical proposition. They rule out behavioral change induced by any nonprice triggers, and they allow a limited role for technological change. This colors their findings.

5. The four propositions were: Proposition 74 (on teacher tenure), Proposition 75 (on union dues for political purposes), Proposition 76 (on spending limits and school funding), and Proposition 77 (on reapportionment).

6. As quoted in the *Los Angeles Times* and the *San Francisco Chronicle* on April 12, 2006.

7. This trend may accelerate given the outcome of the presidential and congressional elections in November 2008.

References

Bailie, Alison, Bill Dougherty, Charlie Heaps, and Michael Lazarus. 2004. *Turning the Corner on Global Warming Emissions: An Analysis of Ten Strategies for California, Oregon, and Washington*. San Francisco: Energy Foundation.

Brown, M. A. 2001. Market Failures and Barriers as a Basis for Clean Energy Policies. *Energy Policy* 29(14): 1197–1207.

California Air Resources Board. 2004. *Staff Proposal Regarding the Maximum Feasible and Cost-Effective Reduction of Greenhouse Gas Emissions From Motor Vehicles*. Sacramento: California Environmental Protection Agency.

California Energy Commission (CEC). 2003. *Integrated Energy Policy Report*. Sacramento: California Energy Commission.

California Energy Commission (CEC). 2005. *2005 Integrated Energy Policy Report*. Sacramento: California Energy Commission.

California Energy Commission (CEC). 2006. *Our Changing Climate: Assessing the Risks to California*. Sacramento: California Energy Commission.

California Energy Commission and California Air Resources Board (CEC/CARB). 2003. *Reducing California's Petroleum Dependence*. Sacramento: California Energy Commission and California Air Resources Board.

California Energy Commission and California Public Utilities Commission. 2003. *California Energy Action Plan*. Sacramento: California Energy Commission.

California Public Utilities Commission (CPUC). 2004. *Order Adopting Long Term Procurement Plans*. San Francisco.

California Public Utilities Commission (CPUC). 2006. *Order Instituting Rulemaking to Promote Policy and Program Coordination and Integration in Electric Utility Resource Planning*. San Francisco: California Public Utilities Commission.

Chinn, Lily N. 1999. Can the Market Be Fair and Efficient? An Environmental Justice Critique of Emissions Trading. *Ecology Law Quarterly* 26(1): 80–125.

Climate Action Team. 2006. *Report to Governor Schwarzenegger and the Legislature*. Sacramento: California Environmental Protection Agency.

Energy Information Administration. 1998. *Impacts of the Kyoto Protocol on U.S. Energy Markets and Economic Activity*. Washington, DC: U.S. Department of Energy.

Farrell, Alexander E., Jill Jaeger, and Stacy D. VanDeveer. 2006. *Overview: Understanding Design Choices*. In *Assessments of Regional and Global Environmental Risks*, edited by Alexander E. Farrell and Jill Jaeger. Washington, DC: RFF Press.

Franco, Guido. 2002. *Inventory of California Greenhouse Gas Emissions and Sinks: 1990–1999*. Sacramento: Public Interest Energy Research Program.

Franco, Guido, Robert Wilkinson, Alan Stanstad, Mark Wilson, and Edward Vine. 2003. *Climate Change Research, Development, and Demonstration Plan*. Sacramento: Public Interest Energy Research Program.

Fried, Jeremy S., Margaret S. Torn, and Evan Mills. 2004. The Impact of Climate Change on Wildfire Severity: A Regional Forecast for Northern California. *Climatic Change* 64(1–2): 169–191.

Gerston, Larry N., and Terry Christensen. 2004. *Recall! California's Political Earthquake*. Armonk, NY: M. E. Sharpe.

Hayhoe, Katharine, Daniel Cayan, Christopher B. Field, Peter C. Frumhoff, Edwin P. Maurer, Norman L. Miller, Susanne C. Moser, Stephen H. Schneider, Kimberly Nicholas Cahill, Elsa E. Cleland, Larry Dale, Ray Drapek, W. Michael Hanemann, Laurence S. Kalkstein, James Lenihan, Claire K. Lunch, Ronald P. Neilson, Scott C. Sheridan, and Julia H. Verville 2004. Emissions Pathways, Climate Change, and Impacts on California. *Proceedings of the National Academy of Sciences of the United States of America* 101(34): 12422–12427.

Howarth, Richard B., Brent M. Haddad, and Bruce Paton. 2000. The Economics of Energy Efficiency: Insights from Voluntary Participation Programs. *Energy Policy* 28(6–7): 477–486.

Krier, James E., and Edmund Ursin. 1977. *Pollution and Policy: A Case Essay on California and Federal Experience with Motor Vehicle Air Pollution 1940–1975*. Berkeley: University of California Press.

Martin, Mark. 2006. Lawmakers Hold Key to Energy Aide's Job: They Won't Hold Hearings to Confirm Governor's Adviser. *San Francisco Chronicle*, April 26.

McCarthy, James E. 2007 *California's Waiver Request to Control Greenhouse Gasses under the Clean Air Act*. Washington, DC: Congressional Research Service Report RL34099.

Nordhaus, William D., and Zili Yang. 1996. A Regional Dynamic General-Equilibrium Model of Alternative Climate Change Strategies. *American Economic Review* 86(4): 741–765.

Rabe, Barry G. 2004. *Statehouse and Greenhouse: The Emerging Politics of American Climate Change Politics*. Washington, DC: Brookings Institution Press.

Roland-Holst, David. 2006. *Economic Growth and Greenhouse Gas Mitigation in California*. Berkeley: Climate Change Center at the University of California.

Roos, Maury. 2003. The Effects of Global Climate Change on California Water Resources. *Public Interest Energy Research Program—Research Development and Demonstration Plan*. Sacramento: California Energy Commission.

Rosenfeld, Arthur H. 1999. The Art of Energy Efficiency: Protecting the Environment with Better Technology. *Annual Review of Energy and the Environment* 24: 33–82.

Schwarzenegger, Arnold. 2005. *Executive Order S-3-05*. Office of the Governor, Sacramento: California.

Shaw, Rebecca. 2002. Ecological Impacts of a Changing Climate. *PIER Environmental Area Report*. Sacramento: California Energy Commission.

Sweeney, James L. 2002. *The California Electricity Crisis*. Stanford, CA: Hoover Institution Press.

6

Climate Leadership in Northeast North America

Henrik Selin and Stacy D. VanDeveer

Introduction

Many of the most ambitious climate change policy efforts in North America are developing in the Northeast, with several state-level initiatives in the region leading the way in climate change policies and regulations that exceed federal requirements. State policymakers have taken action to control greenhouse gas (GHG) emissions from power plants and vehicles; to set standards for energy generation, land use, agriculture, forestry and waste management; and to set green mandates for public sector spending and operations. In Canada, the eastern provinces play similar roles in the context of federal-provincial relations on climate change mitigation and adaptation. A growing number of municipalities, firms, and civil society organizations across the North American Northeast are also at the forefront in expanding climate change policy action and GHG reduction efforts.

Climate change policies in the Northeast are developing in a context of increasingly sophisticated assessments of regional vulnerabilities. Politically and economically important sectors in the region, such as forestry, agriculture, fishing, and tourism, may be greatly affected by temperature and precipitation changes, rising sea levels, and coastal erosion (Daley 2007; National Assessment Synthesis Team 2000; Union of Concerned Scientists 2006). Several states in the region are significant emitters of GHG emissions. For example, Massachusetts, with a population of 6 million, emits GHG emissions roughly equal to those of Austria (population 8 million), Greece (population 10 million), or Egypt (population 70 million) (Hamel 2003). Levels of GHG emissions in New York (population 20 million) are roughly comparable to Thailand (population 65 million), Turkey (population 70 million), or Indonesia (population 230 million) (Rabe 2004).

This chapter analyzes climate change action in Northeast North America. It explores how regional institutions and policy networks with members from public, private, and civil society sectors are taking on important leadership roles in pioneering climate change policymaking and implementation. In addition, the chapter discusses key aspects of major policy developments and their importance for effective climate change mitigation. Political action in the Northeast demonstrates the appropriateness and technical and economic feasibility of more ambitious efforts to reduce GHG emissions, setting important precedents upon which future North American climate change policy may be modeled. However, GHG emissions have continued to grow in the region, albeit slowly, and more aggressive public and private sector efforts are needed to meet short- and long-term GHG emission reduction goals.

First, the chapter discusses the importance of regional institutions and networks in expanding local-level policymaking in the Northeast. This is followed by an analysis of institutions for policy change in the Northeast, including two major state-led cooperative efforts: the 2001 Climate Change Action Plan of the New England Governors Conference and the Eastern Canadian Premiers (NEG-ECP), and the recent establishment of a regional CO_2 cap-and-trade scheme under the Regional Greenhouse Gas Initiative (RGGI). In addition, regional municipal, civil society, and private-sector climate change efforts are examined. The next section assesses possibilities and limitations of Northeastern climate change action in the context of developing regional and national policymaking. The chapter ends with a few concluding remarks on key features of current and future climate change action in northeastern North America.

Institutions and Networks for Climate Change Action

Climate change action and policymaking in Northeast North America are developing under both long-standing and newly created institutions (Selin and VanDeveer 2005, 2006). Many of these overlapping institutions are regional, while others are national or even international in geographical scope. These institutions bring together state and provincial civil servants and policymakers, municipal officials, scientists, technical experts, private sector representatives, and staff members from environmental organizations. This growing web of climate change institutions fosters debate, scientific assessments, and policymaking, facilitating leadership opportunities for actors within networks. As institutionalization of climate change action in the Northeast increases, this expansion sustains attention to climate change.

The creation of the 2001 NEG-ECP Climate Change Action Plan benefited greatly from existing institutions and long-standing regional political and economic cooperation efforts. Governors of the six New England states—Connecticut, Maine, Massachusetts, New Hampshire, Rhode Island, and Vermont—began holding annual meetings in 1937. Regional cooperation was expanded in 1973 to include premiers of the five Eastern Canadian provinces (New Brunswick, Newfoundland and Labrador, Nova Scotia, Prince Edward Island, and Québec). Since then, the governors and premiers have met regularly to develop policies and programs on cross-boundary issues and to promote regional integration in areas of economic development, transportation, energy, environment, and human health. Since the mid-1990s, NEG-ECP environmental cooperation has focused on acid rain, ground-level ozone, mercury pollution, and climate change, with climate change issues growing in significance after 2000.

The activities of the NEG-ECP are coordinated through the secretariat of the New England Governors' Conference, located in Boston, Massachusetts. In Canada, the Council of Atlantic Premiers in Halifax, Nova Scotia acts as a supportive clearinghouse and cooperative mechanism. Most NEG-ECP policy is developed in working groups and steering committees consisting of staff from the governors' and premiers' offices and state/provincial administrative agencies. Climate change issues are mainly addressed in the Environment Committee, which is comprised of the commissioners of the state/provincial departments of environmental protection. The Environment Committee, however, also collaborates with the NEG-ECP energy and transport working groups on related GHG reduction issues. In addition, NEG-ECP staff members work closely with the New England Regional Office of the U.S. Environmental Protection Agency and with officials from Environment Canada.

Northeastern states are furthermore building new collective institutions for CO_2 emissions trading under RGGI. This effort connects the six New England states with four other states in the Northeast not included in NEG-ECP cooperation: New York, New Jersey, Maryland, and Delaware (Canadian provinces are not part of RGGI). The RGGI trading scheme, which became fully operational in 2009, is the first public sector CO_2 emissions trading scheme in North America. This effort draws on U.S. experiences with earlier trading schemes for SO_2 and NO_X emissions, which were originally launched by Northeastern states and later became part of federal policy (Aulisi, Farrell, Pershing, et al. 2005). In another example of a state-led initiative, public clean energy funds were used to create the Clean Energy States Alliance in 2003. This alliance, which by early 2009 involved funds from eighteen states, supports clean energy projects and companies across member states.

In building institutions for climate change action in the Northeast, public officials are taking advantage of the region's many advanced research centers. For example, the Northeast States for Coordinated Air Use Management (NESCAUM), an interstate association of air quality control divisions created in 1967, early on became actively involved in the development of regional emissions inventories and registries linked to RGGI. On the Canadian side of the border, Ouranos was established by several Canadian government agencies and universities in Montreal in 2002 to work on climate change and adaptation. These research organizations—together with many of North America's leading universities located in the region—are engaged in state-of-the-art climate change research that is fed into regional scientific assessments and economic estimates of existing and proposed climate change mitigation and adaptation efforts.

Municipal climate change action in the Northeast also takes place within institutional frameworks connecting small and large municipalities across the region. However, municipal policy typically develops separately from state actions even though municipal actions influence state and regional GHG emissions. Many municipalities draw institutional support from the International Council for Local Environmental Initiatives (ICLEI) and its Cities for Climate Protection (CCP) campaign. The ICLEI was founded in 1993 to aid in the development of local policy across countries (Betsill 2001; Betsill and Bulkeley 2004; Kousky and Schneider 2003). Under the CCP program, municipalities typically set GHG emission reduction goals, develop policies aimed at reaching these goals, and create processes for monitoring and verifying results. In addition, many municipalities in the Northeast are interacting through the Federation of Canadian Municipalities and the U.S. Mayors Climate Protection Center (Gore and Robinson, this volume).

Civil society actors are also increasingly engaged in regional and local institution building on climate change issues. An expanding regional network of environmental nongovernmental organizations (NGOs) has coalesced around climate change action to establish the New England Climate Coalition (NECC). This coalition includes state Public Interest Research Groups (PIRGs), state chapters of Clean Water Action and the Sierra Club, and numerous local environmental and public health groups and newer NGOs focused explicitly on climate change, such as Clean Air–Cool Planet and Environment Northeast. The region's NGOs coordinate much of their research, public awareness, and lobbying through NECC. The Union of Concerned Scientists is also located in Massachusetts, and this group has done extensive work to connect researchers, policymakers, and citizens around climate change issues in the Northeast and across the United States (Union of Concerned Scientists 2006).

Universities in the Northeast are not only involved in cutting-edge climate change research but also engaged in a host of climate change education, outreach, and abatement initiatives. Several of these efforts are institutionalized through a university outreach program sponsored by the New England Board of Higher Education and NEG-ECP. Many universities are the size of towns or small cities and operate as independent entities. As such, campus action can both contribute to regional GHG reductions and serve as models for other public- and private-sector actors considering taking more aggressive climate change actions (Levine, this volume). A growing number of universities in the Northeast, for instance, are setting examples in areas of energy efficiency and renewable energy use. In addition, universities provide a multitude of expertise benefiting local, state, and regional climate change action.

Many climate change efforts throughout the Northeast are facilitated by expanding networks across public, private, and civil society spheres. These climate change networks are often organized around key organizational nodes. State and provincial officials frequently interact through NEG-ECP working groups, RGGI-related organizations, and the Clean Energy States Alliance. Municipal officials utilize ICLEI, the U.S. Mayor Climate Protection Center, and the Federation of Canadian Municipalities to facilitate communication and interaction. NGOs such as Clean Air–Cool Planet and the New England Climate Coalition connect public- and private-sector representatives across the region who work on a host of issues related to climate change and energy.

These organizations and the connections they foster are important for network members, who use them to share information about the best practices, to generate and diffuse new policy ideas, and to gain support for the development and implementation of more ambitious climate change action. These organizations and the networks around them give climate change policy advocates in each Northeastern state, province, municipality, firm, or NGO opportunities to interact and meet with others from within and outside the region who are engaged in climate change initiatives. Network members use these professional and personal connections and the information they gather about GHG mitigation efforts elsewhere to leverage more comprehensive policymaking and action within their own jurisdiction (Selin and VanDeveer 2005, 2006).

Climate Change Action in the Northeast

Joint climate change policy efforts among Northeastern states and provinces include two distinct but overlapping multijurisdictional initiatives. First, the regional Climate Change Action Plan that was signed by the governors of six New England states and the premiers of five Eastern Canadian provinces in 2001 has spurred the development of action plans and a host of policies by states and provinces. Second, RGGI establishes a regional cap-and-trade scheme for CO_2 emissions from power plants in ten states starting in 2009. In addition to these state and provincial actions, a growing number of municipalities in the Northeast are developing more aggressive climate change policy and standards for GHG reductions as civil society actors in the region are becoming increasingly active on climate change issues. This section examines the development and implementation of these multiple efforts.

NEG-ECP and the Development of Action Plans and Policies

The collaborative effort by the New England governors and the Eastern Canadian premiers includes all six New England states and five Eastern Canadian provinces. In 2000, this region's total GHG emissions were higher than those of all but ten industrialized countries reporting under the United Nations Framework Convention on Climate Change (NESCAUM 2004). Under the 2001 Climate Change Action Plan, the states and provinces committed to reducing GHG emissions to 1990 levels by 2010 and to achieve a 10 percent reduction below 1990 levels by 2020. The action plan calls for ultimate reductions in GHG emissions to levels that do not pose a threat to the climate system, which is estimated to require a 75 to 85 percent reduction from 2001 emissions levels. The plan outlines nine general actions and goals pursuant to these emissions reduction targets:

1. Establish a regional standardized GHG emissions inventory.
2. Establish a plan for reducing GHG emissions and conserving energy.
3. Promote public awareness.
4. Encourage state and provincial governments to lead by example.
5. Reduce GHG emissions from the electricity sector.
6. Reduce total energy demand through conservation.
7. Reduce and/or adapt to negative social, economic, and environmental impacts.
8. Decrease the transportation sector's growth in GHG emissions.
9. Create a regional emissions registry and explore a trading mechanism.

To specify policy options for the implementation of these nine actions, the action plan also outlines thirty-four additional policy recommendations. Some recommendations involve building regional institutions for continued policymaking and implementation review, while others call for state and provincial policymaking to support the regional policy goals and emissions reduction targets. In addition, the action plan's recommendations contain provisions for outreach efforts to private- and public-sector groups and for raising public awareness. Since 2001, state and provincial officials have worked to develop and implement state- and provincial-level policies and programs and to develop institutions in support of this action plan (Selin and VanDeveer 2005, 2006). Accomplishments to date are generally modest, however, and also vary significantly among individual states and provinces (David Suzuki Foundation 2006; NECC 2005, 2006, 2008; Stoddard and Murrow 2006).

Initially, state and provincial officials focused their attention on launching relatively small-scale abatement programs. Many such programs are "smart growth" and "no-regrets" measures, which seek to simultaneously reduce energy use, public financial costs, and GHG emissions. Examples of such measures include using more efficient light-emitting diodes in traffic lights, promoting the purchase of Energy Star products in state governments, and switching to more energy-efficient vehicles in state vehicle fleets. Despite the smaller size and ambition of these programs, they can save local budgets millions of dollars in public expenditures annually (Hamel 2003). But state and provincial officials generally recognize that these modest GHG reduction programs will not meet the emission reduction goals for 2020 and beyond formulated in the NEG-ECP action plan.

In 2003, Maine became the first state to include the NEG-ECP emission reduction goals in state law. By 2005, all six New England states had issued relatively broad climate change action plans designed to help meet the NEG-ECP reduction goals and additional state-formulated targets. In 2008, Connecticut became the first New England state to write specific short- and long-term GHG emission targets into state law (10 percent below 1990 levels and 80 percent below 2001 levels by 2050), following similar legislation passed in California, Hawaii, New Jersey, and Washington. The Pew Center for Global Climate Change tracks climate- and energy-related policies and programs among U.S. states. By 2008, the six New England states averaged 16.5 initiatives out of the 20 tracked by the Pew Center—far more than the average U.S. state. Similarly, New York and New Jersey each engaged 17 such policies and programs.

Generally speaking, Canadian provinces and territories have been less active than U.S. states in making and implementing climate change policies. While all measure GHG emissions, fewer than half of the provinces and territories had even completed climate change action plans by 2006 (David Suzuki Foundation 2006). Among the Eastern provinces active in NEG-ECP cooperation, only Newfoundland and Labrador had provincial action plans by early 2006. However, New Brunswick recently completed its action plan as part of a trend in the provinces and territories toward increased climate change action after 2006 (Marshall 2007; Stoett, this volume). By 2008, British Columbia had emerged as a Canadian climate change leader, and several other provincial and territorial authorities began to prioritize policies designed to increase renewable energy investment and production.

By 2007, all the New England states and four out of the five Eastern Canadian provinces had adopted mandatory renewable portfolio standards, some of which have been significantly strengthened over time (see table 6.1). Such standards set minimum percentages for the amount of renewable energy generated and sold within those states' or provinces' overall energy markets. States and provinces have also created public benefits funds designed to support energy efficiency and/or renewable energy development. In addition, some initial regulatory progress can be

Table 6.1
State and provincial renewable portfolio standards

State/Province	Goals (as of April 2008)
Connecticut	House Bill 7432 from 2007 requires that 27 percent of the state's electricity come from renewable sources by 2020.
Maine	Maine's Public Utilities Commission in 1999 mandated that 30 percent of the state's power should come from renewable resources by 2000. In 2007, Maine passed a law requiring an increase in new renewable energy capacity of 10 percent by 2017 (new sources defined as those placed into service after September 1, 2005).
Massachusetts	In 2002, the Massachusetts Division of Energy Resources adopted a renewable portfolio standard requiring that 4 percent of the state's electricity come from new renewable sources by 2009. After 2009, the minimum renewable standard will increase by 1 percent annually.
New Brunswick	The Electricity Act of 2006 mandates that 1 percent of electricity had to come from renewable sources by 2007, to be increased by 1 percent annually up to 10 percent by 2016.
New Hampshire	House Bill 873 from 2007 established a renewable portfolio standard mandating that 25 percent of the state's electricity come from renewable sources by 2025.
Newfoundland and Labrador	None. However the province already derives over 90 percent of its electricity from renewable sources.
Nova Scotia	The renewable energy standards from 2007 mandates renewable energy increases of 5 percent to the total supply by 2010 and 10 percent by 2013. It is estimated that approximately 20 percent of the province's electricity will come from renewable sources by 2013.
Prince Edward Island	The Renewable Energy Act of 2005 requires utilities to acquire at least 15 percent of electrical energy from renewable sources by 2010. An additional goal of 100 percent capacity of electrical energy from renewable energy sources by 2015 has been formulated but has not yet been proclaimed into law.
Québec	No overall percentage goal, but specific targets for increased hydropower and wind power generation have been formulated. The provincial government intends to launch hydroelectric projects through Hydro-Québec to create an additional 4,500 MW of generating capacity by 2015, constituting an increase of almost 10 percent for the province. In addition, the provincial government aims to reach 4,000 MW of wind power by 2015, constituting a provincial increase of over 800 percent.
Rhode Island	The 2004 Clean Energy Act established a requirement that electricity retailers derive at least 3 percent of the electricity they sell in state from renewable sources by 2006. The percentage of renewable energy required increases by 1 percent annually up to 16 percent by 2020.

Table 6.1
(continued)

State/Province	Goals (as of April 2008)
Vermont	The 2005 renewable portfolio standard law requires renewable generation to equal incremental load growth between 2005 and 2012. If utilities do not meet this requirement, the state will set a renewable portfolio standard equal to the percentage of load growth between 2005 and 2012. If by 2012 the state experiences 7 percent load growth, but utilities have not obtained 7 percent of their electricity from renewable sources, the state will adopt a renewable portfolio standard of 7 percent.

Sources: Data for Nova Scotia from http://www.gov.ns.ca/energy/renewables/renewable-energy-standard/. Data for the New England states come from the Pew Center on Global Climate Change while data for the Eastern Canadian provinces come from respective provincial websites and authors' personal communication with provincial officials. Data Vermont from http://www.leg.state.vt.us/docs/legdoc.cfm?URL=/docs/2006/acts/ACT061.HTM.

noted in efforts to cap and reduce CO_2 emissions from power plants prior to RGGI's establishment in, for example, Connecticut, New Hampshire, and Massachusetts.

Despite expanding efforts to increase the generation and use of renewable energy, several older and relatively inefficient oil and coal-fired power plants remain in use across the region. For example, the NGO Environment Northeast (2003, 12) estimates that replacing such facilities with more modern and efficient natural gas plants could reduce Connecticut's GHG emissions by up to 60 percent. Increased use of renewable energy sources would, of course, allow for even more dramatic cuts in GHG emissions. Public- and private-sector actors in New England—home to the country's first large-scale electricity-producing windmill installed in Grandpa's Knob, Vermont in 1941—have shown an increased interest in expanding wind power capacity. Yet, high-profile local and political opposition to Cape Wind—a proposed large-scale wind farm in Nantucket Sound, Massachusetts—demonstrates that the expansion of renewable energy capacity can be highly contentious (Williams and Whitcomb 2007).

Whereas state and provincial policymakers are increasingly attempting to reduce GHG emissions from the utilities sector, policy progress on reducing GHG emissions from the transportation sector to date has been negligible, at best (NECC 2005, 2006). Transportation generates approximately one-third of regional GHG emissions and increases in transportation-related emissions alone make the NEG-ECP emission reduction goal for 2010 difficult, if not impossible, to meet (MASSPIRG 2003; NECC 2005, 2006; NEG-ECP 2006; NESCAUM 2004; Stoddard and Murrow 2006). However, most Northeastern states plan to adopt California's vehicle standards for CO_2 emissions, if these survive ongoing legal challenges. In fact, several states' laws require them to adopt California vehicle emissions standards when they are stricter than federal standards (Farrell and Hanemann, this volume; PEW Center on Global Climate Change 2006).

States in the Northeast have also led legal efforts to force the U.S. federal government to regulate GHG emissions. In 1999, a group of environmental NGOs petitioned the EPA to set CO_2 emission standards for vehicles, but this petition was rejected by the EPA, which argued the Clean Air Act did not provide any authority for regulating CO_2 (Hileman 2006). In 2003, attorneys general from Massachusetts and eleven other states—California, Connecticut, Illinois, Maine, New Jersey, New Mexico, New York, Oregon, Rhode Island, Vermont, and Washington—filed a legal suit challenging this ruling with the support of many state regulatory agencies, city officials, and environmental groups (*Commonwealth of Massachusetts v. U.S. Environmental Protection Agency*). In April 2007, the U.S. Supreme Court (in a 5–4 decision) ruled that the Clean Air Act gives the EPA the authority to regulate CO_2 emissions (Mauro 2007).

This Supreme Court ruling, which constituted a major legal victory for the attorneys general, charged the EPA to decide whether to regulate CO_2 emissions from vehicles under the Clean Air Act (Austin 2007). When EPA officials stalled, the attorneys general of Massachusetts and sixteen other states (together with the corporation counsel for New York City, the city solicitor of Baltimore, and thirteen environmental advocacy groups) filed a Petition for Mandamus with the U.S. Court of Appeals for the District of Columbia in April 2008, requesting the EPA to be required to act (Massachusetts 2008). This legal dispute, together with similar cases involving other states and other areas of GHG policy, demonstrate that political and legal struggles over climate change policy are likely to continue for several years, well into the Obama administration.

The Regional Greenhouse Gas Initiative (RGGI)

In 2003, Governor George Pataki of New York invited states from Maryland to Maine to participate in discussions about a regional cap-and-trade system for CO_2 emissions. The result of these discussions, RGGI, became operational in January 2009 with ten founding state members: Maine, Vermont, New Hampshire, Massachusetts, Connecticut, Rhode Island, New York, New Jersey, Delaware, and Maryland. While no Canadian provinces joined RGGI, several provincial officials attend and observe RGGI meetings. Through the development and implementation of RGGI, leader states in the Northeast are setting important technical and political precedents for North American climate change policymaking. The RGGI process has also aided in the building of regional coalitions for policy change and

GHG reductions among key stakeholders across public, private, and civil society sectors.

The Regional Greenhouse Gas Initiative is intended to help member states to reduce their GHG emission and meet state goals more efficiently. In addition to the statewide GHG reduction goals for the six New England states formulated under the NEG-ECP action plan and Connecticut's even more ambitious state law from 2008, New Jersey and New York have set individual targets. An early New Jersey GHG reduction goal called for a 3.5 percent reduction by 2005, but this goal was missed by a wide margin (Algoso and Rusch 2005). New Jersey's current GHG reduction goal—reaching 1990 levels by 2020 and 80 percent below 2006 levels by 2050—was signed into law in 2007. New York's GHG reduction target, adopted by the State Energy Planning Board in 2002, sets the goal of a 5 percent reduction below 1990 levels by 2010 to be followed by a 10 percent reduction below 1990 levels by 2020. By early 2009, neither Delaware nor Maryland had set any statewide emission reduction goals.

Even though many economists argue that taxes are often more efficient policy tools than a cap-and-trade program for reducing CO_2 emissions, most public- and private-sector actors in the Northeast are committed to emissions trading programs like RGGI rather than new taxation schemes (Cooper 1998; McKibbin and Wilcoxen 2002; Pizer 1997). In part, this is because emissions trading programs create marketable allowances (Betsill, this volume). Also, CO_2 tax schemes establish more visible costs to consumers (and voters) and have fewer identifiable supporters in the public or private sectors. RGGI—which draws heavily on the experiences of the Northeast states from the SO_2 and NOx emissions trading schemes, and on the European Union's Emissions Trading Scheme for CO_2 emissions—established a regional emissions inventory, registry, and trading mechanism for CO_2 emissions from power plants based on several years of negotiations among state officials and extensive debate, data gathering and modeling.

The RGGI rules are outlined in a "model rule" finalized in 2007. Many regional energy producers and advocacy groups, as well as national firms like Wal-Mart, commented on earlier drafts. The model rule is implemented by each state. The RGGI rules initially cover CO_2 emissions and power generators of 25 megawatts (MW) or greater. Each power plant must have enough allowances to cover its CO_2 emissions during each compliance period, if necessary by entering the market to purchase additional allowances (one allowance equals one short ton of CO_2). The RGGI initiative, which operates on the basis of a three-year compliance period, is designed to stabilize CO_2 emissions from the power sector between 2009 and 2014. From 2015 through 2018, each state's annual CO_2 emissions budget will decline by 2.5 percent per year, achieving a total 10 percent reduction by 2019.

The RGGI rules also permit that some emission reductions may be achieved through emissions offset projects outside the regulated power sector. States can review these requirements if a national cap-and-trade program in enacted.

The RGGI states are allotted individual emission caps within a regional ten-state cap of 188 million short tons, or 171 million metric tons, between 2009 and 2014. The initiative mandates that at least 25 percent of state allowances have to be auctioned to companies covered by RGGI, but most participating states decided to auction most or all of their allocated allowances. Proceeds from the sale of allowances are intended for public benefit purposes, including supporting clean energy technology development or investments in renewable energy facilities or energy efficiency programs. The first publicly announced trading of RGGI allowances took place in February 2008 as options were sold for the 2009–2010 period. Although the specific price was not disclosed, it is estimated to have been somewhere in the $5 to $10 range (Point Carbon 2008b, 2).

During the creation of RGGI, much attention was paid to the potential costs of RGGI to the region's firms and households through increased electricity prices. The economic models on which the RGGI estimates are based are rather conservative regarding economic costs and benefits (Brome 2006). Technological innovation may also absorb significant costs of reducing emissions, as witnessed in the case of the federal SO_2 trading scheme established in the early 1990s (Munton 1998). Similarly, regional compliance costs from the NO_X emissions trading scheme created in the late 1990s have been much lower than initially predicted by many models (Aulisi, Farrell, Pershing, et al. 2005). Technological innovation in the area of CO_2 reductions may also mean that companies in the region benefit from these product developments if national and international markets for low emissions technology grow in ways that are widely expected.

The effect of RGGI on residential electricity rates is projected to be less than 1.5 percent through 2021, and energy efficiency components built into the program may even result in a positive economic effect (Brome 2006). The initiative also includes a price safety valve, which expands the compliance period if the allowance price equals or exceeds $10/ton (in 2005 dollars) for twelve months (following an initial fourteen-month "market settling" period at the beginning of each compliance period). Breslow and Goodstein (2005) moreover show that of Massachusetts's twenty-five largest industries (accounting for 81 percent of the state's total economic output), only eight have electricity costs over 1 percent of their total operating costs. This indicates that the economic impact of higher electricity rates would be modest even in the absence of company-based energy savings programs and technological development. In addition, Ruth, Gabriel, Palmer et al. (2008, 2288) argue that "RGGI is predicted to have a positive economic impact" on Maryland's gross state product of about $100 million in 2010 and $200 million in 2015, as a result of energy efficiency savings and new job creation.

Municipal Policy Developments

Legendary Massachusetts native and former congressman and speaker of the U.S. House of Representatives, Tip O'Neill, liked to say that "all politics is local." A growing number of municipalities all over the Northeast are developing a host of climate change–related policies and programs that are designed to reduce GHG emissions, thereby localizing climate change policymaking. Municipal officials cite multiple reasons for initiating such action, including the desire to improve local planning and governance and the responsibility to "contribute to the cumulative solution to climate change" (City of Cambridge 2002). As such, many larger and smaller municipalities across the Northeast are playing important political leadership roles alongside their counterparts all over North America (Gore and Robinson, this volume).

Many mayors in the Northeast are among the well-over-800 mayors from all fifty states who signed Seattle Mayor Greg Nickels's initiative calling on cities to meet or exceed the U.S. commitments under the unratified Kyoto Protocol (7 percent reduction in GHG emissions from 1990 levels by 2012). Many cities also use the U.S. Mayors Climate Protection Center as a mechanism for exchanging best practices and lobbying state and federal governments to enact more stringent GHG legislation (Mayors Climate Protection Center 2007). Similarly, many cities in the Eastern Canadian provinces work on climate change issues through the Federation of Canadian Municipalities. Further evidence that municipal leaders desire more stringent climate change policy can be found in the many Northeastern cities that have joined ICLEI and are in various stages of formulating GHG reduction goals and developing climate change policies pursuant to ICLEI's principles. In the Boston area alone, more than thirty communities are ICLEI members, many of which work closely with ICLEI's Northeast regional coordinator.

Many Northeastern municipalities initiated climate change action before, or around the same time as, more ambitious state-level climate policies were developed in the late 1990s and early 2000s. Municipal officials and climate change action plans are often explicit about their intention to push for more ambitious state, provincial, and federal climate policies (City of Cambridge 2002; Cohen and Murray 2003; Mayors Climate Protection Center 2007). A grassroots campaign in New Hampshire in 2007 moreover resulted in citizens in over 150 town meetings voting in favor of a motion calling on the federal government to adopt more aggressive climate change policies (Laidler 2007; Zezima 2007). In addition, city leaders increasingly engage in international efforts as, for example, when New York City

Mayor Michael Bloomberg hosted the second C40 Large Cities Climate Summit in May 2007 (Cardwell 2007).

Many municipalities, like states and provinces, initially focused on "no-regrets" measures designed to reduce GHG emissions, save energy, and reduce public expenditures (Mayors Climate Protection Center 2007). Large cities in the Northeast are leading the way. For example, Boston is switching to hybrid vehicles in its city vehicle fleet. New York City has adopted several climate change–related standards for public transportation and energy efficiency involving new forms of public-private cofunding (Environment News Service 2007; Revkin and Healy 2007). Many cities mandate green building practices under the LEED (Leadership in Energy and Environmental Design) Green Building Rating System. Practical actions involve installing photovoltaic systems, constructing with recycled materials, minimizing heat/cool air losses, and using sensor lighting systems—all of which are intended to reduce energy use and GHG emissions and to save money in the long term.

Another leadership example in a much smaller community can be found in Hull, Massachusetts (population 11,000). In 2001, Hull commissioned the construction of a municipally owned 660-kilowatt wind turbine to power street lights and traffic lights, with the remaining power up for sale. The turbine saves Hull about $185,000 each year and averts hundreds of tons of CO_2 emissions (Johnson 2006). In 2006, Hull erected a second 1.8-megawatt wind turbine, projected to save the city another half a million dollars annually. Many other regional municipalities and colleges are exploring Hull's successes as they visit Hull, commission local wind studies, and construct their own turbines (Ebbert 2006). Hull and other communities also seek expertise and financial support for wind power investments offered by wind power advocates, state renewable portfolio standards, public benefits funds, and university programs (Ebbert 2006; Johnson 2006; Skolfield 2006).

Civil Society and Private Sector Involvement

Whereas many major national U.S. environmental NGOs have struggled to be effective (Shellenberger and Nordhaus 2004, 2007), a host of new and old environmental advocacy groups in the Northeast have gradually built capacity on climate change advocacy and action since 2000. For example, NGOs such as the New England Climate Coalition, Clean Air–Cool Planet, Environment Northeast, the Union of Concerned Scientists, and state PIRGS are highly active on climate change issues and local and regional policy processes. These civil society organizations play important roles as policy advocates, network members, and supporters of public officials and private sector representatives in favor of more aggressive climate change action.

Regional NGOs prepare well-researched assessments and policy reports; organize stakeholder conferences and workshops; and coordinate lobbying and public awareness campaigns at local, state, and regional levels (Selin and VanDeveer 2005, 2006). In attempts to motivate more active involvement by individual citizens demanding more aggressive climate change policy from elected officials, many of these campaigns are specifically designed to invoke expected negative effects of climate change on iconic aspects of local life in the Northeast, such as fall foliage, maple syrup production, and skiing. In addition, major grant-making foundations with headquarters or offices in the region work closely with both established and newer NGOs on funding research, meetings, and education activities.

NGOs, such as Clean Air–Cool Planet, work with a growing number of businesses in the region on GHG reductions issues. These activities include organizing stakeholder meetings to discuss opportunities and strategies for reducing GHG emissions and working directly with individual firms on company-specific programs targeting buildings, production processes, and supply chains. Major firms engaged in these activities include Timberland, Stonyfield Farms, Verizon Communications, and Bank of America. Many firms in the Northeast are also members of the Pew Center's Business Environmental Leadership Council. In addition, private-sector actors are responding for the development of new state and municipal renewable energy standards. Even the storied Fenway Park baseball stadium has installed solar panels as part of a public-private partnership with the Boston city government designed to increase solar energy production (Ryan 2008).

Many of the region's larger and smaller universities have greatly expanded their climate change initiatives (Levine, this volume; Rappaport 2008). Working with the New England Board of Higher Education, NEG-ECP participates in a university outreach program. To date, over 130 universities in both Canada and the United States have joined this program, which seeks to challenge universities to initiate climate action measures and increase climate change–related research and education efforts on campuses. As part of the program, universities are encouraged to complete and release GHG emission inventories, to stabilize and reduce their GHG emissions, and to share their experiences with each other and with public officials, NGOs, and citizens. In pursuing these goals, universities may use an online "toolkit" supplied by Clean Air–Cool Planet.

Developments in climate change science and politics receive growing coverage in major regional newspapers such as the *New York Times* and *Boston Globe*, together with national magazines and a host of local media outlets. The *New York Times*, in particular, devoted much attention to efforts by the Bush administration to play down or alter scientific data on anthropogenic influences on the global climate system as part of its opposition to mandatory GHG controls (United States House of Representatives 2007). Many of the region's newspapers regularly cover RGGI and other local climate and energy policy initiatives as well as political and legal developments associated with the lawsuits filed by state attorneys general against the federal government. Furthermore, editorials in the region's newspapers tend to support various climate change policy developments at regional, state, and local levels.

Possibilities and Limits of Regional Action

A growing number of political leaders in the Northeast are declaring their belief that human behavior has a discernible influence on the climate and that political measures beyond those mandated by the federal government are needed to more aggressively reduce GHG emissions. Since the late 1990s, policymaking in the Northeast has set a host of important precedents. Yet, it has generated limited environmental impact and rather modest GHG reductions to date.

Developing Climate Change Action in the Northeast

Working together in a host of international, national, regional, and local forums, expanding networks of climate change policy advocates from public, private, and civil society sectors have been important driving forces behind climate policy initiatives in the Northeast (Selin and VanDeveer 2005). These networked policy entrepreneurs frame climate change issues in regional and local terms and exchange scientific, technical, and political information in ways that help to shape policy choices of elected officials and develop more progressive climate change policy across the region.

Northeastern forerunner states, provinces, municipalities, and companies are frequently explicit about their desire to lead by example in the face of lagging federal policymaking (Selin and VanDeveer 2006, 2007). Many forerunners seek to strategically demonstrate the feasibility of more aggressive climate change action. In part, they do this by developing markets. For example, state and provincial renewable portfolio standards expand local and regional markets for renewable energy. Massachusetts's renewable portfolio standard, to take but one example, is driving increased regional investments in renewable energy generation (Commonwealth of Massachusetts 2008), and state energy efficiency programs are slowing aggregate energy demand (Bowles 2008). Furthermore, state/provincial governments, municipalities, and firms promote the purchase of energy-efficient products over more energy-intensive alternatives. All these actions serve to undermine claims about high economic costs of GHG emissions reductions.

Market incentives are also central in the development and implementation of RGGI. Under RGGI, prices of CO_2 emissions in the Northeast are set both through the state-level auctioning of allowances and the subsequent trading of these allowances. Thus RGGI is designed to create direct economic incentives for private- and public-sector utilities to reduce their CO_2 emissions. The creation of RGGI and other efforts mandating GHG reductions also expand markets for consultancy and accounting firms offering their services to private and public organizations that want to participate in credit and/or offset schemes for CO_2 reductions. Moreover, much venture capital is invested in firms in the Northeast and elsewhere in North America that are developing and/or deploying GHG-reducing technologies (Rivlin 2005; Weisman 2006).

There are, of course, obstacles to further expanding climate change policymaking and action in the Northeast. Although skeptics of climate change science are becoming fewer and less vocal, some opponents argue that higher energy prices resulting from GHG controls will be "death to New England" (Reisman 2003). Governors of Massachusetts (Mitt Romney) and Rhode Island (Donald Carcieri) did not initially sign RGGI's Memorandum of Understanding in December 2005 in large part because of private-sector opposition. Massachusetts did not join RGGI until a new governor (Deval Patrick) took office in January 2007. Governor Carcieri announced the same month that Rhode Island would also become a member. In 2008, Maryland legislators voted down the Global Warming Solutions Act in the face of strong opposition from industry groups and trade unions (Point Carbon 2008c, 4–5). This bill would have required Maryland to cut GHG emissions by 25 percent from 2006 levels by 2020 and by 90 percent from 2006 levels by 2050.

In addition, many efforts by municipal leaders to green their cities are not successful. For example, Mayor Bloomberg's administration expended significant resources to make New York City the first U.S. city to introduce a congestion fee. Making the announcement on Earth Day in 2007, Mayor Bloomberg advocated for a charge of $8 for cars and $21 for trucks for driving below 60th Street between 6 a.m. and 6 p.m. Monday through Friday (Newman 2007). This program was proposed as part of a larger initiative to reach the goal of reducing the city's GHG emissions by 30 percent by 2030, with fees to be used to finance subway and other public transport expansions. This proposal was rejected by Albany lawmakers in April 2008 (Confessore 2008). Similarly, many wind energy proposals encounter substantial local opposition. Lastly, questions have been raised about the allocation of RGGI emissions permits, given that 2007 emissions (before RGGI took effect) were below the RGGI cap (Jones and Levy, this volume). An overallocation of permits depresses prices, thereby reducing incentives for firms to cut emissions.

Pushing Actions and Policy beyond the Northeast

In the past, a multitude of early environmental policies developed by leading states, provinces, municipalities, and/or firms have served as models for subsequent initiatives by other such actors and federal policymakers (Rabe, this volume; Rothenberg 2002; Stoett, this volume). The many local initiatives discussed earlier, and the well-networked climate policy advocates associated with them, may be influential well beyond the Northeastern region. Policies enacted in the Northeast, if perceived as even moderately successful, will be invoked by climate change policy advocates across North America as evidence that measured climate change policy is possible while maintaining (and possibly even improving) the competitiveness of local economies. Clean Air–Cool Planet also recently merged with the Climate Policy Center, based in Washington, D.C., because of a shared desire to promote successful programs beyond the Northeast.

The ten RGGI states have agreed to try to recruit additional members that may not be geographically contiguous. Regional groups of states and provinces on the West Coast and around the Great Lakes are increasing their climate change policy efforts, including the development of regional trading schemes (Point Carbon 2008a, 1). Subnational jurisdictions are increasing their collaboration to develop common standards for emissions reporting, accounting, registration, and allowance tracking. The Northeastern states first collaborated on these issues under the Eastern Climate Registry. They subsequently worked with California and other states to form the Climate Registry in 2007. By 2008, members included thirty-nine U.S. states and the District of Columbia, ten Canadian provinces, six Mexican states, and three Native American groups. The Climate Registry tracks, verifies, and publicly reports GHG emissions in a consistent manner across its members, potentially laying some institutional groundwork for future national and/or continental climate change regulations (Reilly, Manion, Weiss, and Coleman 2008).

Local political leaders in the Northeast are increasingly calling directly on federal policymakers in Washington to follow their lead. To this end, they sometimes join forces with counterparts outside the region. In May 2007, Governor Jodi Rell of Connecticut and Governor Arnold Schwarzenegger of California wrote in a joint op-ed in the *Washington Post*, stating that "California, Connecticut and a host of like-minded states are proving that you can protect the environment and the economy simultaneously. It's high time the federal government becomes our partner or gets out of the way" (Schwarzenegger and Rell 2007). Federal lawmakers from the Northeast, building on the many policy developments within their constituencies, have often led efforts in Congress to expand federal climate change legislation, and both the House of Representatives and the Senate have responded more positively to these efforts since 2006 (Motavalli 2007).

Many of the region's senators supported the bills introduced by Senators John McCain (R-AZ) and Joseph Lieberman (D-CT) in 2003 and 2005 proposing mandatory GHG controls. Senators John Kerry (D-MA) and Olympia Snowe (R-ME) in September 2006 cosponsored a bill that sought to halt the increase in national GHG emissions by 2010 to be followed by a gradual reduction of emissions toward the target of a 65 percent reduction below 2000 levels by 2050. Senator Jim Jeffords

(Independent-VT) proposed another piece of national legislation in July 2006 that called for significant cuts in national GHG emissions. Since 2007, a record numbers of bills have been introduced in Congress proposing a variety of GHG reduction targets, national cap-and-trade schemes, vehicle emissions standards, renewable energy requirements, and energy efficiency targets. In general, these measures have had significant support from both Republican and Democratic senators from across the Northeast.

Limited Environmental Impacts

While climate change action is increasingly institutionalized in the Northeast at local, state, and regional levels, GHG emissions reductions remain quite modest. In fact, state actions taken to date are not likely to meet self-imposed short-term GHG emissions reduction targets, which in themselves are relatively modest (Stoddard and Murrow 2006). As a 2006 analysis commissioned by the New England Governors and Eastern Canadian Premiers concludes, "Despite considerable efforts ... emissions continue to rise" (NEG-ECP 2006, 1). In other words, the announced NEG-ECP GHG emissions reduction goals for 2010 and 2020 will simply not be met without additional policy measures (NECC 2006, 2008). There are, however, significant differences in emission trends across states and provinces (see table 6.2).

During the period from 2001 through 2005, only one Eastern Canadian province (New Brunswick) and one New England state (Rhode Island) recorded declines in

Table 6.2
Changes in NEG-ECP state and provincial GHG emissions

State/Province	Change in GHG emissions 2001 to 2005
Connecticut	+4 percent
Maine	+2 percent
Massachusetts	+2 percent
New Brunswick	−6 percent
New Hampshire	+26 percent
Newfoundland and Labrador	+7 percent
Nova Scotia	+10 percent
Prince Edward Island	+5 percent
Québec	+7 percent
Rhode Island	−7 percent
Vermont	+1 percent

Sources: NECC 2008; Environment Canada 2007.

emissions as changes in emissions levels ranged from minus 7 percent (Rhode Island) to plus 26 percent (New Hampshire). Some of these differences are closely tied to developments in regional integrated energy markets. New Hampshire's emissions increased largely as a result of the startup of two of the three largest power plants in the state since 2001. These plants serve the entire New England electric grid, and Rhode Island's emissions fell in large part because power production was reduced instate while import of out-of-state generated power increased (NECC 2008). Other Northeastern states also struggle. For example, New Jersey's GHG emissions increased by over 11 percent between 2000 and 2005 (New Jersey Department of Environmental Protection 2008).

In addition to bringing down GHG emissions from the energy sectors, there is a need to more aggressively reduce emissions from the transportation sector (Stoddard and Murrow 2006). Though state and provincial officials frequently discuss transportation issues, little policy change has been achieved. In part, U.S. state officials are awaiting the outcome of California's current efforts to regulate CO_2 emissions from vehicles and the implementation of the new federal CAFE standards that were passed in 2007. The Canadian federal government and at least three provinces also seem likely to at least meet these CAFE standards in the coming years (AFP 2008; Marshall 2007). However, additional policy measures and technological changes are required to effectively bring down GHG emissions from this sector. At the same time, higher fuel prices in 2007 and 2008 did more to alter U.S. and Canadian driving and vehicle-purchasing habits than any policy initiatives.

Municipalities initiating important policy efforts are also discovering that reducing GHG emissions can be challenging. For example, Cambridge, Massachusetts aims to reduce its emissions to 20 percent below 1990 levels by 2010, which would represent an annual reduction of almost 500,000 tons of CO_2 (City of Cambridge 2002). Yet, data released in 2004 show that Cambridge's GHG emissions were up 27 percent from 1990 levels (Cambridge Climate Action Protection Committee 2004). While emissions from the residential sector in Cambridge were down by 25 percent from 1990 levels, commercial and industrial emissions grew by 63 percent from 1990 levels. Cambridge's transportation emissions furthermore increased by 22 percent between 1990 and 2003. While some of these increases may be exceptional, evidence suggests that many municipalities are similarly struggling to curb their GHG emissions (Gore and Robinson, this volume).

Concluding Remarks

Climate change policy is increasingly institutionalized in Northeast North America. States and provinces have adopted climate change policies regulating CO_2 emissions from oil and coal-fired power plants and setting mandatory renewable portfolio standards. Under RGGI, ten states have launched a CO_2 emissions trading scheme. In addition, municipalities all over the Northeast are developing climate policy and formulating GHGs reduction goals. A growing number of firms in the region are also committing to reducing their GHG emissions, and the local NGO community has greatly intensified its advocacy efforts on climate change. These many actions are driven by a combination of factors, including an acceptance of the science of human-driven climate change, concerns about regional vulnerabilities to a changing climate, efforts to protect the long-term viability of local economies, and a sense of responsibility to act in the face of lagging federal climate policy.

The development of institutions for climate change governance in the Northeast and the emergence of well-connected networks of climate change policy advocates are creating supportive conditions for continued regional political and social action on climate change. As public- and private-sector actors in the Northeast move ahead on climate change action, they have consistently called for more stringent federal political action and economic support for climate change mitigation and adaptation. Attorneys general in the Northeast moreover have led legal efforts to force the U.S. EPA to regulate CO_2 under the Clean Air Act. Local politicians and officials readily acknowledge that supplementary federal policy is necessary to substantially expand on the policy momentum that has been building in the region since the early 2000s and to significantly reduce GHG emissions.

While the effectiveness of policy efforts in the Northeast to date should not be exaggerated, as many GHG emission trends are still going in the wrong direction, it is clear that important political and technical precedents for future climate change action are being set all over the region. Developing climate change action in the Northeast is a critical part of a broader effort by states, provinces, municipalities, firms, and NGOs to develop more aggressive climate change policy and to push federal policymakers and regulators in Washington, D.C., and Ottawa to do the same. In this respect, many public, private, and civil society actors in the Northeast, by formulating more aggressive goals and building new institutions, are critical in shaping future multilevel climate change governance in North America.

References

AFP. 2008. Canada Plans to Reduce Car Fuel Consumption in 2011. *Agence France-Presse*, January 17.

Algoso, Dave, and Emily Rusch. 2005. *Global Warming Pollution in New Jersey: Key Steps to Reduce Emissions from Electricity Generation and Transportation*. Trenton, NJ: NJPIRG Law and Policy Center.

Aulisi, Andrew, Alexander F. Farrell, Jonathan Pershing, and Stacy D. VanDeveer. 2005 *Greenhouse Gas Emissions Trading in U.S. States: Observations and Lessons from the OTC NO_x Budget Program*. Washington, DC: World Resources Institute.

Austin, Jay. 2007. Massachusetts vs. Environmental Protection Agency: Global Warming, Standing and the U.S. Supreme Court. *Review of European Community and International Environmental Law* 16(3): 368–371.

Betsill, Michele M. 2001. Mitigating Climate Change in U.S. Cities: Opportunities and Obstacles. *Local Environment* 6(4): 393–406.

Betsill, Michele M., and Harriett Bulkeley. 2004. Transnational Networks and Global Environmental Governance: The Cities for Climate Protection Program. *International Studies Quarterly* 48(4): 471–493.

Bowles, Ian. 2008. Want to Buy Some Pollution? *New York Times*, March 15.

Breslow, Marc, and Eban Goodstein. 2005. *Impact of the Regional Greenhouse Gas Initiative (RGGI) on Business Operating Costs in Massachusetts*. Boston: Massachusetts Climate Action Network.

Brome, Heather. 2006. *Memorandum on Economic Impact of RGGI*. Boston: Federal Reserve Bank of Boston.

Byrne, John, Kristen Hughes, Wilson Rickerson, and Lado Kudgelashvili. 2007. American Policy Conflict in the Greenhouse: Divergent Trends in Federal, State and Local Green Energy and Climate Change Policy. *Energy Policy* 35(9): 4555–4573.

Cambridge Climate Action Protection Committee. 2004. *2004 Annual Report*. Cambridge, MA: City of Cambridge.

Cardwell, Diane. 2007. At Mayors' Summit in New York, Bloomberg Crusades for Clean Air. *New York Times*, May 16.

City of Cambridge. *Climate Protection Plan*. 2002. Cambridge, MA: City of Cambridge.

Cohen, David, and Timothy P. Murray. 2003. State Must Protect Our Climate Now. *Boston Globe*, September 1.

Commonwealth of Massachusetts. 2008. Massachusetts Renewable Energy Portfolio Standard: Annual RPS Compliance Report for 2006. Boston: Commonwealth of Massachusetts, Division of Energy Resources.

Confessore, Nicholas. 2008. $8 Traffic Fee for Manhattan Gets Nowhere. *New York Times*, April 8.

Cooper, R. 1998. Toward a Real Global Warming Treaty. *Foreign Affairs* 77(2): 66–79.

Daley, B. 2007. Winter Warm-up Costing N.E. Region. *Boston Globe*, January 28.

David Suzuki Foundation. 2006. *All Over the Map*. Vancouver: David Suzuki Foundation.

Ebbert, Stephanie. 2006. Wind Turbines Gaining Power. *Boston Globe*, February 24.

Environment Canada. 2007. *National Inventory Report 1990–2005: Greenhouse Gas Sources and Sinks in Canada*. Gatineau, Québec: Environment Canada.

Environment News Service. 2007. New York Mayor Unveils Multi-Billion Dollar Green City Plan. April 23.

Environment Northeast. 2003. *Climate Change Roadmap for Connecticut: Economic and Environmental Opportunities: Part II: Strategies*. Hartford, CT: Environment Northeast.

Hamel, Sonia. 2003. Climate Change Action Plan: 2003 Update. Presentation given at the 28th Annual Meeting of the New England Governors and Eastern Canadian Premiers, Groton, September 9.

Hileman, B. 2006. Scientists Support Curbing CO_2. *Chemical & Engineering News*, September 11, 10.

Johnson, Carolyn Y. 2006. Catching the Knots for Watts. *Boston Globe*, May 18, 2006.

Kousky, C., and S. H. Schneider. 2003. Global Climate Policy: Will Cities Lead the Way? *Climate Policy* 3(4): 359–372.

Laidler, John. 2007. N.H. Prepares to Vote on Global Warming. *Boston Globe*, February 18.

Marshall, Dale. 2007. *Picking up the Slack: The Provinces' Potential to Act on Climate Change*. Vancouver: David Suzuki Foundation.

Massachusetts Government. 2008. Attorney General Martha Coakley Petitions Court to Require EPA to Comply with Court Order. The Commonwealth of Massachusetts Office of the Attorney General, Press Release, April 2.

MASSPIRG. 2003. *Cars and Global Warming: Policy Options to Reduce Greenhouse Gas Emissions from Massachusetts Cars and Light Trucks*. Boston: MASSPIRG Education Fund.

Mauro, Tony. 2007. High Court Orders EPA to Review Greenhouse-Gas Emissions. *Legal Times*, April 3.

Mayors Climate Protection Center. 2007. *Climate Protection Strategies and Best Practices Guide: 2007 Mayors Climate Protection Summit Edition*. Washington, DC: United States Conference of Mayors.

McKibbin, W. J., and P. Wilcoxen. 2002. *Climate Change Policy after Kyoto*. Washington, DC: Brookings Institution.

Motavalli, Jim. 2007. The Can-Do Congress? *E Magazine*, May/June.

Munton, D. 1998. Dispelling the Myths of the Acid Rain Story. *Environment* 40(6): 4–7, 27–34.

National Assessment Synthesis Team. 2000. Climate Change Impacts on the United States: The Potential Consequences of Climate Variability and Change. Report from the National Assessment Synthesis Team, U.S. Global Change Research Program.

New England Climate Coalition (NECC). 2005. *2005 Report Card on Climate Change Action: Second Annual Assessment of the Region's Progress towards Meeting the Goals of the New England Governors/Eastern Canadian Premiers Climate Change Action Plan of 2001*. Boston: New England Climate Coalition.

New England Climate Coalition (NECC). 2006. *2006 Report Card on Climate Change Action: Third Annual Assessment of the Region's Progress towards Meeting the Goals of the New England Governors/Eastern Canadian Premiers Climate Change Action Plan of 2001*. Boston: New England Climate Coalition.

New England Climate Coalition (NECC). 2008. *Falling Behind: New England Must Act Now to Reduce Global Warming Pollution*. Boston: New England Climate Coalition.

New England Governors/Eastern Canadian Premiers (NEG-ECP). 2006. *Climate Change Action Plan 2006 Discussion Paper*. Boston: New England Governors Conference.

New Jersey Department of Environmental Protection. 2008. *Draft New Jersey Greenhouse Gas Inventory and Reference Case Projections 1990–2020*. Trenton: New Jersey Department of Environmental Protection.

Newman, Maria. 2007. Mayor Proposes a Fee for Driving into Manhattan. *New York Times*, April 22.

NESCAUM. 2004. *Greenhouse Gas Emissions in the New England and Eastern Canadian Region, 1990–2000*.Boston: Northeast States for Coordinated Air Use Management.

PEW Center on Global Climate Change. 2006. *Learning from State Action on Climate Change: March 2006 Update*. Washington, DC: Pew Center on Climate Change.

Pizer, W. 1997. *Prices vs. Quantities Revisited: The Case of Climate Change*. Washington, DC: Resources for the Future.

Point Carbon. 2008a. *Carbon Market North America* 3(1): 1–8, January 16.

Point Carbon. 2008b. *Carbon Market North America* 3(4): 1–8, February 27.

Point Carbon. 2008c. *Carbon Market North America* 3(7): 1–8, April 9.

Rabe, Barry. 2004. *Statehouse and Greenhouse: The Emerging Politics of American Climate Change Policy*. Washington, DC: Brookings Institution Press.

Rappaport, Ann. 2008. Campus Greening: Behind the Headlines. *Environment* 50(1): 7–16.

Reilly, Allison, Michelle Manion, Leah Weiss, and James Coleman. 2008. Laboratories of Innovation: Why States Created a North American Greenhouse Gas Registry. *EM Magazine*, November.

Reisman, Jon. 2003. Governors and Greens Au Groton. *Tech Central Station*, September 5.

Revkin, Andrew C., and Patrick Healy. 2007. Global Coalition Makes Buildings Energy-Efficient. *New York Times*, May 17.

Rivlin, Gary. 2005. Green Tinge Is Attracting Seed Money to Ventures. *New York Times*, June 22.

Rothenberg, L. 2002. *Environmental Choices: Policy Responses to Green Demands*. Washington, DC: CQ Press.

Ruth, Matthias, Steven A. Gabriel, Karen L. Palmer, Dallas Burtraw, Anthony Paul, Yihsu Chen, Benjamin F. Hobbs, Darius Irani, Jeffrey Michael, Kim M. Ross, Russell Conklin, and Julia Miller. 2008. Economic and Energy Impacts from Participation in the Regional Greenhouse Gas Initiative: A Case Study of the State of Maryland. *Energy Policy* 36(6): 2279–2289.

Ryan, Andrew. 2008. Fenway Park Unveils Solar Panels on Roof. *Boston Globe*, May 19.

Schwarzenegger, Arnold, and Jodi Rell. 2007. Lead or Step Aside, EPA: States Can't Wait on Global Warming. *Washington Post*, May 21.

Selin, Henrik, and Stacy D. VanDeveer. 2005. Canadian-U.S. Environmental Cooperation: Climate Change Networks and Regional Action. *American Review of Canadian Studies* 35(4): 353–378.

Selin, Henrik and Stacy D.VanDeveer. 2006. Canadian-U.S. Cooperation: Regional Climate Change Action in the Northeast. In *Bilateral Ecopolitics: Continuity and Change in Canadian-American Environmental Relations*, edited by Philippe Le Prestre and Peter Stoett. Aldershot, UK: Ashgate.

Selin, Henrik and Stacy D. VanDeveer 2007. Political Science and Prediction: What's Next for U.S. Climate Change Policy? *Review of Policy Research* 24(1): 1–27.

Shellenberger, Michael, and Ted Nordhaus. 2004. The Death of Environmentalism: Global Warming Politics in a Post-Environmental World. http://www.thebreakthrough.org/images/Death_of_Environmentalism.pdf.

Shellenberger, Michael and Ted Nordhaus. 2007. *Break Through: From the Death of Environmentalism to the Politics of Possibility*. Boston: Houghton Mifflin.

Skolfield, Karen. 2006. Beyond Bluster. *UMASS Amherst Magazine*, Spring.

Stoddard, Michael D., and Derek K. Murrow. 2006. *Climate Change Roadmap for New England and Eastern Canada*. Hartford, CT: Environment Northeast.

Union of Concerned Scientists 2006. *Climate Change in the U.S. Northeast: A Report of the Northeast Climate Impact Assessment*. Cambridge, MA: Union of Concerned Scientists.

United States House of Representatives. 2007. *Political Interference with Climate Change Science under the Bush Administration*. Washington, DC: House of Representatives, Committee on Oversight and Government Reform.

Weisman, R. 2006. Money Flowing to New Ideas in Energy. *Boston Globe*, August 26.

Williams, Wendy, and Robert Whitcomb, 2007. *Cape Wind: Money, Celebrity, Class, Politics, and the Battle for Our Energy Future on Nantucket Sound*. New York: Public Affairs.

Zezima, Katie. 2007. In New Hampshire, Towns Put Climate on the Agenda. *New York Times*, March 19.

7

Local Government Response to Climate Change: Our Last, Best Hope?

Christopher Gore and Pamela Robinson

Introduction

In the face of ineffective federal responses to climate change in North America, advocacy and action by states, provinces, the private sector, nongovernmental groups, and citizens are building to accelerate a response from the U.S. and Canadian national governments (Selin and VanDeveer 2006, 2007). These subnational initiatives deserve careful attention for their role in serving as models for future national programs and offering complementary approaches to solving problems, as well as for their role in encouraging stronger national action. But attention to the subnational level should not end there. Since the late 1980s, local governments in North America have emerged as leaders in climate change response and have become important actors in a multilevel system of climate change governance.

Municipal greenhouse gas (GHG) reduction efforts challenge the conventional wisdom that climate change is a nonlocal problem and that local governments should be expected "to be policy-takers, not policy-makers" (Sancton 2006, 34). Indeed, despite national inaction, hundreds of municipalities in Canada and the United States have directly embraced the climate change challenge. In Mexico, municipal attention and action in response to climate change is growing as well. While the focus in this chapter is mostly on municipalities in Canada and the United States, it is worth mentioning that a few cities in Mexico are taking important steps to address air quality concerns (Connolly 1999; Pulver, this volume; West et al. 2004). Increasingly throughout North America, municipalities are formulating and implementing GHG reduction strategies, and in a small number of cases, they have achieved measurable and substantial reductions in emissions. These efforts are part of a worldwide trend as local governments become leaders in climate change response (Worldwatch Institute 2007).

This chapter examines municipal climate change responses in North America, identifying the political and practical forces at work in local government actions.

Four broad factors are involved in these municipal actions. First, local governments are members of national, regional, and international networks that promote and motivate climate change response. Second, citizens are looking to local governments to develop progressive environmental action on local and global issues. Third, local governments are concerned about their international reputation and take action to demonstrate leadership. Fourth, local governments are increasingly attentive to the interrelationship among social, economic, and environmental issues when making decisions about the physical form and social and economic quality of urban regions. As a result, they implement practical measures that produce several benefits, including the reduction of GHG emissions.

In this chapter, we provide a brief overview of the federal structure of Canada, the United States, and Mexico and show how municipalities fit within that structure. This leads to an examination of the breadth and depth of municipal climate change action, documenting the extent to which municipal action is institutionalized. We then consider two local governments that have been leaders in their respective countries—Toronto, Ontario, in Canada and Portland, Oregon, in the United States. We explore the actions of each of these local governments and their motivations for taking action. The chapter concludes with a discussion of the reasons for municipal action and a consideration of how municipalities and municipal networks contribute to multilevel climate governance in North America.

Municipal Response to Climate Change: Structural Constraints and Opportunities

In examining the role of North American municipalities in climate governance, it is important to be mindful of the structural constraints and opportunities that permit and restrict action. Canada, the United States, and Mexico are all federal states. Historically, the national constitutions of each country defined only two sovereign orders of government, the national and subnational.

The constitutions of Canada and the United States say almost nothing about local governments, other than that provinces and states have powers to legislate over them. In this respect, the relationship between municipal and subnational governments in both countries is quite similar. This situation has almost uniformly, but somewhat problematically, led to the general argument that municipalities are "creatures of provinces" or "creatures of states," and that legally they have no inherent right to exist (Magnusson 2005). In the United States, this argument was upheld by the 1907 Supreme Court ruling that enshrined the "Dillon's Rule," which gave states preeminence over local governments. In Canada, provincial authority over municipal government is derived from Section 92 of the national constitution, which states that "each province may exclusively make laws in relation to municipal institutions in the province."

An important difference between subnational governments in Canada and the United States is that states have their own written constitutions, which can enumerate duties and responsibilities of lower-level governments, while provinces do not. Nonetheless, in both countries any powers bestowed upon a local government are so granted through the statutory authority of the state or province. Hence, municipalities in Canada and the United States commonly exist in an uneasy tension with higher orders of government, particularly surrounding issues of finance and autonomy. In Mexico by contrast, the constitution has been amended twice in the last twenty-five years (1984 and 1999) to strengthen the authority of municipal governments. As a result, Mexico's constitution now recognizes municipalities as a formal level of government and specifies a list of functions and powers that are under their exclusive jurisdiction and that cannot be limited by state governments.

Local government authority varies by province and state across Canada and the United States. Accordingly, it is not possible to easily create a comprehensive and detailed overview of emission reduction opportunities for municipalities across all provinces and states. In general, however, municipalities in both countries have control over the following functions that may affect GHG emissions:

- Development of building standards and codes to improve energy and water efficiency;
- Facilitating energy and water audits and retrofits of municipal buildings;
- Procuring and installing energy efficient infrastructure (such as street lights, parking meters, and pumping equipment);
- Procurement of low-emission and alternative-fuel municipal fleet vehicles;
- Development of district heating and cooling systems;
- Development of active transportation infrastructure (such as bicycling lanes and pedestrian friendly neighborhoods);
- Public transport;
- Landfill gas capture; and
- Land-use planning.

Despite authority in all these policy areas, North American growth management strategies aimed at reducing GHG emissions continue to be challenged by the proliferation of single-family, single-land-use, low-density, automobile-dependent development, resulting in suburban sprawl. State governments in the United States also exercise less comprehensive control or influence over growth management at the local level, compared to their Canadian provincial counterparts. As a result, municipality by municipality effectiveness in growth management is more widely varied in the United States than in Canada (Scott 1995). Added together, municipal governments have important opportunities to influence a significant volume of GHG emission reductions and to shape and constrain human behavior that is GHG-intensive.

Indeed, in Canada, it has been calculated that municipalities have indirect control or influence over half of national GHG emissions (Municipalities Table 1999; Robinson 2000).

Given this potential for GHG reductions, how widespread is municipal action? On a case-by-case basis, evidence suggests that an increasing number of local authorities and metropolitan regions are actively developing and implementing climate change action plans and emission reduction strategies. In the United States and Canada, municipal action has often been independent of state action and has developed through municipally based activism and leadership. A long list of municipal actions and experiences could be enumerated for large, medium, and small cities throughout North America. However, a more nuanced and informative picture emerges in examining the influence of long-standing international and national municipal networks and programs on climate change actions.

Municipal Response to Climate Change: How Broad, How Deep?

One outcome of increased municipal action is the recognition that local authorities are important to global environmental governance. In large part this is due to the significant role that municipal transnational networks play in climate action. Since the early 1990s, these networks have grown, with some estimates that there are at least twenty-eight in Europe alone (Betsill and Bulkeley 2006, 143). One of the most important networks is the International Council for Local Environmental Initiatives (ICLEI). Headquartered in Toronto, ICLEI was founded in 1990 as an international association of local governments and national and regional local government organizations committed to sustainable development. The council provides informational and technical services and support to local governments, and it also plays a significant role in knowledge exchange.

Shortly after its establishment, ICLEI engaged local authorities in discussions about GHG emission reductions. In 1991, ICLEI sponsored the Urban CO_2 Reduction Project—a project funded by the U.S. EPA, several private foundations, and the City of Toronto (Bulkeley and Betsill 2003, 51). The initial project was deemed successful, and subsequently evolved into a program called the Cities for Climate Protection (CCP) Campaign. The ICLEI CCP program pioneered a five-step process for addressing climate change: (1) creating a GHG emissions inventory and forecast; (2) setting an emissions reductions plan; (3) developing a local action plan; (4) implementing the local action plan; and (5) monitoring progress and reporting results. These milestones need not be implemented sequentially, though they often are. In February 2007, the CCP had 674 participants from 30 different countries. In North America, there were 269 CCP participants—152 in the United

States, 109 in Canada, and 8 in Mexico. In October 2007, an ICLEI office was established in Mexico. Relative to the approximate number of various local governments in the United States (85,000 plus), Canada (5,000 plus), and Mexico (2,400 plus), these levels of participation may not seem impressive.

Under ICLEI's CCP program, hundreds of local authorities have formally adopted resolutions to act in the absence of national leadership. Evidence from the United States, Europe, and Australia, however, shows that only minor emission reductions by CCP members have been achieved and also that cities have been motivated by cobenefits rather than a strict concern about climate change (Lindseth 2004, 330–332). Another concern with the CCP program is that by constructing climate change as a local issue, it might give the impression that climate-related problems can be solved locally (Lindseth 2004, 334). However, with the potential to control upward of half of national GHG emissions, municipalities have the potential to play a leading role in reducing GHG emissions relating to infrastructure, energy use, waste management, and transit and transportation, leaving higher-order governments the opportunity to regulate large emitters, establish and enforce emission standards, and to address emissions crossing national boundaries.

Other international networks connecting municipalities and municipal leaders in efforts to respond to climate change include the Clinton Climate Initiative's "C40 Cities," which aims to leverage financial resources from governments and the private sector to support forty large urban centers committed to climate change response. Mexico City is one of the C40 cities and is actively participating in several C40 initiatives, such as the foundation's Energy Efficiency Building Retrofit Program and a taxi replacement program that aims to remove 10,000 old taxis by 2012.[1] As of April 2008, Mexico City had replaced 3,000 old taxis with more fuel-efficient models. The municipal government is subsidizing the purchase of new taxis, with a local bank offering loans and credits for drivers to pay off the remaining cost.

Other networks that facilitate knowledge exchange relating to climate change include associations of municipal professionals such as engineers, planners, and architects. Together, these international and domestic networks play an important role in facilitating knowledge exchange and leveraging financial resources to assist in GHG reduction measures. Hence, despite suggestions that ICLEI's CCP program has not produced significant emission reductions from participating cities, its early role in facilitating national networks of municipal climate change action remains significant. In the United States and Canada, ICLEI has partnered with two national municipal networks to address environmental concerns generally and climate change specifically: the U.S. Conference of Mayors and the Federation of Canadian Municipalities (FCM).

The U.S. Conference of Mayors

The U.S. Conference of Mayors began in 1932. Its origins rest with national mayoral advocacy for federal assistance to cities in the wake of the Great Depression. Congress responded with a $300 million municipal assistance bill, and shortly afterward the Charter of the Conference of Mayors was written. The Conference of Mayors is a nonpartisan organization of cities with populations of 30,000 people or more. In February 2007, the organization had 1,139 members. Each member city is represented by its mayor—its chief elected official. Although there are other discreet local initiatives occurring throughout the United States, which are embedded within this and other national networks (Vogel 2007), the Conference of Mayors is the dominant U.S. national network of municipal actors engaged in GHG emission reduction efforts.

The conference first articulated a formal position on climate change in 1998. The conference had earlier passed resolutions on energy and air pollution, but none specific to climate change. In the wake of U.S. Senate Resolution 98 (also known as the Byrd-Hagel Resolution), which outlined conditions under which the United States would participate in an international climate agreement, the conference urged the U.S. president to adhere to Resolution 98 and to include local government representation in international meetings where decisions might have a direct financial impact on local governments. In 2000 the conference, however, adopted a resolution on global warming, stating that mayors are uniquely positioned to provide leadership and take action to reduce GHG emissions causing global warming. The resolution also called upon the Congress and the White House to make global warming a priority and to implement policies and programs working with local communities. In 2002, the conference passed the Municipal and National Commitment to Reduce Greenhouse Gases resolution, noting that 125 municipalities had committed to setting specific GHG emission reduction targets.

By 2005, the mayors' advocacy coalesced into an even stronger movement. On the same day that the Kyoto Protocol entered into force—February 16, 2005—Seattle mayor Greg Nickels challenged mayors across the United States to join Seattle in taking action to reduce GHG emissions. Elected in 2001, Mayor Nickels championed urban environmental improvements in Seattle through his Environmental Action Agenda. A month after Nickels's invitation, ten mayors, representing 3 million U.S. citizens, came together under the banner of a draft resolution: the U.S. Mayors Climate Protection Agreement. The mayors included Greg Nickels, Seattle, Washington; Peter Clavelle, Burlington, Vermont; Rocky Anderson, Salt Lake City, Utah; Rosemarie Ives, Redmond, Washington; Gavin Newsom, San Francisco, California; Pam O'Conner, Santa Monica, California; Tom Potter, Portland, Oregon; Mark Ruzzin, Boulder, Colorado; and R. T. Rybak, Minneapolis, Minnesota.

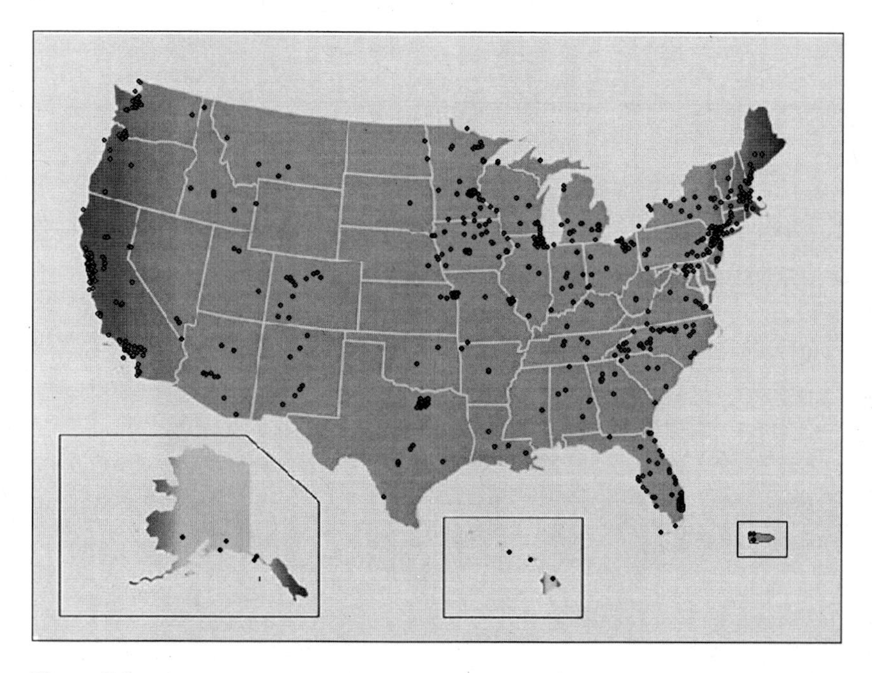

Figure 7.1
Map of U.S. cities participating in U.S. Climate Protection Agreement as of March 1, 2009. Source: http://www.usmayors.org/climateprotection/ClimateChange.asp, accessed March 1, 2009.

In March 2005, in a lead-up to the conference's meeting in June 2005, the ten mayors sent a "call to action" to 400 other mayors encouraging them to endorse and sign the U.S. Mayors Climate Protection Agreement. Recognizing that climate protection resolutions had been adopted at previous meetings, this time they urged "specific actions." At the June meeting in 2005, the U.S. Mayors Climate Protection Agreement passed unanimously. A short time later, 187 mayors representing 40 million people in thirty-eight states had endorsed this agreement. By 2009, 916 mayors, representing a total population over 80 million citizens, had signed the U.S. Mayors Climate Protection Agreement. Figure 7.1 is a map of the U.S. cities that had endorsed the agreement as of March 1, 2009.

Two factors should be considered when assessing the strength of this agreement: its support and the content of the agreement. With respect to support, in three years, over two-thirds of the members of the U.S. Conference of Mayors have become participants. Thus, municipal leaders of over one-quarter of the total U.S. population (300 million) have formally committed to reducing GHG emissions. Given that at least one city from every U.S. state is participating, and that the number of cities participating has continued to increase, support for the agreement is strong. Furthermore, at their meeting in January 2007 the mayors called for a $2 billion annual Energy and Environmental Block Grant to help cities combat global warming (U.S. Conference of Mayors 2007a). This proposal was one component of a larger ten-point plan that was aimed at the 110th Congress, the Bush administration, and the 2008 Presidential candidates (U.S. Conference of Mayors 2007b). This grant was included in the 2007 U.S. Energy Independence and Security Act.

With respect to content, the U.S. Mayors Climate Protection Agreement is written in strong and unequivocal language. It identifies the urgent need for response owing to scientific evidence of human-induced climate change disruptions. The agreement also notes the failure of the U.S. federal government to respond to the climate change threat, while the United States is "responsible for producing approximately 25 percent of the world's global warming pollutants" while having "less than five percent of the world's population." It further acknowledges that local and state governments throughout the Untied States are adopting emission reduction targets, and "that this leadership is bipartisan, coming from Republican and Democratic governors and mayors alike."

The U.S. Mayors Climate Protection Agreement also specifies the actions mayors will take in signing on. Actions include urging the federal and state governments to enact policies and programs to meet or beat the U.S. national target under the Kyoto Protocol (if it had become a party) of reducing GHG emissions 7 percent below 1990 levels by 2012; reducing dependence on fossil fuels and accelerating development of clean, renewable energy technology; and passing bipartisan GHG reduction legislation that includes timetables, emission limits, and a market-based system of tradable allowances. In articulating these goals and actions, the Conference of Mayors has challenged the prevailing wisdom that has dominated federal climate change politics and asserted the prominence of local governments in climate action and advocacy.

It remains, however, that the impact of the Conference of Mayors in terms of GHG emission reductions will be uncertain until a comprehensive survey of achievements by participating cities is compiled. In February 2007, the Conference of Mayors launched the Mayors Climate Protection Center to support mayors in their climate protection efforts. One of the first activities undertaken by the center was a survey of 134 mayoral efforts to reduce GHG emissions. The survey shows that a strong majority of mayors surveyed were engaged in many actions focused on improving municipal and community energy efficiency, transportation, green building practices, and outreach (U.S. Conference of Mayors 2007c). The survey also showed that 92 percent of cities surveyed consider efforts to reduce GHG emissions part of broader efforts to address public health concerns, relating to such things as air quality.

As promising as these actions and initiatives are, the survey also shows that cities are largely focused on corporate operational practices, outreach efforts, and creating new business opportunities. Corporate emissions are tied to things a city owns or controls, including municipal buildings, street lighting, water and wastewater treatment facilities, and municipal vehicle fleets. In contrast, cities have been much less active targeting community emissions within municipal boundaries, which are emissions that do not originate from municipal facilities and operations, but from residential, commercial, or industrial sectors. As such, most municipalities have not yet attempted to target or restrict intensive GHG-emitting activities that are tied directly to patterns of human behavior or social preferences, such as automobile use. Nonetheless, of the cities surveyed, half said they would use additional federal funds from the Energy and Environmental Block Grant to improve community energy efficiency.

Federation of Canadian Municipalities

Turning to Canada, there are important parallels and differences with U.S. municipal advocacy. The Federation of Canadian Municipalities (FCM) has been a national voice for Canadian municipalities since the early 1900s. In February 2007, FCM had 1,427 members from Canada's ten provinces and three territories. The organization promotes strong, effective, and accountable municipal government, as well as a high quality of life for Canadian communities. While FCM is a professional association serving elected municipal officials, it also promotes policy issues. For over a decade, "sustainable communities" has been one of FCM's central policy interests. In partnership with ICLEI, one of its most prominent initiatives is the Partners for Climate Protection (PCP) program. Prior to partnering with ICLEI, FCM had its own emissions reduction program called the "20% Club," which started in 1995. Members of this club committed to reduce their corporate GHG emissions 20 percent below 1994 levels.

The PCP is the Canadian version of ICLEI's CCP campaign. It is based on a similar five-step approach to climate change action pioneered by ICLEI in the form of CCP milestones. To join the PCP program, Canadian municipalities must pass a council resolution. The FCM resolution has many elements similar to the U.S. Mayors Climate Protection Agreement, but its tone is weaker and it does not advocate for federal or provincial government action. This may be explained partly by Canada's ratification of the Kyoto Protocol in December 2002, thus giving the (false) impression that concerted federal action was forthcoming (Stoett, this volume). Municipalities in Canada are also not known for having a strong influence in national politics, and FCM is not an outspoken critic of the national government.

Table 7.1
Canadian municipal participation in Partners for Climate Protection program

Province/Territory	Number of participants
Alberta	15
British Columbia	58
Manitoba	13
New Brunswick	16
Newfoundland	7
Northwest Territories	3
Nova Scotia	10
Nunavut	1
Ontario	43
Prince Edward Island	2
Québec	7
Saskatchewan	2
Yukon	1
Total	**178**

Source: http://www.sustainablecommunities.fcm.ca/Partners-for-Climate-Protection.

Given that provinces have constitutional authority over municipalities, FCM is in the challenging position of trying to advocate nationally while not usurping provincial authority.

Participants in the FCM PCP program commit to working toward reducing GHG emissions in municipal operations by 20 percent below 1994 levels and 6 percent below 1994 levels throughout the community within ten years of joining. In 1998, 53 municipalities had registered with the program. By 2005 the number had increased to 122 municipalities. By March 2009, 178 municipalities were participating in the program, or about 10 percent of FCM's 1,400 members (see table 7.1). Thus, there has been a steady, slow growth in the number of participating municipalities.

When the total number of municipalities participating in the PCP program is compared to the total number of FCM members or Canadian municipalities, the result is not impressive. However, when the total number of people living in PCP member municipalities is compared to the national population, the result is far more significant. By 2007, the sum of all the municipal populations that are PCP members was approximately 21 million out of a national population of 32 million.[2] In other words, roughly 65 percent of Canadians live in municipalities that have committed to GHG reductions. Two factors help explain this: about 80 percent of Canadians live in urban areas, and most of Canada's largest cities are members of the PCP program. It is also notable how early municipalities began to take action. Of 142 participating municipalities in 2007, 47 had joined the PCP in 1998 or earlier (FCM 2007). The ICLEI international headquarters were also located in Toronto in 1990.

As in the United States, intent to reduce emissions is different than achieving reductions. One important indicator of municipal success in emission reductions is the number of Canadian municipalities that have successfully achieved some or all of the PCP milestones and/or begun to implement a formal plan for emission reductions. The FCM data shows that only a small handful of Canadian municipalities have achieved three of five milestones, while the majority of municipalities have completed two or fewer milestones. Hence, many PCP member municipalities are struggling to substantively cut GHG emissions, but there are a few success stories.[3]

For example, one of FCM's most celebrated cities is Calgary, Alberta. Calgary's corporate climate change plan includes the first wind-powered light-rail transit system, methane capture from landfills and wastewater treatment plants, and a building retrofit program. These measures have cut the city's GHG emissions by more than 80,000 tons (FCM 2006, 18–19). As of early 2007, an 80-megawatt wind farm south of the city was expected to fulfill 75 percent of its electricity requirements. The wind farm will further cut the city's emissions by more than 200,000 tons. In light of Calgary's success, and past research highlighting the barriers to and reasons for municipal response (Betsill and Bulkeley 2003; Robinson 2000; Robinson and Gore 2005), future research in Canada needs to build on U.S. research (see Zahran, Grover, Brody, et al. 2008), which examines and characterizes the social and political conditions that promote municipal action.

In addition to the PCP program, FCM manages several other innovative programs intended to reduce energy use and improve energy efficiency. One of its most long-standing programs in this area is the Green Municipal Fund, which was started in 2000 following a federal government endowment of $125 million. Increases in the endowment now put the fund at $550 million (FCM 2006). In 2007, the FCM also launched the Green Municipal Corporation. The fee-for-service Green Municipal Corporation will purchase, aggregate, and sell carbon-offset credits earned by Canadian municipal governments in an anticipated future federal carbon-offset market. It is moreover designed to help municipal governments overcome regulatory, technical, and financial barriers to entering that market.

The preceding discussion of transnational and national municipal networks indicates the breadth of municipal climate action in North America. It also provides some sense of the range of actions these networks have achieved and continue to fulfill. The next section highlights two cities that have been early leaders in municipal climate change action in North America—Toronto, Ontario, and Portland, Oregon.

Early Leaders in Municipal Response: Toronto and Portland

Toronto and Portland have been at the forefront of international and North American municipal climate change response. Although many other cities are worthy of detailed consideration, efforts undertaken by these two cities demonstrate some of the major opportunities and challenges municipalities confront when trying to make a substantive commitment to reduce GHG emissions.

Toronto

Toronto is the largest city in Canada, located in the province of Ontario on the northwest shores of Lake Ontario. The population of Toronto proper is approximately 2.5 million, but the greater Toronto region is home to about 5 million Canadians, or about one-sixth of Canada's total population.

Following the 1988 Toronto Conference on the Changing Atmosphere, Toronto adopted the conference's goal of reducing per capita emissions to 20 percent below 1988 levels by 2005. It also formed a Special Advisory Committee on the Environment (SACE) to develop an emission reductions plan. In 1990, SACE delivered a final report to the council that made three recommendations: (1) reduce carbon dioxide (CO_2) emissions to 20 percent below 1988 levels by 2005; (2) create an energy efficiency office; and (3) establish an atmospheric fund to promote measures to stabilize greenhouse gas emissions (City of Toronto 1993; Robinson 2000). In 1997, the Toronto Board of Health reported that the city was on track to achieve GHG emissions that were 3 percent lower than those released in 1990 by 2005.[4] This was despite a projected 10 percent increase in population, a 13 percent increase in the number of households, and a 14.5 percent increase in employment. Indeed, in 1998, at the Kyoto summit, ICLEI recognized Toronto as an international leader in emission reductions.

By 2003, Toronto exceeded its Toronto Target goal, reducing emissions from its own corporate facilities and operations to 42 percent below 1990 levels. In 2004, Toronto announced a further $29 million plan to reduce emissions 60 percent below 1990 levels by 2010 (The Climate Group 2005). In practice, Toronto's emissions reductions are due to ongoing support for and from the Toronto Atmospheric Fund (TAF), landfill gas capture, the Better Buildings Partnership, and deep lake cooling. In 1991, the city council established the TAF as a source of funding for initiatives to improve local air quality and to reduce GHG emissions. The $23 million that the council originally allocated to the TAF came from the sale of a city-owned rural property for $110 million, of which the city council allocated $23 million to establish the TAF (the property was purchased by the city as a jail farm in 1912 for $184,000). Today, through a series of annual grants and loans, grantees from the community and the City of Toronto receive funding resulting in real and "significant" emissions reductions. Annually, the TAF distributes approximately $1.2 million in grants and up to $8 million in loans.

In 1990, 72 percent of Toronto's GHG emissions were from methane gas generated by landfill sites and the waste collection system (Climate Group 2005). Toronto's Keele Valley Landfill Gas to Energy Project is one of the largest landfill gas-capture programs in Canada, providing 30 MW of electricity for the city using a combined-cycle process and resulting in an annual GHG reduction of over 1 million tons CO_2 equivalent (Environment Canada 1999). Toronto's landfill gas-capture program produces approximately 44 MW of electricity (equivalent to the electricity needed to power 34,000 residences) and these projects provide the city with annual royalties in excess of $2 million (Climate Group 2005).

Toronto's Better Building Partnership (BBP) is a public-private venture that dates back to 1996. The BBP initiative mirrors the practices of a vibrant energy service contractor industry in Canada whereby the costs of energy and water retrofits are fronted by the BBP and the client repays the cost at a premium over a fixed duration of time, by the reduced costs of their energy and water bills. In 2006, the BBP had retrofitted 39 million square feet in the institutional, commercial, and industrial building sectors resulting in an emissions reduction of 172,000 tons of CO_2 (Toronto Environmental Alliance 2006). In 2008, London, England established its own BBP based on Toronto's approach, and the Clinton Climate Initiatives Energy Efficiency Building Retrofit Program similarly follows the approach pioneered in Toronto.

Deep lake cooling is anticipated to reduce GHG emissions by 40,000 tons CO_2. Costing $180 million, this system was developed in partnership with the private company Enwave Energy. It uses cold water drawn from the bottom of Lake Ontario as the cooling source for large-scale commercial and institutional clients in the downtown core of Toronto. At full capacity, this system is expected to reduce energy consumption by 75 percent and reduce grid demand by 35MW (Climate Group 2005).

Municipal leaders in Toronto continue to develop and implement new programs. In December 2006, David Miller was reelected mayor. In his inaugural address, he stated: "Where National Governments have failed to lead, cities must lead. And as Canada's biggest city we have an opportunity to lead by tremendous example. Reducing green house gases is THE issue of our time. Maybe of all time" (City of Toronto 2006). Reinforcing these words, the mayor appointed a Special Advisor on Climate Change—the first senior political appointment of a climate change staff member in the city's history, and the only one in Canada. With funding from TAF, the city is undertaking a Greenhouse Gas Analysis project that will inventory energy consumption and pollutants emitted by corporations and the community in order to determine the city's progress in achieving its goal of a 20 percent reduction in corporate emissions below 1990 levels. At the same time, the urban area target remains a reduction of 6 percent below 1990 levels by 2012 (City of Toronto 2007).

With the Greenhouse Gas Analysis project, Toronto is moving beyond corporate emissions reductions to produce a more robust inventory that includes all activities within its municipal boundaries. This tool will allow for a more integrated approach to emissions reductions and local air quality improvement (City of Toronto 2007). Recently, the powers of the city have increased due to new provincial legislation. Despite this, Toronto remains strongly influenced and challenged by provincial and federal government oversight. In 2006 and 2007, for example, city-approved transit improvements were challenged in court and in provincial land-use planning tribunals. Despite these political challenges, Toronto's historic and present leadership in climate change action, along with its physical size, population, and contribution to the Canadian economy, suggests that it will continue to strive to be both an international and a Canadian leader in climate change response.

Portland

Portland was an early municipal leader in climate action. In 1993, it became the first American local government to develop a climate change action plan. Portland is also viewed as a U.S. leader in urban sustainability initiatives (Portney 2003). Located in the state of Oregon in the northwest United States, roughly 550,000 people lived in Portland proper in 2007, while the population in the metropolitan area was approximately 2 million. In 2001, Portland's Local Government Action Plan on Global Warming set a target to reduce emissions 10 percent below 1990 levels by 2010. This plan was in partnership with Multnomah County in which Portland is located. This target included both corporate and community emissions. To date, Portland has achieved per capita emission reductions of 13 percent while also experiencing strong economic growth (Bailey 2007). The city's and county's total GHG emissions are roughly the same as those in 1990, while national emissions have increased by about 13 percent over the same period (City of Portland 2007a).

Portland's climate change strategy includes 150 different activities that are intended to provide direct local benefits (de Steffey, Rojo, and Sten 2005). The city has relied upon land use planning, new transit initiatives, renewable energy, recycling, tree planting, and building retrofits to achieve these reductions. Some specific achievements include:

- 75 percent growth in public transit use since 1990 in part through the addition of two light-rail lines;
- 10 percent of city's electricity derived from renewable sources;

- New funding for energy efficiency and renewable energy initiatives;
- A local recycling rate of 53 percent;
- Forty new high-performance green buildings;
- The establishment of the Energy Trust of Oregon;
- 750,000 trees and shrubs planted since 1996; and,
- Since 2005, 10,000 multifamily units and over 800 homes retrofitted for energy efficiency (City of Portland 2007b).

The Energy Trust of Oregon was founded in 2002. It invests in cost-effective energy conservation, helping to pay the above-market costs of renewable energy resources, and encouraging energy market transformation in Oregon (Energy Trust of Oregon 2007). In June 2005, the city and county jointly released the *Progress Report on the City of Portland and Multnomah County Local Action Plan on Global Warming*. The report outlined progress in meeting GHG reduction goals set earlier, and also identified new priorities for activities for the period 2006–2008. These included supplying all city facilities with 100 percent renewable energy through a wind power project; increasing local recycling rates; making the city's climate change commitments legally binding through participation in the Chicago Climate Exchange; and implementing the West Coast Governor's Global Warming Initiative and Governor Ted Kulongoski's Global Warming Strategy (de Steffey, Rojo, and Sten 2005). Whether Portland will meet its target by 2010 remains uncertain. By early 2008, the 2005 *Progress Report* was the most recent update on Portland's emission reduction efforts. The time lag between action and assessment reinforces the need for municipalities to have human and technical capacity to track emission reduction efforts in an accurate and timely manner.

A variety of stakeholders are targeted in the climate change work at the city and county levels. Both levels of government have logged success with inhouse emission reduction efforts focusing on municipal operations through technology change, education campaigns, and financial incentives aimed at individual behavior change related to transportation and consumption patterns. Through participation in other networks, Portland is sharing its success stories as well as enhancing its own capacities to act on climate change issues. The private sector has also been engaged through green-procurement policies to facilitate the market transformation required to meet Portland's green building standards and energy conservation programs. Collaboration with the state government is also continuing to reduce barriers to emission reductions on complex issues that neither the city nor the county fully controls.

Ten "Kyoto Cities" have adopted local-level emission reduction goals similar to those in the Kyoto Protocol. A recent study of these cities showed that while community-wide emissions have risen since 1990, and that few if any will reduce emissions 7 percent below 1990 levels by 2012, Portland is an exception (Bailey 2007, 3). Portland is reporting a 0.7 percent increase above its 1990 baseline, at the same time that Oregon's emissions have increased between 18 and 30 percent, depending on whose calculations are used and what assumptions are made in completing the calculations (Bailey 2007, 12). Portland's early leadership is credited with sparking action in over 400 municipal governments worldwide (City of Portland 2007b). At the same time, Portland's municipal emission reduction efforts, promoted by its participation in ICLEI's CCP program, allowed it to learn from the experiences of Stockholm, Sweden, and Copenhagen, Denmark (EPA 2007). Specifically, Portland's green fleets and inventory approach were directly informed by exchanges between council members and city staff in these locations (EPA 2007). These experiences reinforce the significance of international knowledge networks in initiating, supporting, and advancing local efforts.

Comparing Efforts and Achievements by Toronto and Portland

When comparing fifteen years of emission reduction efforts in Toronto and Portland, a few lessons can be seen. The approaches employed by each city remind other municipalities and higher-order governments that to achieve real emission reductions, a variety of tools—carrots, sticks, and education—are required. The achievements in Portland and Toronto are also the result of careful, prolonged actions, with sustained political and civic commitment—they were not achieved quickly. Civic involvement and participation are also important. Recent research shows that U.S. metropolitan areas most at risk from the adverse affects of climate change are the least likely to adopt policy reforms to control CO_2. In contrast, metropolitan areas that already have high levels of civic and environmental capacity are most likely to engage in CO2 reduction efforts (Zahran, Grover, Brody, et al. 2008).

At the same time, both Portland and Toronto have been challenged by, and entangled in, a complex jurisdictional web, receiving varying degrees of support. Experience in both cities also highlights the importance of sharing municipal success stories. Portland's 2005 Global Warming Progress Report provides an excellent overview of specific activities that facilitated inquiry by other interested municipalities. In contrast, Toronto has not communicated its initiatives in a coordinated manner, making knowledge exchange more complicated and uncertain, despite its successes. The ICLEI and FCM have played important roles in facilitating information exchange for both cities, yet the roles of both networks are hampered when municipalities do not regularly report their progress in detail. Nevertheless, both cities show that meaningful emission stabilization and reductions can occur at the municipal level, even in the face of economic and population growth and federal governments antagonistic or reluctant to take meaningful action on climate change.

Local Government Action and Influence

Globally, municipalities that have stabilized and/or reduced their emissions below 1990 levels have accomplished something that few other political jurisdictions have attained. In Canada and the United States, municipalities have responded to a global problem in the face of inaction by their national governments, disproving assumptions that the global character of climate change presents too many barriers for municipal response. Based on the high percentage of U.S. and Canadian populations residing in municipalities that are formally participating in climate change reduction programs, municipal action to reduce GHG emissions is unique in the history of social and environmental movements in North America. It is difficult to identify another environmental issue that has locally elected governments of over 100 million people actively supporting it. Municipalities may not be "our last, best hope," but they are part of a promising movement, which shows capacity and potential to reduce GHG emissions, while at the same time improving the quality of life in cities and promoting greater citizen participation in the response to climate change (Moser, this volume).

The argument can be made that it is easy to make commitments to reduce GHG emissions without taking action. Equally, while municipal emission reduction efforts are to be congratulated, they cannot be relied upon to fix the climate change problem, and even those municipalities that are taking action are pursuing emission reduction goals far short of the reductions that are ultimately needed. Indeed, municipalities will have to make practical and ideological changes if they are to make substantial reductions in emissions. Canadian municipalities have achieved emission reduction success with initiatives such as building retrofits and landfill gas capture, but continue to be vexed by the demand for low-density, automobile-oriented, residential land uses (Robinson 2006). Similarly, American cities typically pick the low-hanging fruit in the form of landfill gas capture, LED traffic lights, and renewable energy procurement. They too must wrestle with the relationship between land use and emission reductions to make further progress (Bailey 2007).

All of these observations, however, do not get to the heart of a central question: Why are municipalities taking action? Independent municipal climate change action challenges orthodox assumptions about collective action problems. Why in the face of a problem replete with free-riding are municipalities taking actions to reduce GHG emissions? Revisiting the reasons advanced in the introduction, municipal action rests on the interrelationship of four broad factors.

First, this chapter describes the central role of international and national networks in municipal climate change action. Given the recent and ongoing strengthening of the U.S. Mayors Climate Protection Agreement, and the longstanding presence of the Federation of Canadian Municipalities in national and international climate action, there is a need to understand the longevity of these networks, how they are institutionalized, and their potential to influence policy. Such networks use specific pathways of change to influence local and national decision-making (Selin and Van-Deveer 2007). Municipalities are demonstrating that it is possible to make strategic interventions that reduce GHG emissions; municipal demand for energy savings encourages market expansion for these products and services; municipalities are engaged in policy diffusion and learning through their networks; and, owing to the scale of municipal attention to climate change, taking action to reduce emissions is becoming the "right thing to do."

Second, municipal leaders recognize that the imperative to respond to the global problem of climate change resonates locally. In expressing reasons for taking action, mayors and municipal staff committed to reducing GHG emissions raise attention to collective responsibility, children's futures, quality of life, equity, and environmental quality. Hence, these commitments respond directly to public opinion and genuine citizen support for more action on climate change. At the same time, and thirdly, municipal action also results from moral and strategic obligations to respond to global problems at the local level. For two decades, through the international system and international networks, municipalities have been encouraged to join the local movement toward global environmental improvement. From the early establishment of ICLEI, to the creation of Local Agenda 21 at the "Rio Summit" in 1992, municipalities have been persuaded to act on climate change out of moral and collective obligation. It follows that municipalities are also concerned about their international reputation and pursue policy goals and actions that demonstrate municipal leadership.

Finally, more tangible reasons for action are also prominent. Local governments have not just recently recognized the relationship between the quality of human settlements and the quality of the local and global environment. The 1972 United Nations Conference on the Human Environment focused centrally on the relationships among human settlements, environmental quality, and human well-being. Today, most municipalities in North America are highly cognizant of broader principles of urban sustainability, and they are taking practical and institutional steps to integrate these principles into municipal practices. Through more participatory and inclusive processes of decision making, municipalities are achieving tangible local benefits with respect to costs and resource use, at the same time producing positive global outcomes. The achievement of these benefits is a function of several factors, including leadership from elected officials and support from municipal bureaucracies

(Bulkelely and Betsill 2003). Past and potential extreme weather events in North America also encourage municipalities to consider their operations and policies more holistically, bringing near, medium, and long-term timeframes into decision making.

Conclusion

Whether municipal responses to climate change are a result of moral obligation, the ability to implement policies that achieve cobenefits, or due to the prominence and influence of national and international networks, many North American municipalities are well positioned to play a central role in national and international climate change mitigation and adaptation. More specifically, national and international municipal networks focused on climate change represent prominent and growing pathways of knowledge exchange that have diffused and continue to diffuse across jurisdictions. In turn, municipalities are central actors in emerging multilevel climate change governance in North America. Reducing GHG emissions in North America, however, will require a more concerted, multilevel effort.

It remains unclear whether provincial, state, and national governments will embrace and promote local government enthusiasm for climate change response and build on municipal innovation. Federal institutions have not directly impeded municipal actions, but have only provided limited and largely inconsistent support. Similarly, state and provincial officials have generally been slow to support municipal climate change policymaking (Selin and VanDeveer, this volume). Municipalities should certainly not be relied upon as the solution to climate change in North America, but they are independently and collectively showing that they are policy leaders, and should figure prominently in any future climate governance regime in North America.

Notes

1. See http://www.c40cities.org/bestpractices/transport/mexicocity_taxi.jsp.

2. Personal communication, Amy Seabrooke, FCM.

3. Data on milestones achieved are derived from municipal reporting. Therefore, if a municipality has not updated its progress in milestone implementation or completion, FCM would not present this. For example, FCM data indicate that Toronto has only completed two corporate and two community milestones. However, Toronto's climate change action plans indicate that the city has achieved several other milestones.

4. Following the baseline year established under the Kyoto Protocol, Toronto changed its original baseline year from 1988 to 1990.

References

Bailey, John. 2007. *Lessons from the Pioneers: Tackling Global Warming at the Local Level.* Washington, DC: Institute for Local Self-Reliance.

Betsill, Michele M. 2001. Acting Locally, Does It Matter Globally? The Contributions of U.S. Cities to Global Climate Change Mitigation. Paper presented at the Open Meeting of the Human Dimensions of Global Environmental Change Research Community, Rio de Janeiro, Brazil, October 6–8.

Bulkeley, Harriet and Michele M. Betsill. 2003. *Cities and Climate Change: Urban Sustainability and Global Environmental Governance.* London: Routledge.

Betsill, Michele M., and Harriet Bulkeley. 2004. Transnational Networks and Global Environmental Governance: The Cities for Climate Protection Program. *International Studies Quarterly* 48(2): 471–493.

Betsill, Michele M., and Harriet Bulkeley. 2006. Cities and Multilevel Governance of Global Climate Change. *Global Governance* 12(2): 141–159.

City of Portland. 2007a. Community Wide Actions. Portland, OR: City of Portland. http://www.portlandonline.com/osd/index.cfm?&a=111833&c=32927.

City of Portland. 2007b. Local Actions Against Global Warming. Portland, OR: City of Portland. http://www.portlandonline.com/osd/index.cfm?c=41896&a=111833.

City of Portland. 2007c. What Is Local Government Doing? Portland, OR: City of Portland. http://www.portlandonline.com/osd/index.cfm?c=41896&a=111834.

City of Toronto. 1993. *Draft Strategic Action Plan for the Reduction of Carbon Dioxide Emissions from the City of Toronto.* Toronto: International Council for Local Environmental Initiatives.

City of Toronto. 1997. Global Climate Change: Toronto—Contribution and Commitment. Toronto: City of Toronto.

City of Toronto. 2006. Mayor David Miller's inaugural address. Toronto: City of Toronto. http://www.toronto.ca/mayor_miller/speeches/inaugural_address06.htm.

City of Toronto. 2007. *Greenhouse Gases and Air Pollutants in the City of Toronto: Toward a Harmonized Strategy for Reducing Emissions.* Toronto: City of Toronto. Retrieved from: http://www.toronto.ca/taf/pdf/ghginventory_jun07.pdf.

The Climate Group. 2005. Low Carbon Leader: Canada 2005. London: Climate Group. http://www.theclimategroup.org/assets/resources/low_carbon_leader_canada.pdf.

Connolly, Priscilla. 1999. Mexico City: Our Common Future? *Environment and Urbanization* 11(1): 53–78.

de Steffey, Maria Rojo, and Erik Sten. 2005. Global Warming Progress Report: A Progress Report on the City of Portland and Multnomah County Local Action Plan on Global Warming. Portland, OR: City of Portland and Multnomah County. http://www.portlandonline.com/shared/cfm/image.cfm?id=112118.

Energy Trust of Oregon. 2007. Who We Are. Portland, OR: Energy Trust of Oregon. Portland: Energy Trust of Oregon. http://www.energytrust.org/who/index.html.

Environment Canada. 1999. Six Successful Landfill Gas Utilization Projects Demonstrate Early Actions to Reduce Greenhouse Gases. Ottawa: Environment Canada.

Environmental Protection Agency (EPA). 2007. Portland's Local Action Plan Learned from Stockholm and Copenhagen. Washington, DC: United States Environmental Protection Agency. http://www.epa.gov/NCEI/international/airclimate.htm.

Federation of Canadian Municipalities (FCM). 2006. *Our Communities, Our Future. Annual Report 2005–2006*. Ottawa: FCM.

Federation of Canadian Municipalities (FCM). 2007. Members and Milestones. Ottawa: FCM.

IPCC. 2007. *Climate Change 2007: The Physical Science Basis. Summary for Policymakers*. Geneva: IPCC.

Larsen, Janet. 2006. U.S. Mayors Respond to Washington Leadership Vacuum on Climate Change. Washington, DC: Earth Policy Institute.

Lindseth, Gard. 2004. The Cities for Climate Protection Campaign (CCPC) and the Framing of Local Climate Policy. *Local Environment* 9(4): 325–336.

Magnusson, Warren. 2005. Protecting the Right of Local Self-Government. *Canadian Journal of Political Science* 38(4): 897–922.

Municipalities Issue Table. 1999. *Options Paper*. Ottawa: Government of Canada, National Climate Change Program.

Portney, Kent E. 2003. *Taking Sustainable Cities Seriously: Economic Development, the Environment, and Quality of Life in American Cities*. Cambridge MA: MIT Press.

Robinson, Pamela. 2000. Canadian Municipal Response to Climate Change: A Framework for Understanding Barriers. Ph.D. dissertation, University of Toronto.

Robinson, Pamela J. 2006. Canadian Municipal Response to Climate Change: Measurable Progress and Persistent Challenges for Planners. *Planning Theory and Practice Interface* 7(2): 218–223.

Robinson, Pamela, and Christopher D. Gore. 2005. Barriers to Canadian Municipal Response to Climate Change. *Canadian Journal of Urban Research* 14(1): 102–120.

Sancton, Andrew. 2006. Cities and Climate Change: Policy-Takers Not Policy-Makers. *Policy Options* 27(8): 32–34.

Scott, Mel. 1995. *American City Planning since 1890*. Chicago: American Planning Association.

Selin, Henrik, and Stacy D. VanDeveer. 2006. Canadian-U.S. Cooperation: Regional Climate Change Action in the Northeast. In *Bilateral Ecopolitics: Continuity and Change in Canadian-American Environmental Relations*, edited by Philippe Le Prestre and Peter Stoett. Aldershot, UK: Ashgate.

Selin, Henrik, and Stacy D. VanDeveer. 2007. Political Science and Prediction: What's Next for U.S. Climate Change Policy? *Review of Policy Research* 24(1): 1–27.

Toronto Environmental Alliance. 2006. Toronto Smog Report Card 2006. Toronto: Toronto Environmental Alliance. http://www.torontoenvironment.org/sites/tea/files/TEASmogReport2006.pdf.

U.S. Conference of Mayors. 2007a. Mayors Call for $4 Billion in Energy and Environmental Block Grant at 75th Winter Meeting of the U.S. Conference of Mayors. Press Release. Washington, DC: U.S. Conference of Mayors. http://usmayors.org/75thWinterMeeting/eebg_012507.pdf.

U.S. Conference of Mayors. 2007b. Mayors Release 10–Point Legislative Agenda on Issues Impacting Cities and Families at 75th Winter Meeting of the U.S. Conference of Mayors. Press Release. Washington, DC: U.S. Conference of Mayors. http://usmayors.org/75thWinterMeeting/10pointplanadvisory_012407.pdf.

U.S. Conference of Mayors. 2007c. *Survey on Mayoral Leadership on Climate Protection*. Washington, DC: Mayors Climate Protection Center.

Vogel, Ronald K. 2007. Multilevel Governance in the United States. In *Spheres of Governance: Comparative Studies of Cities in Multilevel Governance*, edited by Harvey Lazar and Christian Leuprecht. Montreal: McGill-Queen's University Press.

West, J. Jason, Patricia Osnaya, Israel Laguna, Julia Martinez, and Adrián Fernández. 2004. Co-control of Urban Air Pollutants and Greenhouse Gases in Mexico City. *Environmental Science and Technology* 38(13): 3474–3481.

Worldwatch Institute. 2007. *State of the World: Our Urban Future*. New York: W. W. Norton.

Zahran, Sammy, Himanshu Grover, Samuel D. Brody, and Arnold Vedlitz. 2008. Risk, Stress and Capacity: Explaining Metropolitan Commitment to Climate Protection. *Urban Affairs Review* 43(4): 447–474.

III

Continental Politics

8

NAFTA as a Forum for CO_2 Permit Trading?

Michele M. Betsill

Introduction

This chapter explores emerging policies and practices related to global climate change within the North American Free Trade Agreement (NAFTA), with a particular focus on the possibility of establishing a North American greenhouse gas (GHG) emissions trading system. Emissions trading is emerging as a central policy response to the problem of climate change. Systems for emissions trading are being considered and/or are in operation in more than thirty different venues, ranging from local to global levels and involving both public and private sectors, many of which are located in North America (Betsill and Hoffman 2008; Farrell and Hanemann, this volume; Selin and VanDeveer, this volume). These discussions of emissions trading reflect a trend toward multilevel governance on the issue of climate change, which in turn raises questions about what level of social organization is most appropriate for particular governance tasks (Betsill and Bulkeley 2006; Young 2002). Since 2000, there has been some preliminary discussion within the North American Commission for Environmental Cooperation (CEC), NAFTA's environmental organ, about establishing a carbon dioxide (CO_2) permit trading system to mitigate the environmental impacts of electricity generation. This chapter draws on this discussion in order to evaluate whether the CEC is an appropriate venue for GHG emissions trading in North America.

The chapter begins with a brief introduction to emissions trading and an overview of the CEC discussions related to climate change in general and GHG emissions trading more specifically. The chapter then addresses three sets of issues related to establishing a CEC-based CO_2 permit trading system in North America, with particular focus on its implications for climate protection: the institutional context, design elements, and interplay with other trading systems. This discussion highlights the complex system of multilevel governance that has developed around climate change in North America and the challenges and opportunities related to creating a regional emissions trading system in the face of a diverse set of interests and policy approaches. In the final section, I question the wisdom of establishing a CO_2 permit trading system within the NAFTA regime to address the problem of climate change.

Emissions Trading as a Mechanism for Climate Governance

Over the past decade, there has been a proliferation of interest in emissions trading as a mechanism for governing global climate change. Allowance trading systems—or cap-and-trade systems—involve setting a cap on emissions levels, distributing emissions allowances among participants, then permitting participants to trade allowances among themselves in order to meet their respective commitments. Credit (or project-based) trading systems engage in the purchase and transfer of emissions credits derived from specific projects (Hasselknippe 2003; Rosenzweig, Varilek, and Janssen 2002). The carbon market is one of the world's fastest-growing markets, with trade volume increasing from 94 million metric tons in 2004 to 1.2 billion metric tons in the first six months of 2007 at an approximate value of €16 billion (Roine and Tvinnereim 2007). This chapter focuses on the development of an allowance trading system within the NAFTA regime.

Emissions trading was a divisive issue during the Kyoto Protocol negotiations (Betsill 2004). The United States, supported by Canada and several other industrialized countries as well as industry representatives, argued that states should be given a great deal of flexibility in deciding how to meet their emissions obligations and pushed strongly for emissions trading to be included in the protocol. In contrast, the European Union (EU), many developing countries, and most environmental groups stressed that extensive reliance on such mechanisms would allow rich countries to buy their way out of making any meaningful domestic commitments. They believed this further violated the "polluter pays" principle enshrined in the United Nations Framework Convention on Climate Change (UNFCCC). Language allowing emissions trading was nevertheless included in the Kyoto Protocol as the result of a desire to appease the United States in recognition of its significant role in producing GHG emissions. However, the Kyoto Protocol did not specify the rules for how a trading system would be implemented, and in subsequent negotiations, the EU sought to impose restrictions on the extent to which countries could use trading and other flexible mechanisms to meet their obligations to reduce GHG emissions.

In an interesting turnaround, European decision makers and civil servants have become much more supportive of emissions trading as a tool for controlling GHG emissions in the decade following the adoption of the Kyoto Protocol in 1997. As they began reviewing options for achieving their own Kyoto commitments, trading became a much more attractive option, particularly once a proposed EU carbon tax was defeated. Trading was seen as an economically efficient way to achieve the EU Kyoto target. Moreover, several EU member states began developing national and bilateral trading systems, and EU policymakers had an interest in ensuring such schemes were compatible (Christiansen and Wettestad 2003; Skjærseth and Wettestad 2008). The EU Emissions Trading Scheme (ETS) became operational in January 2005, and it is meant to serve as the model for a Kyoto-wide trading system.

Interest in emissions trading as a mechanism for climate governance involves actors operating at a variety of levels of social organization (from the local to the global) in both the public and private spheres. In the addition to the EU trading system, the government of Switzerland launched an allowance trading system in 2008, and several other national governments, including Canada, Japan, and New Zealand, are considering domestic trading schemes.[1] In the United States, a number of states are in the process of developing statewide trading systems and/or regional trading schemes (Rabe, this volume; Selin and VanDeveer, this volume). In the private sector, the Chicago Climate Exchange is a voluntary trading program whose members include subnational governments, universities, and firms primarily in North America. In short, this broad interest in emissions trading illustrates the fact that global climate governance has evolved into a complex system of multilevel governance characterized by the vertical and horizontal fragmentation of authority (Betsill and Bulkeley 2006).

The proliferation of interest in emissions trading as a preferred policy option for addressing climate change is not surprising given that there has been a general growth trend in the use of market-based mechanisms for achieving environmental objectives in the industrialized world since the 1970s (Jordan, Wurzel, and Zito 2005). In contrast to command-and-control approaches to environmental management, market-based instruments such as emissions trading give agents flexibility in meeting environmental goals and are seen to lower the cost of action. Moreover, market-based instruments fit easily within the metadiscourses of ecological modernization and liberal environmentalism that dominate discussions of global environmental governance today and emphasize the compatibility of economic growth and environmental protection (Bäckstrand and Lövbrand 2006; Bernstein 2001; Mol 2001).

There are, however, also critiques of permit trading as a means of controlling GHG emissions. Some emphasize that the international system lacks sufficient enforcement mechanisms or argue that taxes are more efficient (Bell 2006; Nordhaus 2006). Other critics contend that emissions trading schemes are not sufficient to stimulate fundamental and necessary social changes such as moving societies away from a dependence on fossil fuels and reorganization of the global economy.

Moreover, emissions trading schemes institutionalize a right to pollute for those like the United States and other industrialized countries who have historically contributed most to the problem (Lohmann 2006). In the following sections, I situate the CEC discussion on climate change in this wider discussion on emissions trading as a possible tool for climate change governance.

The CEC and Climate Change

The CEC is the primary mechanism for addressing environmental concerns within the NAFTA regime. It was created by the North American Agreement on Environmental Cooperation (NAAEC), a side agreement negotiated at the time NAFTA was created in order to appease environmental groups concerned about the ecological impacts of increased trade in North America. The NAAEC, and its related institutions, seek to manage the economic growth associated with trade liberalization in an environmentally sustainable way; it also allows for action on the environment beyond trade.

The CEC consists of three bodies: the Council, the Secretariat, and the Joint Public Advisory Committee (CEC 2005). The Council is an intergovernmental body made up of the environmental ministers from all three NAFTA member states. It meets annually and is responsible for initiating inquiries and releasing environmental information. The Secretariat is located in Montreal, Canada and is the support structure for the NAAEC. It produces several types of reports and administers the CEC's four program areas: (1) Environment, Economy, and Trade; (2) Conservation of Biodiversity; (3) Pollutants and Health; and (4) Law and Policy. Finally, the CEC includes the Joint Public Advisory Committee, which consists of fifteen individuals appointed by the Parties. Members act in their capacity as citizens to provide advice to the Council and the Secretariat.

The CEC's discussions related to mitigating CO_2 emissions are in their infancy, and it is important to acknowledge that climate change is by no means at the top of the CEC agenda. While climate change has been addressed directly on a limited basis, climate-related issues have been taken up in several CEC program areas (Betsill 2007). The Environment, Economy, and Trade program includes projects on renewable energy in the context of greening trade, green procurement programs, and identifying the environmental implications of the North American electricity market (discussed in greater detail below). One project under the Pollutants and Health program aims to facilitate coordination among air quality management agencies in North America, including in the monitoring of CO_2. The CEC Council has passed two climate-related resolutions calling for coordination of methodologies for GHG emissions inventories and forecasts (CEC 1995, 2001). In 2008, the CEC released a report on opportunities and challenges related to green building in North America (CEC 2008).

Between 2000 and 2004, CEC discussions on climate change were linked to concerns about the environmental impacts, especially related to air quality, of an increasingly integrated North American electricity market (CEC 2002). Of course, discussions of air quality and the electricity sector are not divorced from the problem of climate change since electricity generation is a significant (and growing) source of GHG emissions in each NAFTA member state, accounting for 39 percent of all CO_2 emissions in the United States, 22 percent in Canada, and 30 percent in Mexico (Miller and Van Atten 2004). The North American electricity market has experienced rapid change over the past decade largely due to trade liberalization and regional convergence in competitiveness and trade policy (CEC 2002; Dukert 2002; McKinney 2000). Market integration is likely to continue and generation capacity is expected to increase to meet rising demand (CEC 2002; Ferretti 2002). North America already accounts for half of all electricity produced and consumed in the industrialized world (Dukert 2002; Rowlands, this volume). Ultimately, a number of factors will determine the implication of market integration and increased generation capacity for North American GHG emissions, including location, fuel choice, price, infrastructure, market access, grid access, and regulations (CEC 2002; Dukert 2002; McKinney 2000; Mumme and Lybecker 2002; Rowlands, this volume).

In 2002, the CEC Secretariat released a special report entitled *Environmental Challenges and Opportunities of the Evolving North American Electricity Market* (CEC 2002). This report was the culmination of a process launched in 2000 under Article 13 of the NAAEC, which states that the Secretariat can prepare a report on "any matter within the scope of its annual work program." The process was overseen by an advisory board consisting of academics, NGO representatives, and members of the energy sector from each member state and involved the production of several working papers and three public events. The initiative sought to address the challenge of ensuring "an affordable and abundant supply of electricity without compromising environmental and health objectives" (CEC 2002, v). The CEC examined the environmental aspects of the regional electricity market and prospects for green electricity. In the 2002 CEC report, the initiative's advisory board made four recommendations specific to climate change and emissions trading:

- Develop a regional GHG emissions inventory;
- Establish a framework for a regional GHG emissions trading regime;
- Demonstrate that carbon trading can generate resources for developing countries (such as Mexico); and
- Develop programs to stimulate investment in clean and renewable energy production (especially in the United States).

In 2002, the CEC Council agreed to include some items from the electricity report in the 2003 CEC Air Quality Work Plan, including comparative studies of North American air quality management standards, regulation and planning, compatibility of standards for construction and operation of electricity generation facilities, and opportunities for emissions trading (JPAC 2003). In 2004, the CEC issued a report detailing 2002 sulfur dioxide (SO_2), nitrogen oxides (NO_x), mercury, and CO_2 emissions from North American power plants as a result of a 2001 Council Resolution and the *Environmental Challenges* report (Miller and Van Atten 2004). The detailed pollutant information in this report was envisioned to support decision-making related to energy needs and environmental protection, including "the development of robust and viable trinational emissions trading programs" (Miller and Van Atten 2004, iv).

The CEC and Emissions Trading

This limited consideration of CO_2 mitigation and climate change sets the stage for thinking about the desirability of establishing a North American emissions trading system within the NAFTA regime. First, it should be noted that a North American trading system need not be situated in the CEC. An alternative model would be to link national and subnational permit trading systems. However, many of the same concerns about the institutional context, design elements and institutional interplay would apply. In a situation of multilevel governance, a key challenge is to allocate specific governance tasks to the "appropriate" level of social organization. Young (2002) cautions against trying to find the "right" scale of governance. Rather, he argues, "In most cases, the key to success lies in allocating specific tasks to the appropriate level of social organization and then taking steps to ensure that cross-scale interactions produce complementary rather than conflicting actions" (Young 2002, 266). In that spirit, this section considers the appropriateness of developing a CEC-based CO_2 permit trading system by examining three sets of issues: the institutional context in which such a system would be developed; how the system would be designed; and how such a system would interact with other trading systems in North America.

The Institutional Context

Emissions trading systems develop in a particular institutional context, which in turn shapes the form of each system. One important aspect of the institutional context has to do with the allocation of authority, and specifically, the ability to facilitate political agreement to control emissions, which is seen to be a key factor in the success of any trading system (Aulisi, Farrell, Pershing, et al. 2005; Hasselknippe

2003). In considering the ability of the CEC to facilitate agreement among its member states to control CO_2 emissions, I examine the specific rules and structures of the CEC and the broader NAFTA regime, and conclude that reaching political agreement on controlling CO_2 emissions will be difficult given the CEC's limited authority over its member states. As a result, agreement depends on linking CO_2 to broader air quality and energy concerns (rather than climate change) in a way that is consistent with NAFTA's core commitment to trade liberalization.

The CEC's capacity to facilitate political agreement on controlling CO_2 emissions is constrained foremost by the fact that it is an intergovernmental body with limited authority over its member states. The CEC can facilitate political agreement only when the member states have a common interest in doing so (Stevis and Mumme 2000). Notably, NAFTA member states do not have common interests on climate change; each has a distinct domestic approach to climate change, which has developed independently of the other member states, and in the case of the United States, outside the multilateral regime. This stands in stark contrast to the situation in the European Union where a supranational body has authority over member states on the issue of climate change and where regional and national climate governance systems have coevolved since the 1990s.

Moreover, Canada, the United States, and Mexico have different levels of economic development, which has also shaped their respective responses to climate change. Under the George W. Bush Administration, the United States focused on reducing the carbon intensity of its economy 18 percent below 2002 levels by 2012 through voluntary programs (U.S. Department of State 2006). Prior to 2006, the Canadian approach consisted of government regulations and strategic investments designed to achieve its Kyoto target (reduction of GHG emissions 6 percent below 1990 levels in the period 2008–2012) and to become a leader in clean technology (Government of Canada 2005). The 2006 election of a minority Conservative government resulted in a shift away from a Kyoto-focused strategy toward a "made-in-Canada" approach aimed at reducing total GHG emissions 20 percent below 2006 levels by 2020 while maintaining a strong economy (Government of Canada 2008; Stoett, this volume). Mexican climate policy has focused on developing an emissions inventory, introducing mitigation projects in the forestry and energy sectors, and attracting investment through the Clean Development Mechanism program under the Kyoto Protocol (Instituto Nacional de Ecología 2001; Pulver, this volume).

Moreover, Canada, the United States, and Mexico have distinctly different relationships to the international climate change regime. Mexico participates in the negotiations as a developing country, while the United States and Canada have often worked together in specialized negotiating blocs with other industrialized countries. However, it should be noted that membership in these negotiating blocs was based on common concerns about the economic implications of climate mitigation rather than any regional affinities. Today, Canada and Mexico are both parties to the Kyoto Protocol, although only Canada has a binding commitment to reduce its GHG emissions in the first commitment period. The Bush administration rejected the Kyoto Protocol in March 2001, but the United States remains a party to the UNFCCC.

The member states of NAFTA do appear to have common interests related to air quality and energy supply as reflected in the CEC's programs and projects. In this context, the CEC could facilitate political agreement on controlling CO_2 emissions to the extent that CO_2 is linked to air quality and energy issues rather than climate change. Regional organizations, such as the CEC, typically address many issues simultaneously, and thus they can promote the development of shared interests by linking a particular issue to other issues addressed within the organization that may be of greater concern to one or more member states (Axelrod and Keohane 1986; Levy, Keohane, and Haas 1995). Indeed, we see this sort of issue linkage in the CEC, where CO_2 is often treated like other air pollutants produced by utilities rather than as a contributor to climate change. For example, in the CEC's 2004 power plant emissions report, CO_2 is identified as "an important greenhouse gas," but the entire report contains only one mention of climate change or global warming (Miller and Van Atten 2004). This occurs in the "Note from the Executive Director," which notes that some North American companies are partnering with developing countries through the Kyoto Protocol's Clean Development Mechanism to "help address the looming threat of global warming" (Miller and Van Atten 2004, iv). The *Environmental Challenges* report notes the transboundary impacts of electricity generation and even identifies the specific problems associated with the medium- and long-range transport of NO_x, SO_2, mercury, and persistent organic pollutants. Of CO_2 emissions, the report states only that they "are of global concern wherever they are emitted" without specifying the exact nature of that concern (CEC 2002, 7). However, as discussed below, identifying CO_2 as an air pollutant has implications for the design of an emissions trading system. Specifically, such a system might include other non-GHGs such as SO_2, NO_x, and mercury, which may in turn limit its impact on the problem of global climate change.

Another important component of the institutional context concerns the relationship between a CEC-based CO_2 permit trading system and the broader NAFTA regime, where trade liberalization is the primary objective. Consistent with the NAFTA treaty, the CEC rests upon a core set of neoliberal economic assumptions —trade will increase prosperity, environmental protection is an important part of prosperity, and trade will create greater resources for environmental protection. The CEC works to promote environmental sustainability in ways that are consistent with NAFTA's goal of trade liberalization (Ferretti 2002). Stevis and Mumme (2000) contend that in practice, the environment plays a secondary role in the NAFTA context, reflecting "weak ecological modernization" where environmental protection is assumed to have potentially negative implications for economic growth that need to be mitigated (Mol and Spaargaren 2000).

On the issue of climate change, much attention has been focused on the potential tensions between trade and climate policy (Brewer 2003; Frankel 2005). Charnovitz (2003) notes that economic growth linked to trade liberalization will likely lead to increased GHG emissions while also potentially providing incentives for the innovation and diffusion of climate-friendly technology. In addition, climate policies may have implications for economic competitiveness. The need to support NAFTA's trade liberalization goal will shape the types of policies and standards considered within the CEC to reduce GHG emissions, with likely preference for technological management of emissions rather than policies that more directly require changes in the underlying processes that give rise to GHG emissions (such as a reliance on fossil fuels) (CEC 2002). It is therefore not surprising that emissions trading was seen as a promising policy instrument for controlling CO_2 emissions within the CEC.

At the same time, the fact that a CEC-based CO_2 permit trading system would be nested within the broader NAFTA regime means that it would need to be consistent with NAFTA's trading rules in general as well as rules specific to the electricity sector in particular (Horlick, Schuchhardt, and Mann 2002). Russell (2002) identifies a number of potential conflicts between a NAFTA-wide emissions-trading system and trade and investment provisions in the NAFTA treaty (which also arise in the development of national or subnational permit trading systems in NAFTA member states). For example, would tradable emissions units (TEUs) be treated as goods and therefore subject to treaty provisions on national treatment and market access? Would purchasing TEUs from entities in other countries fall under treaty rules regarding investment? Emissions allowance units could be considered as subsidies subject to extra duty when transferred between countries. Is trade in TEUs an activity linked to the procurement of energy goods and services? Finally, could activities involved in the trading system be viewed as trade restrictions? These critical issues would require careful consideration in any future discussions of a CEC-based trading system.

Design Issues

The design of an emissions trading system affects its environmental integrity and economic efficiency. This section argues that framing the problem of CO_2 emissions as an issue of air quality and energy has implications for the design of a CEC-based permit trading system. I analyze key design elements involving coverage (gases and sectors) and targets (caps and allocation), and I also identify issues that may arise in the context of designing a CEC-based CO_2 permit trading system. In several instances the goals of environmental integrity and economic efficiency are likely to come into conflict, and given NAFTA's focus on trade liberalization, the CEC can be expected to resolve such conflicts by giving priority to economic efficiency. As such, some design choices might render a CEC-based CO_2 permit trading system meaningless for addressing the problem of climate change.

Trading systems that include a variety of gases give participating installations the flexibility to reduce emissions where costs are lowest. Roughly half of all CO_2 permit trading systems include a basket of six GHGs, which makes sense in addressing the problem of climate change since each has a global warming potential (Hasselknippe 2003). At the same time, it can be very difficult to establish monitoring and verification systems for emissions reductions for several GHGs. As a result, many systems such as the EU's ETS only include CO_2. A CEC-based trading system would also likely include only CO_2 since it is the only GHG that has been monitored to date on a cross-national basis in North America. In addition, reaching agreement among member states to include a broader range of GHGs could be difficult given the preference for viewing CO_2 as an air pollutant (rather than a contributor to climate change) within the CEC (Miller and Van Atten 2004).

Of course, this framing also makes it likely that a CEC-based system would include other air pollutants that are monitored cross-nationally, such as SO_2, NO_x, and mercury. As noted above, including several gases is economically desirable because facilities can choose to reduce emissions of gases where the costs are lowest. However, by including non-GHGs, a CEC-based system may have little impact on the problem of climate change if participating facilities routinely choose to reduce SO_2, NO_x, or mercury emissions rather than CO_2. One way to achieve environmental integrity in terms of climate protection is to set emissions caps on a gas-by-gas basis. However, this would reduce the flexibility that comes with multigas coverage, potentially raising compliance costs.

Ideally, a CO_2 permit trading system should have broad participation from a variety of sectors and emissions sources, since sources are highly diffuse across the economy and broad participation allows for greater opportunity to identify low-cost reduction options (Aulisi, Ferrell, Pershing, et al. 2005). At the same time, broad participation may be administratively or politically prohibitive. The vast majority of CO_2 permit trading systems focus on large final emitters (such as power plants), which tends to keep the number of facilities at a manageable level while also covering a relatively high percentage of emissions (Christiansen and Wettestad 2003; European Commission 2004; Hasselknippe 2003). Discussions within the CEC have focused on the electricity generation sector, which seems to be appropriate given its central role in producing CO_2 emissions as well as other air pollutants.

Many allowance trading systems allow participants to purchase credits from offset projects (Hasselknippe 2003). Allowing offset credits is one way of encouraging developing countries to participate in trading schemes since they are most likely to attract investment in offset projects. In North America, allowing project credits to some extent could be a particularly attractive option for Mexico, whose domestic approach to climate change has included situating itself to provide this service under the Kyoto Protocol. However, the problem with offset credits is that they fall outside the emissions cap and initial allocation formula. When market participants rely on offset credits rather than allowances to meet their commitments, they may introduce additional emissions into the trading system thereby jeopardizing its environmental integrity (Aulisi, Ferrell, Pershing, et al. 2005).

Two key tasks in establishing an allowance trading system are (1) setting the overall cap and (2) allocating an emissions allowance among facilities in the selected sector(s). Caps can be expressed in a variety of ways: absolute tons of emissions, percentage of emissions from a base year, or intensity-based emissions standards (Hasselknippe 2003). In systems that include a number of gases, targets may reflect an overall cap on the emissions of all gases combined or they can be set on a gas-by-gas basis. As mentioned above, if a CEC-based permit trading system is to be climate-relevant, it must have a gas-by-gas cap. However, setting a CO_2 cap in the North American context is likely to be difficult as there is no clear basis for doing so. Member states of NAFTA differ in their domestic approaches to climate change, which has resulted in distinct preferences for setting caps. As noted above, the United States has an intensity cap while Canada has a percentage from-base-year cap under its Kyoto Protocol commitment. The CEC, with its limited authority, has no capacity to reconcile these differences.

Once a target is set, emissions allowances must be allocated among participating facilities, which can be done by a central authority (such as the CEC) or by jurisdictions within the trading system (such as national governments). The latter option is most likely in the North American context given the CEC's weak authority over member states. Each jurisdiction could then decide how to allocate allowances. Under grandfathering, allowances are distributed to participating facilities based on historical emissions and/or production levels. Alternatively, facilities can be required to purchase allowances through the auctioning of allowances (Aulisi, Ferrell, Pershing, et al. 2005; Christiansen and Wettestad 2003). Auctioning is more economically efficient and can be useful for early price discovery. At the same time, auctioning can be politically contentious because participating installations incur an upfront cost.

In the initial stages, grandfathering is less likely to mobilize opposition from covered facilities and may enhance prospects of getting a system up and running (Christiansen and Wettestad 2003). However, a potential weakness of the grandfathering approach became apparent in April 2006 when carbon credits within the European system lost 50 percent of their value. Several EU countries announced that their 2005 emissions were smaller than expected, which in turn reduced future demand for credits. According to the *Economist* (May 6, 2006), this reflects "industry's success in getting itself allocated more permits than actual emissions warranted when the scheme was launched." The EU's ETS experience has resulted in greater political support for the auctioning of allowances, both in the EU and in developing markets such as the Regional Greenhouse Gas Initiative (Selin and VanDeveer, this volume).

Interplay with Other CO_2 Permit Trading Systems

The design of a CEC-based trading system also has implications for its relationship with other CO_2 permit trading systems. This section considers the nature of institutional interplay between a CEC-based CO_2 permit trading system and five permit trading systems in operation, under development, or proposed in North America as of April 2008. Table 8.1 lists several major GHG emissions trading schemes that are proposed or operational. In addition, California and the Western Climate Initiative (including several U.S. states and Canadian provinces) are developing ideas and plans for GHG emissions trading schemes. The schemes listed in table 8.1 can be used to analyze critical interplay issues.

The literature on institutional interplay points to the futility of analyzing governance arrangements in isolation (Aggarwal 1998; Alter and Meunier 2006; Oberthür and Gehring 2006). This becomes all the more important in situations of multilevel governance. As discussed above, interest in CO_2 permit trading has emerged at a variety of levels of social organization in both the public and private spheres. In any situation of multilevel governance, governance arrangements may interact both horizontally (across space) and vertically (across levels of social organization). Bressers and Rosenbaum (2003, 14) observe that such interactions constitute multiple "games" that affect "the overall incentive structure of those affected by the governance." Similarly, synergies between institutions cannot be assumed and institutional interplay is a key determinant of the effectiveness of any given governance arrangement (Berkes 2002; Young 2002).

In considering the possible outcomes when governance arrangements interact, it is useful to distinguish between forms of institutional interplay, as each has its own political dynamic and means for dispute resolution. "Nested" institutions are linked through a hierarchy of rule systems, where rules at higher levels of social organization shape the development and design of institutions at lower levels (Aggarwal 1998; Alter and Meunier 2006). Conflicts may be less likely to occur in such situations since lower-level governance schemes must take higher-level rules and norms into account when they are created. Moreover, when disputes do occur, nested

Table 8.1
GHG allowance trading systems in North America

Trading system	Status	Description	Sectors covered
America's Climate Security Act of 2007	Proposed (2007)	Plan to reduce emissions from major emitters 70 percent below current levels by 2050; trading permitted to meet regulatory commitments.	Electric power plants; oil companies (producers and importers); large industrial emitters (more than 10,000 CO_2e/year)
Canada's "Turning the Corner" Plan	Proposed (2007)	Part of Canadian government's comprehensive plan for reducing GHG emissions 20 percent below 2006 levels by 2020; domestic trading permitted to meet regulatory commitments.	Electricity generation produced by combustion; oil and gas; pulp and paper; iron and steel; iron ore pelletizing; smelting and refining; cement; lime; potash; and chemicals and fertilizers
Chicago Climate Exchange	Operational (2006)	Voluntary trading program for companies, municipalities, and universities, primarily in North America.	Multiple depending on members who join
New Hampshire	Operational (2002)	Mandatory caps on CO_2, SO_2, NOx, and mercury emissions for state's power plants; trading allowed to meet CO_2, SO_2, and NOx targets.	Fossil-fuel-fired power plants
Regional Greenhouse Gas Initiative (RGGI)	Operational (2009)	Establishes a common CO_2 permits trading system from Maryland to Maine covering power plants.	Fossil-fuel-fired power plants

Sources: Chicago Climate Exchange 2004; Environment Canada 2008; Pew Center on Global Climate Change 2006; RGGI 2005; Selin and VanDeveer 2005, this volume; U.S. Senate 2007.

arrangements are likely to have a clear mechanism for resolution. In contrast, "overlapping" institutions have similar mandates but are not necessarily nested within a common higher-level rule system. Conflicts occur when overlapping institutions with common membership establish contradictory obligations. Resolving such conflicts can be contentious since there is no clear authority with jurisdiction over the dispute (Aggarwal 1998; Raustiala and Victor 2004).

The CO_2 permit trading systems in table 8.1 and a possible CEC-based trading system as discussed above vary in terms of the economic sectors covered. Despite this variation, it is notable that the electricity sector is subject to regulation in all but the Chicago Climate Exchange system. In the case of the New Hampshire and RGGI systems, this interplay produces complementarity because the RGGI is explicitly designed to help states meet their specific goals for controlling CO_2 emissions. In other words, the New Hampshire system is nested within the RGGI system (although it could be argued that in this case authority is in the hands of lower-level institutions). However, these systems would overlap with a CEC-based trading system, which could result in conflict if power-generation facilities in Canada and the United States find themselves subject to competing regulations.

When institutional interplay creates conflict, there may be incentives for actors to "venue shop" or seek to shift authority to the political arena most likely to promote a favorable policy (Alter and Meunier 2006; Gerber and Kollman 2004). In the face of conflicting regulations, owners of North American power-generation facilities may prefer to shift primary authority to a CEC-based trading system for two reasons. First, the CEC jurisdiction would cover all power plants in North America, which would lessen the risk of some facilities gaining a competitive advantage because they face less restrictive regulations or none at all. Second, it is possible that a CEC-based program rationalized in terms of air quality would set a less stringent CO_2 reduction target than the other systems, which are justified in terms of mitigating the threat of climate change, thereby reducing the cost of compliance.

Conclusion

This chapter has examined several issues related to the possibility of establishing a NAFTA-wide CO_2 permit trading system within the CEC, particularly in the context of the increasingly multilevel nature of North American climate change governance. The political foundation for such a system depends on framing CO_2 as an issue related to air quality and energy rather than climate change and relying on policies consistent with NAFTA's overarching goal of trade liberalization. In addition, the design of such a system could give rise to conflicts between economic efficiency and environmental integrity, with economic efficiency likely to prevail given

NAFTA's trade liberalization goal. Institutional interplay between a CEC-based CO_2 permit trading system and other trading systems in North America could also result in conflicts over the regulation of the electricity generation sector.

In the end, there seems to be little value-added benefit in establishing a North American CO_2 permit trading system under NAFTA. Because of its intergovernmental nature, the CEC has virtually no authority to promote harmonization of climate policy among member states without their consent. Instead, it must address climate change indirectly by linking the problem to broader issues of air quality and energy. While this may be a politically useful strategy to achieve agreement about the need to reduce emissions, the trading system's impact on climate change could be diluted if CO_2 emissions are not considered separately from other air pollutants. In addition, the fact that the CEC is nested within the broader NAFTA regime means that conflicts between environmental integrity and economic efficiency are likely to be resolved in favor of economic efficiency so as to be consistent with the goal of trade liberalization.

Finally, there is danger that a CEC-based CO_2 permit trading system could undermine the effectiveness of trading systems at other levels of social organization by creating incentives for venue shopping. This is not to say that NAFTA has no role in governing climate change. As cross-border climate change and energy-related activities (such as electricity generation and trade) intensify (Rowlands, this volume), NAFTA institutions are likely to face greater pressure to address climate change in the future. The ongoing challenge will be to identify appropriate tasks that will not undermine efforts to mitigate GHG emissions in other spheres and tiers of the multilevel system of climate governance.

Note

1. Point Carbon (www.pointcarbon.com) and the International Emissions Trading Association (www.ieta.com) are useful sources for updated information on developments in this rapidly evolving policy domain.

References

Aggarwal, Vinod K. 1998. Reconciling Multiple Institutions: Bargaining, Linkages, and Nesting. In *Institutional Designs for a Complex World*, edited by V. K. Aggarwal. Ithaca, NY: Cornell University Press.

Alter, Karen J., and Sophie Meunier. 2006. Nested and Overlapping Regimes in the Transatlantic Banana Trade Dispute. *Journal of European Public Policy* 13(3): 362–382.

Aulisi, Andrew, Alexander F. Farrell, Jonathan Pershing, and Stacy D. VanDeveer. 2005. *Greenhouse Gas Emissions Trading in U.S. States: Observations and Lessons from the OTC NO_x Budget Program*. Washington, DC: World Resources Institute.

Axelrod, Robert, and Robert O. Keohane. 1986. Achieving Cooperation under Anarchy: Strategies and Institutions. In *Cooperation under Anarchy*, edited by Kenneth A. Oye. Princeton, NJ: Princeton University Press.

Bäckstrand, Karin, and Eva Lövbrand. 2006. Planting Trees to Mitigate Climate Change: Contested Discourses of Ecological Modernization, Green Governmentality and Civic Environmentalism. *Global Environmental Politics* 6(1): 50–76.

Bell, Ruth Greenspan. 2006. What To Do about Climate Change. *Foreign Affairs* 85(3): 105–113.

Berkes, Fikret. 2002. Cross-Scale Institutional Linkages: Perspectives from the Bottom Up. In *The Drama of the Commons*, edited by Elinor Ostrom, Thomas Dietz, Nives Dolsak, Paul C. Stern, Susan Stonich, and Elke U. Weber. Washington, DC: National Academy Press.

Bernstein, Steven. 2001. *The Compromise of Liberal Environmentalism*. New York: Columbia University Press.

Betsill, Michele M. 2004. Global Climate Change Policy: Making Progress or Spinning Wheels? In *The Global Environment: Institutions, Law and Policy*, edited by Regina S. Axelrod, David Downie, and Norman J. Vig. 2d ed. Washington, DC: CQ Press.

Betsill, Michele M. 2007. Regional Governance of Global Climate Change: The North American Commission for Environmental Cooperation. *Global Environmental Politics* 7(2): 11–27.

Betsill, Michele M., and Harriet Bulkeley. 2006. Cities and the Multilevel Governance of Global Climate Change. *Global Governance* 12(2): 141–159.

Betsill, Michele M., and Matthew J. Hoffman. 2008. The Evolution of Emissions Trading Systems for Greenhouse Gases. Paper presented at the annual meeting of the International Studies Association, San Francisco, California, March 26–29.

Bressers, Hans T. A., and Walter A. Rosenbaum. 2003. Social Scales, Sustainability and Governance: An Introduction. In *Achieving Sustainable Development: The Challenge of Governance across Social Scales*, edited by Hans T. A. Bressers and Walter A. Rosenbaum. Westport, CT: Praeger.

Brewer, Thomas L. 2003. The Trade Regime and the Climate Regime: Institutional Evolution and Adaptation. *Climate Policy* 3(4): 329–341.

CEC. 1995. *Statement of Intent to Cooperate on Climate Change and Joint Implementation*. Montreal: Commission for Environmental Cooperation.

CEC. 2001. *Council Resolution: 01–05. Promoting Comparability of Air Emissions Inventories*. Montreal: Commission for Environmental Cooperation.

CEC. 2002. *Environmental Challenges and Opportunities of the Evolving North American Electricity Market: Secretariat Report to the Council under Article 13 of the North American Agreement on Environmental Cooperation*. Montreal: North American Commission for Environmental Cooperation.

CEC. 2005. *Our Programs and Projects*. Montreal: North American Commission for Environmental Cooperation.

CEC. 2008. *Green Building in North America: Opportunities and Challenges*. Montreal: North American Commission for Environmental Cooperation.

Charnovitz, Steve. 2003. Trade and Climate: Potential Conflicts and Synergies. In *Beyond Kyoto: Advancing the International Effort against Climate Change*. Arlington, VA: Pew Center on Global Climate Change.

Chicago Climate Exchange. 2004. *About CCX*. Chicago: Chicago Climate Exchange.

Christiansen, Atle C., and Jørgen Wettestad. 2003. The EU as a Frontrunner on Greenhouse Gas Emissions Trading: How Did It Happen and Will the EU Succeed? *Climate Policy* 3(1): 3–18.

Dukert, Joseph M. 2002. *A Review: Environmental Challenges and Opportunities of the North American Electricity Market*. Paper presented at a symposium organized by the Commission on Environmental Cooperation in North America, June.

European Commission. 2004. *EU Emissions Trading: An Open Scheme Promoting Global Innovation to Combat Climate Change*. Brussels: European Commission.

Ferretti, Janine. 2002. NAFTA and the Environment: An Update. *Canada-United States Law Journal* 28: 81–89.

Frankel, Jeffery. 2005. Climate and Trade: Links between the Kyoto Protocol and WTO. *Environment* 47(7): 8–19.

Gerber, Elisabeth R., and Ken Kollman. 2004. Authority Migration: Defining an Emerging Research Agenda. *PS: Political Science and Politics* 37(3): 397–410.

Government of Canada. 2005. *Moving Forward on Climate Change: A Plan for Honouring Our Kyoto Commitment*. Ottawa: Government of Canada.

Government of Canada. 2008. *Turning the Corner: Regulatory Framework for Industrial Greenhouse Gas Emissions*. Ottawa: Government of Canada.

Hasselknippe, Henrik. 2003. Systems for Carbon Trading: An Overview. *Climate Policy* 3(2): 43–57.

Horlick, Gary, Christiane Schuchhardt, and Howard Mann. 2002. *NAFTA Provisions and the Electricity Sector*. Paper presented at a symposium organized by the Commission on Environmental Cooperation in North America, June

Instituto Nacional de Ecología. 2001. *México: Segundo Comunicación Nacional ante la Convención Marco de las Naciones Unidas sobre el Cambio Climático*. D. F. México: Secretaría de Medio Ambiente y Recursos Naturales, Instituto Nacional de Ecología.

Jordan, Andrew, Rudiger K. W. Wurzel, and Anthony Zito. 2005. The Rise of "New" Policy Instruments in Comparative Perspective: Has Governance Eclipsed Government? *Political Studies* 53(3): 477–496.

JPAC. 2003. *Discussion Questions on CEC Air Quality Work Plan: Assessments of Transboundary Air Issues under the 2002 CEC Council Final Communique*. Montreal: CEC.

Levy, Marc, Robert O. Keohane, and Peter M. Haas. 1995. Improving the Effectiveness of International Environmental Institutions. In *Institutions for the Earth: Sources of Effective International Environmental Protection*, edited by Peter M. Haas, Robert O. Keohane, and Marc Levy. Cambridge, MA: MIT Press.

Lohmann, Larry. 2006. *Climate Politics after Montreal: Time for a Change*. Washington, DC: Foreign Policy in Focus.

McKinney, Joseph A. 2000. *Created from NAFTA: The Structure, Function and Significance of the Treaty's Related Institutions*. Armonk, NY: M. E. Sharpe.

Miller, Paul J., and Chris Van Atten. 2004. *North American Power Plant Air Emissions*. Montreal: North American Commission for Environmental Cooperation.

Mol, Arthur P. J. 2001. *Globalization and Environmental Reform: The Ecological Modernization of the Global Economy*. Cambridge, MA: MIT Press.

Mol, Arthur P. J., and Gert Spaargaren. 2000. Ecolotical Modernisation Theory in Debate: A Review. *Environmental Politics* 9(2): 17–49.

Mumme, Stephen P., and Donna Lybecker. 2002. North American Free Trade Association and the Environment. In *Encyclopedia of Life Support Systems, Volume II*, edited by Gabriella Kütting. Oxford: EOLSS Publishers.

Nordhaus, William D. 2006. *After Kyoto: Alternative Mechanisms to Control Global Warming*. Washington, DC: Foreign Policy in Focus.

Oberthür, Sebastian, and Thomas Gehring, eds. 2006. *Institutional Interaction in Global Environmental Governance: Synergy and Conflict among International and EU Policies*. Cambridge: Cambridge University Press.

Pew Center on Global Climate Change. 2006. *State and Local Net Greenhouse Gas Reduction Programs: Multiple Pollutant Reduction Program (New Hampshire)*. Washington, DC: Pew Center on Global Climate Change.

Raustiala, Kal, and David Victor. 2004. The Regime Complex for Plant Genetic Resources. *International Organization* 58(2): 277–309.

RGGI. 2005. *Regional Greenhouse Gas Initiative-Overview*. http://www.rggi.org/docs/mou_rggi_overview_12_20_05.pdf.

Roine, Kjetil, and Endre M. Tvinnereim. 2007. The Global Carbon Market in 2007. In *Greenhouse Gas Market 2007*, edited by David Lunsford. Geneva: International Emissions Trading Association.

Rosenzwieg, Richard, Matthew Varilek, and Josef Janssen. 2002. *The Emerging International Greenhouse Gas Market*. Washington, DC: Pew Center on Global Climate Change.

Russell, Douglas. 2002. *Design and Legal Considerations for North American Emissions Trading*. Montreal: Commission for Economic Cooperation.

Selin, Henrik, and Stacy D. VanDeveer. 2005. Canadian-U.S. Environmental Cooperation: Climate Change Networks and Regional Action. *American Review of Canadian Studies* 35(2): 353–378.

Skjærseth, Jon Birger, and Jørgen Wettestad. 2008. *EU Emissions Trading: Initiation, Decision-Making and Implementation*. Aldershot, UK: Ashgate.

Stevis, Dimitris, and Stephen P. Mumme. 2000. Rules and Politics in International Integration: Environmental Regulation in NAFTA and the EU. *Environmental Politics* 9(4): 20–42.

U.S. Department of State. 2006. *U.S. Climate Action Report–2006. Fourth National Communication of the United States of America under the United Nations Framework Convention on Climate Change*. Washington, DC: U.S. Department of State.

U.S. Senate. 2007. S. 2191 America's Climate Security Act of 2007. 110th Congress, 2d session, October 10.

Young, Oran R. 2002. Institutional Interplay: The Environmental Consequences of Cross-Scale Interactions. In *The Drama of the Commons*, edited by Elinor Ostrom, Thomas Dietz, Nives Dolsak, Paul C. Stern, Susan Stonich, and Elke U. Weber. Washington, DC: National Academy Press.

Zhang, Zhong Xiang. 2003. Open Trade with the U.S. without Compromising Canada's Ability to Comply with Its Kyoto Target. Paper presented at Second North American Symposium on Assessing the Environmental Effects of Trade, March 24-28, Mexico City. http://www.cec.org/files/PDF/ECONOMY/Zhang-final_en.pdf.

9

Renewable Electricity Politics across Borders

Ian H. Rowlands

Introduction

Many of the world's systems of generating electricity are unsustainable, even though they have contributed to unprecedented levels of economic wealth for some.[1] The system of large, centralized power stations, mainly fueled by fossil fuels and uranium, connected to a web of transmission and distribution lines, has a number of negative environmental consequences (Holdren and Smith 2000). One of the most significant of the sustainability impacts of fossil fuels is the effect that systems of electricity supply have on global climate change. With 67 percent of the world's commercial electricity generated by fossil fuels in 2006 (REN21 2007), conventional methods to generate power are serving to increase concentrations of carbon dioxide and other greenhouse gases in the atmosphere.

Public electricity and heat production accounted for 5,449 megatons of carbon dioxide equivalent per year, or approximately 30 percent of global greenhouse gas emissions (excluding land use) in 2005 (UNFCCC 2008). As such, a reconsideration of how electricity services are provided (and the extent to which they are needed, or might be provided, by nonelectrical means) is critical. While an effective response consists of many different approaches—including energy efficiency and conservation—it is often argued that greater use of renewable resources in electricity supply is needed to mitigate climate change and advance sustainability. This sentiment has been expressed by a variety of international organizations and national governments.

Working Group III of the Intergovernmental Panel on Climate Change, for example, devotes considerable attention to renewable electricity issues. In particular, it is often argued that it is over the longer term—twenty or more years—that renewable electricity could play a large role, since key elements of electricity systems have a useful lifetime of twenty, thirty, or forty years or more (IPCC 2007). Furthermore, the leaders of the G8, in the 2007 Heiligendamm Statement, committed "to increase the use of cleaner and renewable energy sources" (G8 2007a, 2). Energy issues were also placed within a broader climate change context (G8 2007b). National governments have often done the same: the United Kingdom's March 2006 Climate Change Programme, for example, placed the topic of energy supply—with renewable energy resources playing a key role—as the first of six chapters that, together, outlined the central elements of the national plan (Government of the United Kingdom 2006).

This chapter investigates current and potential cross-border issues between Canada and the United States with respect to efforts to increase the use of renewable electricity. In the first section, a brief review of Canada-U.S. electricity exchanges and relations outlines the ways in which the two countries already interact on power issues. In the next section, issues that have already arisen with respect to renewable electricity between the two countries are examined with a particular focus on electricity generated by large-scale hydropower facilities. This discussion leads into a subsequent exploration of additional concerns that could arise between the two countries, including issues related to cross-border investment, green procurement, subsidies, and tradable certificates. The last section summarizes the main arguments and highlights some areas for further investigation.

Canada-U.S. Electricity Exchanges and Relations

Why should a book analyzing North American climate change politics include a chapter focusing upon electricity exchanges across an international border? After all, the very nature of electricity—that it is economically unfeasible to store and that physical and economic losses are associated with its long-distance transmission—means that it is a commodity largely contained within local systems. Little is usually made of the international trade in electricity compared to many other goods and services. Granted, energy use is consequential to climate change, but why should there be climate change–related interest in cross-border electricity exchanges between Canada and the United States?

Table 9.1 provides basic information about electricity generation in Canada and the United States. Traditionally in these two countries, electricity has been predominantly a national or subnational concern. Relatively little of the electricity generated in either Canada or the United States is usually exported to the other country. In 2007, an estimated 7.3 percent of the electricity generated in Canada was exported to the United States, while an estimated 0.4 percent of the electricity generated in the United States was exported to Canada (National Energy Board 2008). Nevertheless, given the significant total value of the electricity supply industry in both Canada and the United States, this still represents a substantial figure in absolute terms—in 2007, it totaled $4.2 billion in Canadian dollars (National Energy Board 2008).

Table 9.1
Electricity generation by source for Canada (2005) and the United States (2006)

	Coal	Natural gas	Other fossil fuels	Nuclear	Hydro-power	Other renew-ables	Wood and other*	Total generation (gigawatt-hours)
Canada	16.8%	5.6%	2.6%	14.4%	59.6%		1.2%	604,131
United States	49.0%	20.0%	2.0%	19.4%	7.1%**	2.4%	0.3%	4,064,702

Notes: * Includes, for Canada, wood and spent pulping liquor, and for the United States, nonbiogenic municipal solid waste, batteries, chemicals, hydrogen, pitch, purchased steam, sulfur, tire-derived fuels, and miscellaneous technologies. ** Conventional hydropower only.
Sources: EIA 2007; Statistics Canada 2006.

Table 9.2 highlights the largest exchanges of electricity that occurred between the two countries in 2007.

It is important to recognize that there are a variety of institutional and physical electricity links between Canada and the United States (Gattinger 2005). For one, governance of different parts of these countries' interconnected power systems is international. The North American Electricity Reliability Council—and its various committees including the Western Electricity Coordinating Council, the Mid-Continent Area Power Pool, and the Northeast Power Coordinating Council—was formed in 1968 following a major power blackout in 1965. This council is charged with ensuring that the bulk electric system in North America is reliable, adequate, and secure. Moreover, the August 14, 2003, blackout demonstrated how closely the two countries' electricity systems are physically linked. At that time, accidents and errors in the state of Ohio cascaded out of control, eventually affecting 50 million people in eight U.S. states and the province of Ontario (U.S.-Canada Power System Outage Task Force 2003). With fifty-one electricity grid connections that cross the Canada-U.S. border, electricity is, every minute of every day, a transnational issue (Wilson 2006).

Current Issues in Renewable Electricity

Before turning to cross-border issues in renewable electricity, it is useful to reflect, initially, upon the motivations for such international cooperation. It is often assumed that a government's pursuit of renewable electricity is driven by its desire to preserve the environment and to counter the negative effects on local ecology from energy produced through fossil fuels. Indeed, much has been made—at least

Table 9.2
Electricity trade between Canada and the United States (>1,000 GWh for 2007)

6,330				2,812	2,887		2,150		1,203	
British Columbia	British Columbia	British Columbia	Manitoba	Ontario	Ontario	Ontario	Québec	Québec	Québec	New Brunswick
↓↑	↓	↓	↓	↑	↓↑	↓	↓↑	↓	↓↑	↓
Washington	Oregon	California	North Dakota/ Minnesota	Minnesota	Michigan	New York	New York	Vermont	New England	Maine
3,403	1,440	4,618	11,063		1,681	7,416	6,815	2,200	6,898	1,463

Source: National Energy Board 2008.

at the rhetorical level—as to how synergies between the two countries' electricity systems (often a winter peak driven by lighting and heating demands in Canada, complemented by a summer peak driven by cooling demands in the United States) could lead to increased efficiencies and thus result in environmental, and particularly climate change, benefits. These efficiencies might be realized, for example, by avoiding the use of fossil fuel generators to meet spikes in demand. However, if we turn to recently developed electricity policies in Canadian provinces and American states, it soon becomes apparent that international environmental aspirations can quickly be displaced by national economic priorities.

Consider, first, the renewable electricity policies of the northern U.S. states. The importance of economic issues is evident when governors, such as New York State Governor George Pataki, announce their renewable energy plan as "a blueprint to encourage additional private sector development of alternative energy sources, attract jobs and investments in clean energy, and help to diversify our fuel supplies" (New York State 2006, 1). Another example is from Ohio, where it is argued that the state should pursue renewable energy because of Ohio's strong manufacturing infrastructure and workforce. Many in the state believe this will allow them to make the components for renewable energy, bringing an additional $3.64 billion into the economy (Hanauer 2005). In Minnesota, a similar argument is made by the American Wind Energy Association, an industry association of wind energy companies. They believe that Minnesota has 5,000 megawatts (MW) of untapped wind energy, which could bring millions of dollars into the local economy (MEAN 2006).

Leaders in Canadian provinces are not particularly different in their promotion of renewable electricity for economic reasons. For example, Premier Dalton McGuinty, in a speech announcing his "balanced plan" for Ontario's electricity future, emphasized energy reliability, stable prices, supporting businesses, and creating investment (Ontario Ministry of Energy 2006). Concern for the environment received little mention. Moreover, renewable energy is also promoted as a means of employment for the First Nations, and the benefits that major capital investment would bring to Ontario businesses and employees are often trumpeted (Ontario Ministry of Natural Resources 2006). Québec is perhaps the one province that has tended to highlight the protection of the environment as the primary reason for its pursuit of renewable energy. This promotion, however, may well be a defensive reaction to groups opposing large-scale hydroelectric projects. So while the transnational frame is often dominated by environmental (and reliability) images, the national frame is usually economic.

The most contentious Canada-U.S. issue in the area of renewable electricity involves the appropriate role of hydropower in the pursuit of sustainable electricity goals. More specifically, there are disagreements regarding the role of large-scale hydropower. On the one hand, proponents of large-scale hydropower argue that it is a renewable resource with low atmospheric emissions. Hence, they continue, because large-scale hydropower can contribute to a variety of clean air goals including climate change mitigation, it should not be shut out of any market where renewable electricity is being encouraged. On the other hand, opponents argue that large-scale hydropower presents a number of associated challenges: environmental problems include habitat destruction and associated biodiversity loss, and social difficulties include the displacement of settlements (World Commission on Dams 2000).

Moving from the general to the specific, debates have emerged between Hydro-Québec and the northeastern United States (markets in which Hydro-Québec is active) and Manitoba Hydro and Minnesota (similarly, a market in which Manitoba Hydro is active). In the case of Québec, discussions date back to the late 1980s and early 1990s. At that time, citizens of the northeastern U.S. states adamantly opposed the importation of electrical energy from Hydro-Québec's Great Whale hydroelectric project in the James Bay area of northern Québec. When public pressure increased—catalyzed by environmental organizations, including Greenpeace—New York Governor Mario Cuomo eventually found it irresistible, and he canceled New York's billion dollar contract with Hydro-Québec (Morrison and Nitsch 1993).

Such debates persist. A number of states in the United States have explicitly excluded large-scale hydropower from policy tools that encourage increased use of renewable electricity. In Rhode Island, for example, only hydropower under 30 MW can qualify for the state's Renewable Portfolio Standard.[2] Hydro-Québec has responded vigorously, advancing its case in state-level deliberations. For example, it intervened in New York State discussions about renewable policy options (H. Q. EnergyServices 2003) and prepared a submission to the NAFTA body investigating "environmental challenges and opportunities of the evolving North American electricity market" (Hydro-Québec 2000, 1). The Québec government has similarly contributed to the debates, arguing that the development of its hydroelectric potential should, once again, be a top priority (Government of Québec 2006).

Nevertheless, the use of large-scale hydropower continues to be opposed by many in northeastern United States. One example comes from Vermont, which has long-term contracts with Hydro-Québec, which will expire between 2012 and 2020. Some nongovernmental organizations, such as the Vermont Public Interest Research Group, are using the debate surrounding the desirability of renewing the contract as an opportunity to encourage state-developed wind power (and other renewables) in its place (VPIRG 2006). Of course, during the debates regarding the construction of new facilities in the 1980s and early 1990s, there was also extensive opposition from several groups, including transnational links between U.S. activist organisations and Canadian First Nations groups (McRae 2004).

In addition to the expected assertion that these restrictions by individual states primarily serve to contribute to local goals, there are also claims that some actors in Québec hold double standards in their international dealings on renewable electricity. As evidence, note that in May 2003, Hydro-Québec issued a call for tenders for 1,000 MW of locally produced wind power. This request was doubled by the Québec government, which hoped these projects could help revive the economy of the Gaspé Peninsula. The government also insisted that the turbine nacelles be produced in the local region (Lewis and Wiser 2005). Another call for tenders was issued in 2005, requesting an additional 2,000 MW of wind power. It stipulated that 30 percent of the equipment cost must be spent in the Gaspé region and 60 percent of the entire project cost must be spent within the Province of Québec (Lewis and Wiser 2005). In reaction to calls like these, the U.S.-based independent energy development company Energy Management argued that "the proposal that New York give RPS credit for all renewable energy that is either generated and/or delivered into New York is by no means burdensome" compared to the arrangement in Québec (Duffy 2003, 7).

In Minnesota, a number of citizens' group have been active in challenging the social and environmental attributes of electricity imports generated by large-scale hydropower in Manitoba. Many have claimed that Manitoba Hydro has not traditionally integrated the full environmental and human rights costs of its dam projects; as a consequence its electricity has been unjustifiably inexpensive and has unfairly competed with Minnesota renewable energy sources. (This position was historically advanced by the group JustEnergy; more recently, the group Fresh Energy has worked to keep many of these themes in the public debate (Fresh Energy n.d.). For its part, Manitoba Hydro has repeatedly refuted the accusations put forward groups like this one (Manitoba Hydro n.d.).

The debate also plays out between national governments. The Canadian federal government has continued to note its unease about the way in which hydropower was being defined in U.S. legislation. In 2003, for example, Canada expressed concern "over proposals in recent U.S. federal and state legislation to exclude Canadian-origin renewable-energy resources and hydroelectric power from U.S. renewable-energy programs. Canadian advocacy in this sector has raised U.S. awareness of a North American electricity market and the impact that discriminatory measures could have on this market. Canada continues to monitor developments in U.S. renewable energy standards" (International Trade Canada 2003, 45).

In 2003, the Canadian ambassador to the United States told U.S. legislators, "Canada notes the Senate proposal to mandate a renewable portfolio standard

(RPS) for electricity generation. All hydroelectricity, not just incremental hydroelectric generation, is renewable energy. Should an RPS emerge in your legislation, we would request that hydroelectricity not be disadvantaged. We wish to point out that given NAFTA and WTO obligations, any RPS must be non-discriminatory vis-à-vis Canadian and U.S. generated electricity" (Kergin 2003, 1). In 2005, the subsequent Canadian ambassador argued: "We should note that hydroelectricity is clearly a renewable energy. As it represents 57 percent of the electricity generated in Canada, there is no need for Canada to establish a five to ten percent renewable portfolio standard as many states in New England and elsewhere in the United States have done.... However, to Canada, hydro-generated electricity, whether produced or purchased, should count for any RPS" (McKenna 2005, 1).

For its part, the U.S. government continues to develop its own federal policies that involve definitions of renewable electricity. Similarly, it appears supportive of those being developed at the state level as well. In 2002, in response to NAFTA work on this issue, the assistant administrator of the U.S. EPA commented: "The [NAFTA] report suggests that U.S. state renewable-energy programs may be viewed as possible barriers to international trade.... We have not encountered any trade disputes related to differing renewable-energy standards or definitions, and we see no indication of any trade barrier arising from differing definitions" (Ayres 2002, 16). It remains, however, a point of debate between the two countries.

Recent work by the Renewable Energy and International Law Project—an international collection of academics, lawyers, and others, whose work was catalyzed by the International Conference on Renewable Energies in Bonn, Germany in June 2004—and the ways in which connections between renewable electricity and global climate change are being made therein, effectively demonstrate that interest in this area continues to exist (REIL 2006). Again, while calls for international cooperation to promote climate change mitigation are brought front and center when advantageous to proponents—often Canadian provinces seeking export markets in the United States—they are challenged when seen to hinder opportunities for local economic development.

Potential Issues in Renewable Electricity

There exist different views across North America with respect to how renewables (or, alternatively, green power or green energy) should be defined. Indeed, a 2003 report from the Commission on Environmental Cooperation—the international organization created by Canada, Mexico, and the United States under the North American Agreement on Environmental Cooperation (NAAEC), which was, in turn, a product of the North American Free Trade Agreement negotiations—revealed a variety of perspectives across the continent. While there was "the most unanimous and unqualified support" for solar energy (thermal or photovoltaic) and wind, others received more varied reaction: in particular, "biomass and hydropower [were] both important sources that are widely considered renewable, but which are generally included with other restrictions that vary widely from jurisdiction to jurisdiction" (CEC 2003, 1). Such differences are likely to continue: the Database of State Incentives for Renewable Energy identifies, in the United States, 299 different rules, regulations, and policies for renewable energy. Continuing developments regarding renewable electricity policy in Canada lend further support to this observation (Whitmore and Bramley 2004).

Why might this be problematic? By introducing legislation that effectively restricts part of an electricity market only to renewable forms of electricity, critics could maintain that a particular government has introduced an unjustifiable restriction on trade. After all, it is largely accepted that electricity is a good and that all electrons look the same. They could well demand similar treatment for their (nonrenewable) form of electricity. Indeed, Horlick, Schuchhardt, and Mann (2002) put forth this argument, which suggests that all kinds of renewable portfolio standards could be challengeable under trade law. This argument, however, has been challenged by others who maintain that nonrenewable and renewable electricity are, in fact, different.

Hempling and Radar (2002) contend that because there is a much higher public appetite for renewable electricity than conventional forms of electricity, it is effectively shown that the two kinds of electricity are different (because people feel differently about them). Howse (2005) argues that electricity cannot be viewed in isolation from the way it is generated: physics teaches us that electricity is not produced but instead is simply another form of energy that has been transformed. As such, it cannot be divorced from its generation process. As Howse (2005, 11) states: "Put simply, energy *is* a process. Thus, in considering 'physical characteristics' in the context of determining whether renewable energy is like or unlike non-renewable energy, the WTO adjudicator would almost necessarily, on the basis of sound science, be required to consider the physical nature of a *process*." Put another way: "Energy is inherently dynamic—it *is* a process of transformation. The product *is* the process" (Howse 2005, 14). Finally, recent case law judgments regarding Turtle/Shrimp, EC-Asbestos, and Japan-Alcohol may open the door for looking at the production and process methods behind a good. Proponents maintain that this provides additional support for the argument that renewable electricity is different.

Does the discussion change when we restrict attention to renewable electricity itself, and consider different kinds of electricity within this more restricted subset? As noted above, there is already such a discussion surrounding hydropower (large-scale versus low-impact, for example). Additionally, there are debates regarding whether restrictions should be based upon the geographical location of facilities that generate renewable electricity (particularly whether they are inside or outside of the particular jurisdiction enacting the legislation) or based upon the age of these facilities (with some programs favoring new renewables, with "new" defined differently in different places). Some RPS programs currently in place in the United States limit renewable electricity on the basis of one or both of these characteristics. Critics of such policies maintain that these RPS policies have little to do with any environmental goals (including climate change goals) that may be laid out in the preamble of the relevant piece of legislation, but instead are about protecting and/or developing local industries (DSF 2005; Rabe 2004).

In these cases, it would seem that the particular goal of the legislation and the extent to which the goal is defensible would be key. If the aim of the renewable electricity legislation is to meet the challenges of global climate change, then it might be that even nuclear power could be included—certainly the location of the generating facility would not seem to matter. Alternatively, if the aim is about local air quality, then the proximity of the generators would seem to be particularly important. In cases such as this, the goal of the legislation and how the legislation could potentially be protected by General Exceptions in international trade law (GATT's Article XX and NAFTA's Article 2101) would be critical.

Similarly, investment disputes could also lead to cross-border discussions and/or conflicts (Rowlands 2006). It is possible that certain provisions of NAFTA's Chapter 11 (the investment chapter) could apply even in the absence of "true deprivations of property" (that is, one's traditional view of corporate expropriation by host governments "nationalizing" foreign companies and taking over their assets). Instead, there might need to be "compensation for any government action which has a significant impact on the profit-making ability of an investment" (Horlick, Schuchhardt, and Mann 2002). Horlick and colleagues (2002, 24–25) go on to argue that: "If the approach set out there [in the Metalclad case] is maintained, then any post-investment environmental measure applied in the electricity generation and distribution sectors that impact on the profitability of a foreign investor will require compensation to be paid."[3]

An example of such a challenge, involving renewable electricity in Canada and the United States, can be envisaged. Consider fictional jurisdictions called "A" and "B." The government of A had traditionally taken a laissez-faire attitude toward renewable electricity. As a result, a company from jurisdiction B sets up a landfill gas recovery-to-electricity unit in jurisdiction A, and it markets the resulting power using green power language and images. Further imagine that the leadership in jurisdiction A then has a change of heart and decides to actively advance renewable electricity by introducing its own support scheme (rather than using the default national one, which is the emerging norm developed by industry's practices). Legislators there decide to introduce an RPS. Following the results of local polling, these legislators decide that "renewable" consists exclusively of solar and wind. As a result, the company from jurisdiction B can no longer market its biomass-sourced electricity as a premium (environmental) product. That company's officials may then argue that because biomass is just as renewable as the privileged sources (solar and wind), the legislation is unfair. They proceed to argue that the introduction of the RPS amounts to a "de facto expropriation of assets" and they demand compensation for lost revenues. Although it is hard to anticipate the outcome of such a case, it is reasonable to state that the case put forward by the company from jurisdiction B could be viewed sympathetically by a NAFTA panel. Indeed, what makes the potential challenge all the more intriguing is that it would not need to be instituted by (or even supported by) the government in jurisdiction B; private companies have standing in such cases under NAFTA.

Staying with the point about varying definitions of renewable energy, issues related to government procurement also have the potential to become prominent. The government purchase of green products in a systematic manner has been a relatively popular form of encouraging uptake of renewable electricity to encourage "learning by doing" and to stimulate the market for these kinds of electricity (Levine, this volume; Selin and VanDeveer, this volume). Key examples include Natural Resources Canada, which began purchasing green power for some of its facilities in 1997 (CEC n.d.). In North America, green procurement is affected by the terms of NAFTA's chapter 10, which applies to listed federal government entities and enterprises of NAFTA parties. It obliges relevant bodies to follow particular rules, to ensure transparency and to adhere to a "national treatment" obligation.

Given that last point, there is—as discussed earlier—the potential of conflict, for opponents could argue that the electrons are providing government services (lighting, heat, ventilation, and so forth) and that it does not matter how they were created. This would, as noted above, open discussions related to the production and process methods of electricity generation. Perhaps more significantly, if any such program were to favor in-jurisdiction green power, then a challenge on "national treatment" grounds might achieve greater traction. International lawyer Barry Appleton reviewed Ontario's Renewables Request for Proposals (a mechanism whereby it was seeking bids for long-term supplies of renewable electricity) for the Society of Professional Engineers. He argued that, by stipulating that generation had to be in Ontario, the province was setting itself up for an international trade challenge (Spears 2005).

The differences in the definition of green power could affect international relations not only with respect to the kind of resources used to generate the electricity, but also with respect to the standards that apply to the manufacture and/or use of renewable energy technology. For example, one jurisdiction might prohibit the use of hazardous products in photovoltaic cells, or it might have particular levels of noise performance for wind turbines. The debate with respect to differences in definition could also manifest itself in discussions about labeling. This might include not only differing perspectives with respect to what qualifies as green power but also whether goods made with green power could qualify for an environmental label or not.

Another area where politics across the border may arise is with respect to subsidies. Generally, the global trade and investment regime frowns on subsidies. Historically, however, subsidies have been central to energy activities, with fossil fuels and nuclear power, in particular, receiving millions of dollars in support in many countries, Canada and the United States included. While challenges to these subsidies, in order to promote the increased use of renewable energy, could conceivably arise, the well-entrenched (and universal) nature of these subsidies may mean that they do not attract such attention. Instead, subsidies (or, at least, claimed subsidies) to encourage renewable electricity may be the ones that come under scrutiny. In Europe, there is an oft-cited debate about "prices versus quantities" with respect to supporting renewable electricity. In other words, should there be explicit prices for renewable electricity with the market determining the quantity provided, or explicit quantities for renewable electricity with the market determining the price. While North America has often favored the latter in the form of an RPS, Europe has adopted both approaches.

The European mix of both approaches, however, appears to be increasingly moving to North America. In March 2006, some observers claimed that Ontario took the lead with respect to an approach representative of the former, feed-in tariffs (what are sometimes called "standard offer contracts" in North America). At that time, the government announced that the Ontario Power Authority would purchase electricity produced by wind, biomass, or small hydroelectric at a base price of 11 Canadian cents per kilowatt-hour. The fixed price for solar would be 42 Canadian cents per kilowatt-hour (Ontario Ministry of Energy 2006). Other jurisdictions in North America seem set to follow suit. Paul Gipe (2008) identifies California, Florida, Hawaii, Illinois, Michigan, Minnesota, Oregon, Rhode Island, Washington, and Wisconsin as being in the forefront in this regard.

A key precedent for determining the relationship between a feed-in tariff and international economic law comes from Germany in the case of PreussenElektra versus Schleswag. PreussenElektra, a German electricity supplier, complained that it was paying too much for renewable electricity under the German feed-in tariff law, which requires suppliers to purchase renewable electricity within their area of supply at a set (premium) price. PreussenElektra maintained that the law violated European rules on subsidies because it was in effect state aid. The European Court of Justice, however, disagreed and declared that this was not problematic because it did not constitute aid granted through state resources. Instead, it was the private grid operators that were obliged to make the payments (Kreutzmann and Schmela 2001). In contrast, payments in Ontario are made by the Ontario Power Authority, a public entity. As such, it would be interesting to see how the Ontario government would react if someone in Buffalo, New York put solar panels on their roof and arranged for the electricity to be submitted to the Ontario grid, demanding payment for it.

Finally, the emerging market for renewable-electricity certificates is another area for continued investigation. In different schemes around North America and elsewhere, systems of tradable emission credits have been established to address environmental concerns. Renewable-electricity certificates are closely related. It is worth noting that renewable-electricity certificates may be interpreted as financial services instead of electricity as a good, and may fall under the General Agreement on Trade in Services rather than the General Agreement on Tariffs and Trade (Howse and van Bork 2006). As renewable electricity in place of fossil fuel-powered electricity can serve to meet environmental challenges at a variety of different scales—reduced emissions of nitrogen oxides ameliorate smog challenges, fewer sulfur emissions lessen acid precipitation, and lower carbon dioxide emissions serve to mitigate global climate change—there might be a range of legislative obligations to which the act of encouraging renewable electricity is contributing. The fact that air sheds are often international simply adds another layer of complexity.

Concluding Remarks

Recognizing that binational cooperation on renewable electricity issues is already a feature of international climate change politics (if not yet part of the explicit policy), this chapter examines existing cross-border issues between Canada and the United States as well as issues that may arise with respect to the generation and use of renewable electricity. It is suggested that many more issues could emerge than are currently part of the political agenda, which is now dominated by the debate about large-scale hydropower.

These debates may increasingly interact with climate change politics. Remember that Canada and the United States use carbon-intensive resources to generate much of their electricity (see table 9.1). Therefore, as efforts to develop climate change mitigation policies and programs continue at national and subnational levels, development of renewable electricity sources will no doubt continue to be part of the discussions. This chapter suggested many links between the two: an additional example includes the work of the New England governors and Eastern Canadian premiers, where they view the promotion of renewable energy as part of their climate change goals (Selin and VanDeveer, this volume). As such, the ways in which the kinds of potential debates could arise will be, at least in part, linked to the development of climate change policies both within Canada and the United States, as well as between the two countries.

The potential for international cooperation to assist the sustainable development of renewable energy, and thus climate change mitigation efforts, is great (Rowlands 2005). As such, it is important to anticipate and respond to disputes that could arise between countries, Canada and the United States included, as different players pursue the establishment of sustainable energy systems.

Three key areas for ongoing attention and examination stand out. First, a number of organizations are becoming more concerned with the intersections among these kinds of issues. This includes organizations that are traditionally concerned with electricity issues, such as reliability councils and system operators, as well as organizations primarily focused on international trade and diplomacy, such as trade agencies and investment bodies. These organizations, however, increasingly see their traditional mandate being extended into new areas. As a result, the engineers who have historically run electricity bodies must now be concerned with international and environmental dimensions to an unprecedented degree; similarly, the lawyers who have focused upon trade in conventional goods and services must now bring electrons onto their agendas. Either these organizations will have to rise to the challenges, or it is likely that calls elsewhere that led to the recently established International Renewable Energy Agency—will similarly emerge also at the continental scale in North America.

Second, a growing number of actors are involved in renewable electricity issues. Environmental issues, by their very nature, encourage citizen participation. Additionally, the restructuring of the electricity industry in North America over the past ten to twenty years (which has served to introduce competition and consumer choice, among other characteristics) has increased the number and kinds of actors involved. Businesses can now more easily be involved on the supply-side of the electricity system (investing in new generating facilities, for example); they also have more choice in their demand-side actions (choosing among a variety of energy supply products, for example). Indeed, with the increased popularity of feed-in tariffs across North America, these kinds of choices are now available not only to those in the private sector but also to households.

Third, the ways in which the renewable electricity issue is framed will continue to be critical. In Ontario, for example, the extent to which the electricity issue was tied to public concerns about health and economic development was pivotal to its unfolding (Rowlands 2007). Similarly, this chapter has revealed that the ways in which different constituencies prioritize particular advantages and/or disadvantages of different renewable electricity technologies are influential in the outcome of transborder debates. What has traditionally been a technical issue of reliability of electricity supply is increasingly having more and more issues attached to it—in particular, elements of environmental quality and economic development. Accordingly, analysts and activists must continue to be cognizant of the ways in which the links among renewable electricity and global climate change are made. Similarly, greater Canada–United States cooperation on this set of issues appears critical. As such, the ways in which the discourse unfolds regarding renewable electricity will be an important influence upon the outcome of developments within the climate change agenda throughout the continent.

Notes

1. The author would like to acknowledge the research assistance of Jeremy Schembri, and the financial support of the Social Sciences and Humanities Research Council of Canada through the grant "Business and Green Power in Electricity Transformation: Markets and Policies."

2. Unless otherwise noted, information about specific renewable electricity policies in the United States is taken from the Database of State Incentives for Renewable Energy: http://www.dsireusa.org.

3. In the Metalclad case, a U.S. waste management company challenged "decisions by Mexican local government to refuse it a permit to operate a hazardous waste landfill ... and by state government to create an ecological preserve in the area" (CCPA 2004, 8). A NAFTA tribunal found in favor of the U.S. company based on chapter 11 of NAFTA.

References

Ayres, Judith E. 2002. Letter to Janine Ferretti, Executive Director, Commission for Environmental Cooperation, from Judith E. Ayres, Assistant Administrator, United States Environmental Protection Agency. In *Appendix: Government Comments on Environmental Challenges and Opportunities of the Evolving North American Electricity Market: Secretariat Report to Council under Article 13 of the North American Agreement on Environmental Cooperation*. Montreal: Commission for Environmental Cooperation of North America, June. http://www.cec.org/files/PDF//10-govcomments-e.pdf.

CCPA. 2004. *NAFTA Chapter 11 Investor-State Disputes*. Ottawa: Canadian Centre for Policy Alternatives.

CEC. n.d. *Existing Green Procurement Initiatives*. Montreal: Commission for Environmental Cooperation of North America. http://www.cec.org/files/PDF/ECONOMY/Green-Procurement_Initiatives_en.pdf.

CEC. 2003. *What Is Renewable? A Summary of Eligibility Criteria across 27 Renewable Portfolio Standards*. Montreal: Commission for Environmental Cooperation of North America, June.

DSF. 2005. *All Over the Map: A Comparison of Provincial Climate Change Plans.* Vancouver: David Suzuki Foundation.

Duffy, Denis. 2003. *Reply Comments by Energy Management Inc: Case No. 03–E-0188: Proceeding on Motion of the Commission Regarding a Retail Renewable Portfolio Standard.* Albany: State of New York Public Service Commission.

EIA. 2007. *Electric Power Annual 2006.* Washington, DC: Energy Information Administration.

Fresh Energy. n.d. *Minnesota and Hydropower from Manitoba Hydro.* St. Paul, MN: Fresh Energy. http://www.fresh-energy.org/resources/hydropower.htm.

G8. 2007a. Joint Statement by the German G8 Presidency and the Heads of State and/or Government of Brazil, China, India, Mexico and South Africa on the Occasion of the G8 Summit in Heiligendamm, Germany, June 8.

G8. 2007b. Chair's Summary, Heiligendamm, June 8.

Gattinger, Monica. 2005. Canada–United States Electricity Relations: Policy Coordination and Multi-Level Associative Governance. In *How Ottawa Spends 2005–06: Managing the Minority*, edited by G. Bruce Doern. Montreal: McGill-Queen's University Press.

Gipe, Paul. 2008. *Renewable Tariffs and Standard Offer Contracts in the U.S.A.* Available at http://www.wind-works.org/FeedLaws/USA/USAList.html.

Government of Québec. 2006. Summary. In *Using Energy to Build the Québec of Tomorrow: Québec Energy Strategy 2006–2015.* Québec City: Government of Québec.

Government of the United Kingdom. 2006. *Climate Change: The UK Programme 2006.* London: Government of the United Kingdom.

Hanauer, Amy. 2005. National Renewable Energy Investment Could Spark 22,000 New Ohio Jobs. Columbus: Policy Matters Ohio, Press Release, October 25. http://www.policymattersohio.org/pdf/generating_press_release.pdf.

Hempling, S., and N. Rader. 2002. *Comments of the Union of Concerned Scientists to the Commission for Environmental Cooperation.* Cambridge, MA: Union of Concerned Scientists.

Holdren, John P., and Kirk R. Smith. 2000. Energy, the Environment, and Health. In *World Energy Assessment: Energy and the Challenge of Sustainability.* New York: United Nations Development Programme.

Horlick, G., C. Schuchhardt, and H. Mann. 2002. *NAFTA Provisions and the Electricity Sector, Background Paper 4, Environmental Challenges and Opportunities of the Evolving North American Electricity Market.* Montreal: North American Commission for Environmental Cooperation.

Howse, Robert L. 2005. Post-Hearing Submission to the International Trade Commission: World Trade Law and Renewable Energy: The Case of Non-Tariff Measures. May 5.

Howse, Robert L., and Petrus van Bork. 2006. *Opportunities and Barriers for Renewable Energy in NAFTA.* Montreal: Third North American Symposium on Assessing the Environmental Effects of Trade.

H. Q. EnergyServices. 2003. Letter to Honorable Janet Hand Deixler, New York State Public Service Commission, from Gilles Favreau, H. Q. EnergyServices (U.S.) Inc. Regarding Case

03–E-0188—Proceeding on Motion of the Commission Regarding a Retail Renewable Portfolio Standard, March 28. http://www.dps.state.ny.us/rps/3-28-03_comments/HQ.pdf.

Hydro-Québec. 2000. Environment and Electricity Restructuring in North America. Paper presented to the North American Commission for Environmental Cooperation, Hydro-Québec, External Regulatory Affairs. June. http://www.cec.org/files/PDF/HydroQuebec-e_EN.PDF.

International Trade Canada. 2003. *Opening Doors to the World: Canada's International Market Access Priorities 2003*. Ottawa: International Trade Canada.

IPCC. 2007. *Climate Change 2007: Mitigation of Climate Change*. Geneva: Intergovernmental Panel on Climate Change.

Kergin, Michael. 2003. Letter to The Honorable W. J. "Billy" Tauzin, United States House of Representatives, Chairman, House Committee on Energy and Commerce (and others) from Michael Kergin, Ambassador, September 12.

Kreutzmann, Anne, and Michael Schmela. 2001. Neither State Aid Nor Trade. *Photon International: The Photovoltaic Magazine* 4 (April): 12.

Lewis, Joanna, and Ryan Wiser. 2005. *Fostering Renewable Energy Technology Industry: An International Comparison of Wind Industry Policy Support Mechanisms*. Berkeley: Ernst Orlando Lawrence Berkeley National Laboratory.

Manitoba Hydro. n.d. *Welcome to Manitoba Water Power*. Winnipeg, MB: Manitoba Hydro. http://www.manitobawaterpower.com/index.html.

McKenna, Frank. 2005. Canadian Ambassador to the United States, Speech to the New England-Canada Business Council Energy Conference, Boston, November 4.

McRae, Glenn. 2004. Grassroots Transnationalism and Life Projects of Vermonters in the Great Whale Campaign. In *In the Way of Development: Indigenous Peoples, Life Projects and Globalization*, edited by Mario Blaser, Harvey A. Feit, and Glenn McRae. London: Zed Books.

MEAN. 2006. *Support Fair Trade for MN Wind!* Minnesota Environmental Action Network. http://www.mnaction.org/showalert.asp?aaid=438.

Morrison, Allen J., and Detlev Nitsch. 1993. *Hydro-Québec and the Great Whale Project*. Glendale, AZ: Thunderbird, The Garvin School of International Management.

National Energy Board. 2008. *Electricity Exports and Imports, Monthly Statistics for December 2007*. Calgary: National Energy Board.

New York State. 2006. *Governor Unveils Plan to Cut Dependence on Imported Energy*. New York: New York State Department of Environmental Conservation.

Ontario Ministry of Energy. 2006. *Expanding Opportunities for Renewable Energy in Ontario*. Toronto: Ontario Ministry of Energy. News Release, March 21.

Ontario Ministry of Natural Resources. 2006. *Renewable Energy*. Toronto: Ontario Ministry of Natural Resources.

Rabe, Barry G. 2004. *Statehouse and Greenhouse: The Emerging Politics of American Climate Change Policy*. Washington, DC: Brookings Institution Press.

REIL. n.d. *The Renewable Energy and International Law Project*. http://www.reilproject.org/.

REN21. 2007. *Renewables 2007: Global Status Report.* Paris: Renewable Energy Policy Network for the 21st Century.

Rowlands, Ian H. 2005. Renewable Energy and International Politics. In *Handbook of Global Environmental Politics*, edited by Peter Dauvergne. Cheltenham, UK: Edward Elgar.

Rowlands, Ian H. 2006. North American Integration and "Green" Electricity. In *Bilateral Ecopolitics: Continuity and Change in Canadian-American Environmental Relations*, edited by Philippe Le Prestre and Peter Stoett. Aldershot, UK: Ashgate.

Rowlands, Ian H. 2007. The Development of Renewable Electricity Policy in the Province of Ontario: The Influence of Ideas and Timing. *Review of Policy Research* 24(3): 185–207.

Spears, John. 2005. Electricity Laws May Break Trade Rules, Lawyer Says. *Toronto Star*, February 15, D06.

Statistics Canada. 2007. *Report on Energy Supply-Demand in Canada, 2005.* Ottawa: Statistics Canada.

UNFCCC. 2008. *Greenhouse Gas Inventory Data.*

U.S.-Canada Power System Outage Task Force. 2003. *Interim Report: Causes of the August 14th Blackout in the United States and Canada.* Ottawa: Natural Resources Canada.

VPIRG. 2006. *Let's Focus on Clean, Local Energy.* Montpelier: Vermont Public Interest Research Group.

White House. 2005. President Bush Signs into Law a National Energy Plan. News Release, Office of the Press Secretary, August 6.

Whitmore, Johanne, and Matthew Bramley. 2004. *Green Power Programs in Canada—2003.* Drayton Valley, AB: Pembina Institute.

Wilson, Michael. 2006. Canada's Commitment to the Continental and Global Agenda. Canadian Ambassador to the United States, Address to the University of Buffalo, Buffalo, NY, October 26.

World Commission on Dams. 2000. *Dams and Development.* London: Earthscan.

10

Arctic Climate Change: North American Actors in Circumpolar Knowledge Production and Policymaking

Annika E. Nilsson

Introduction

Arctic images have been prominent in North American media coverage of climate change.[1] A *Time Magazine* special report on climate change in April 2006 was illustrated by a cover picture of a polar bear on melting sea ice accompanied by the headline that stated, "Be worried. Be very worried." Statements on a *60 Minutes* special on climate change the same year stressed the importance of climatic changes in the Arctic region: "This is a bellwether, a barometer. Some people call it the canary in the mine.... In 10 years here in the Arctic, we see what the rest of the planet will see in 25 or 35 years from now" (Corell 2006). Large declines in Arctic sea ice and an open Northwest Passage in the summer of 2007 were extensively covered in the media and provided further compelling images of rapid climatic change with economic, political, and security implications for the entire region and beyond. While the pivotal role of the Arctic in climate change has been recognized in international scientific assessments since at least the mid-1980s (Bolin, Döös, Jäger, et al. 1986), North American political interest in the Arctic has grown sharply in recent years.

Increasing North American popular and political interest in Arctic climate change since 2000 coincided with the Arctic Climate Impact Assessment (ACIA), which was conducted between 2000 and 2004. The ACIA has been described as a "tide-turner" in the U.S. climate debate (Warriors and Heroes, 2005), and an online search of *Washington Post* articles for the keywords "Arctic" and "climate change" gives twice as many hits in both 2004 and 2005 compared to 2003. The ACIA made it clear that significant climatic changes in the circumpolar north were already happening (ACIA 2005). These changes were further documented in the latest IPCC assessment, which was released in 2007 (IPCC 2007). Observed effects include coastal erosion that is forcing the relocation of indigenous communities, thawing permafrost that is damaging infrastructure, and reduced sea ice extent. Projected changes include shifting vegetation zones, the migration of new species into the Arctic, and that Arctic warming is likely to have global consequences by affecting ocean currents and natural sources and sinks of greenhouse gases (GHGs).

This chapter analyzes the role of the Arctic in North American climate change debate and policymaking. Specifically, it investigates the role of the Arctic Council in linking scientific and policy concerns and in focusing attention on the impacts of climate change on indigenous peoples. The chapter begins with a discussion on important characteristics of scientific assessments to inform the following examination of the ACIA. A key feature of the ACIA was a prominent emphasis on climate change impacts on indigenous peoples. The chapter argues that this focus was a result of dominating norms and structures of regional cooperation under the Arctic Council. Moreover, this focus helped make the ACIA salient, credible, and legitimate to a broad set of actors in international climate politics in a dynamic that was intimately connected to U.S. and Canadian politics. The chapter discusses linkages between the Arctic Council and the global climate regime, including the limitations of a regional effort in changing well-established power dynamics in the global arena.

Institutions and Scientific Assessments

In the study of global environmental governance, there is a growing interest in how the structure of international collaboration can shape scientific knowledge production. Based on a number of case studies, Young (2002, 2004) has pointed to how environmental institutions (defined as collections of rights, rules, and decision-making procedures governing human action in specific issue areas) can frame research agendas by channeling resources to specific research areas, privilege certain types of knowledge claims, and guide the application of knowledge to specific policy concerns. An activity that can be important in all these mechanisms of institutions guiding knowledge production is scientific assessments, which are formal efforts to assemble and evaluate knowledge in a form that makes it useful for decision making (Mitchell, Clark, Cash, et al. 2006). To this end, assessments bring together resources and people, produce written documents collecting previously disparate information, and raise expectations about policy action.

Scientific assessments have been studied in relation to policymaking, with the Global Environmental Assessment (GEA) Project providing a systematic analysis on how and why these assessments influence policymaking. As an issue becomes established, the most likely policy impact of scientific assessments is in promoting alternative framings (Mitchell, Clark, Cash, et al. 2006). Scientific assessments can thus have a role both in knowledge production and in the creation of specific framings guiding policymaking. Based on the GEA project's work, Mitchell, Clark, Cash et al. (2006,15) highlight three critical factors that affect when and why scientific assessments are influential: salience, credibility, and legitimacy. Salience refers to whether the assessment is seen as relevant, credibility deals with whether it is judged to be reliable, and legitimacy refers to whether the process takes account of concerns and insights of all stakeholders.

In short, the more salient, credible, and legitimate a scientific assessment is to a particular user, the more likely it is to shape the behavior and decisions of that user. In other words, the ability of the ACIA to influence and shape Arctic debates and policymaking in large part would depend on it being perceived as salient, credible, and legitimate by a host of different stakeholders across the Arctic region. The next sections explore the extent to which this was achieved by the ACIA, along with a discussion of the influence of key actors throughout the ACIA process.

The Arctic Climate Impact Assessment

The ACIA was formally launched by a decision by the Arctic Council in 2000, but the process leading to the initiation of the ACIA started several years earlier. The Intergovernmental Panel on Climate Change (IPCC) beginning in the mid-1990s identified a need to focus more on different regions as a way to enhance understanding of global climate change (IPCC 1997, 2004). The call by the IPCC for more regionally focused assessments coincided with interests of the International Arctic Science Committee (IASC), a nongovernmental organization that coordinates Arctic research. The IASC had previously initiated climate impact assessments in the Barents and Bering sea regions, and it also served as a bridge between the IPCC and ongoing initiatives in the Arctic Council, where climate change was gaining increased political attention.

The Arctic Council is a high-level forum that was created in 1996 as a continuation of the Arctic Environmental Protection Strategy (AEPS) that was launched in 1991. Its niche has been to produce scientific assessments, and its role in knowledge production has made it a "cognitive forerunner" in international environmental governance (Schram Stokke 2006). One of the more visible activities of the AEPS and the Arctic Council has been assessments of pollution issues, including persistent organic pollutants (Downie and Fenge 2003). Climate change initially had low priority in the AEPS. However, a smaller initial assessment was published in 1997–1998 (AMAP 1997, 1998), which led governments to call for more in-depth assessments of the potential impacts of climate change (AEPS 1997).

The climate change concerns of the Arctic Council connected with ongoing initiatives in IASC, resulting in a joint process between the Arctic Council and IASC. In October 2000, representatives from all eight Arctic member states and six

Permanent Participants of the Arctic Council adopted an implementation plan for the ACIA with the aim of producing three related documents: an extensive scientific assessment, a popular science synthesis, and a policy document (Arctic Council 2000). The Arctic Council member countries are Canada, Denmark, Finland, Iceland, Norway, Russia, Sweden, and the United States. The Permanent Participants are the Aleut International Association, the Arctic Athabascan Council, the Gwich'in Council International, the Inuit Circumpolar Conference, the Saami Council, and the Russian Association of Indigenous Peoples of the North.

The ability to work through the Arctic Council, which provided political legitimacy to the regional assessment process, has been identified as a major reason for conducting the first comprehensive regional climate impact assessment in the Arctic, rather than somewhere else that may be just as valuable from a climate science perspective (Corell 2004). Key actors in the ACIA included the scientific community (initially through IASC and later through nationally nominated experts), indigenous peoples' organizations through the Permanent Participants in the Arctic Council, and government representatives of the eight member states in the Arctic Council. Only member states have formal voting rights in the Arctic Council, but the Permanent Participants take part in all discussions and can put forward their own proposals (Heininen 2004; Selin and Selin 2008). The IASC and two Arctic Council working groups—the Arctic Monitoring and Assessment Programme (AMAP) and the working group on the Conservation of Arctic Flora and Fauna (CAFF)—were at the center of much of the ACIA work.

Creating a Scientific Framing

Three factors were critical in creating media and policy attention for the ACIA: efficient scientific synthesis of knowledge, the inclusion of indigenous perspectives, and the political dynamics of the policy process. The synthesis of knowledge was mainly linked to the scientific report and the overview document. The scientific assessment was carried out by people who were identified as experts in their respective fields. In that sense, the ACIA was no different from the scientific assessments conducted by the IPCC. However, the regional context created a potential for deeper insights into issues that were specific to the Arctic region, such as the role of changing ice and snow conditions. Personal interviews with lead authors indicate that the ACIA helped build a scientific community with a wider knowledge base than scientists who had previously been involved in Arctic climate science (Nilsson 2007), which in turn created a scientifically credible consensus about a message that the Arctic was warming rapidly with global consequences (Hoel 2006). In addition, the task of assessing impacts led to a complementary focus on the local level, in particular on the importance of climate for indigenous peoples' livelihoods and identities (Nilsson 2007). This local focus was based mainly on case studies and gave climate change a human face illustrated by quotations from personal observations.

The United States (under the Clinton administration) accepted primary responsibility for carrying out the scientific assessments. The United States also by far had the most lead and contributing authors, followed by Canada. This large participation by North American scientists in the ACIA helped make the assessment salient to North American audiences as it ensured that changes that had been observed in Canada and Alaska were prominently featured. In the overview document, ACIA's scientific knowledge base was presented as ten key findings.[2] This structure resulted from a conscious effort to create a message that was short and concise enough to be conveyed to a policymaker during an elevator ride. Although the scientific authors signed off on the scientific content of the overview report, ensuring that it matched what was in the scientific report, the overview document was designed for policy impact. However, about a year before the formal release of the ACIA, U.S. officials from the Bush administration started to actively discredit the overview document. As discussed in more detail below, U.S. officials alleged that its scientific base had not been available for sufficient scrutiny.

If an effective synthesis of science was one reason for ACIA's visibility, a second reason was the inclusion of indigenous peoples' observations of climate change. For indigenous peoples, this inclusion made the factual basis of the assessment more credible than an assessment based only on scientific observations. The ACIA moreover became salient because it addressed issues of concern to indigenous peoples, including the complexity of local impacts and connections to indigenous cultures. In addition, indigenous peoples regarded the ACIA as legitimate because of their inclusion in the assessment process. Using indigenous information that was not published in the peer-reviewed scientific literature could have decreased the scientific credibility, but in this case evidence from all knowledge sources pointed in the same direction: climate change was happening. The ACIA also introduced a discourse of indigenous peoples having a special relationship to the environment, which has played an important role for indigenous peoples gaining political influence (Nuttall 2000).

The Inuit Circumpolar Conference (ICC) was particularly active in bringing ACIA's message into the public debate (Watt-Cloutier, Fenge, and Crowley 2006). The ICC chair at the time of the ACIA was a frequent speaker at climate policy events in 2004, including at U.S. congressional hearings and meetings organized under the UN Framework Convention on Climate Change. In its climate change work, the ICC drew from experiences gained from previous involvement on scientific assessment of persistent organic pollutants (Selin and Selin 2008; Watt-Cloutier,

Fenge, and Crowley 2006). The engagement with pollution issues widened their experience in international environmental politics, to which they could sometimes gain access as part of the Canadian government delegation. The Arctic Council by design also provided a political setting that facilitated the participation of indigenous groups. Broader norms about the importance of indigenous participation that were officially recognized by the Arctic Council gave indigenous groups full access to the process. In addition, the ACIA chair Robert Corell explicitly stressed the importance of including indigenous peoples in all stages of the assessments (Nilsson 2007).

A third factor for the high visibility of ACIA in North America and elsewhere was the very public political controversy that surfaced toward the end of the assessment process, during the political negotiations for a policy summary document. The ACIA policy process, more than anything else, highlighted diverging interests and roles of North American actors in the ACIA and Arctic governance more broadly.

The ACIA Policy Process: Arctic Cooperation at Stake

The ACIA's policy document was initially developed by a group appointed by the Arctic countries and the Permanent Participants. While some Nordic countries appointed their chief climate negotiators to the UN Framework Convention on Climate Change to this group, the United States and Canada chose people who were closer to the scientific community. Several Nordic countries saw the ACIA and the Arctic Council work as an opportunity to push global climate politics forward, making a strong policy document from the ACIA process a political priority for the Nordic governments (Nilsson 2007). Indigenous peoples' representatives shared many of these interests, but they also wanted to showcase the Arctic as had been done in the 2001 Stockholm Convention on Persistent Organic Pollutants, where the preamble acknowledges that Arctic ecosystems and indigenous communities are particularly at risk from hazardous chemicals (Selin and Selin 2008).

The policy drafting team met with the ACIA scientists in the spring of 2003 to get a picture of the science and to scope out a policy document. The ICC took the lead in presenting an outline, which became a basis for drafting the text. As the Arctic Council normally works via consensus, this was also the norm within this policy drafting team. As the process proceeded, it became increasingly clear that the Senior Arctic Officials (government representatives to the Arctic Council directly below the ministerial level) wanted to stay apprised of events. Norway invited the Senior Arctic Officials and their experts for an informal meeting on the island Svalbard in August 2003. This appears to have been a wake-up call for the Bush administration that the ACIA results might be contentious. When the policy-drafting team met again in October 2003, the U.S. delegate delivered a paper, which came to set the tone for the rest of the policy process. After an expression of appreciation of the work of the policy drafting team, it states:

> As we have sought to review the draft Policy Document, we have come increasingly to the conclusion that there is a fundamental flaw in the process we are following—a process that is significantly different from that we have developed in the IPCC and other such efforts. Specifically, we are seeking here to develop the scientific assessment and its summary in tandem with the policy recommendations that logically should flow from them. Moreover, these policy recommendations should be developed only after governments have had an opportunity to consider the Scientific Document and the Synthesis Document on which they are based and draw their conclusions. In effect, we are putting the cart alongside the horse with the risk that neither cart nor horse will arrive at the destination. (U.S. Statement on Policy Document 2003, distributed at the Policy Drafting Group Meeting in London October 2003)

Several ACIA participants and observers expected the U.S. paper to be formally tabled at the upcoming meeting of Senior Arctic Officials, which set in motion counteractions from the indigenous peoples' representatives. The ICC Alaska wrote a letter to the U.S. Senior Arctic Official stating that "it will surely appear to the Arctic Council that the United States intends to delay preparation and presentation of the policy document until after the presidential election. In doing so, the United States is opening itself to criticism that domestic political and electoral considerations override agreed ministerial direction" (Letter to Sally K. Brandel, Senior Arctic Official, U.S. Department of State from Chuck Greene, ICC Alaska, October 17, 2003). The letter made an explicit connection to the upcoming presidential election of 2004, which many people in the ACIA process were speculating about informally.

When the Senior Arctic Officials met in Svartsengi, Iceland a few weeks after the London meeting, there was no longer any doubt about the U.S. intention to delay (or even stop) the policy process. Shortly after the Svartsengi meeting, the policy drafting team formally handed over their draft to the Senior Arctic Officials. The Permanent Participants were not happy with developments. From their perspective, not only the policy document was at stake, but also the political credibility of the Arctic Council (Watt-Cloutier, Fenge, and Crowley 2006). The Arctic Council records are silent about what happened in the policy process after the Senior Arctic Officials took over. However, the U.S. position soon became public when the newsletter *Inside EPA* reported that "some foreign officials and Native American groups suspect the Bush administration is seeking the delay to avoid addressing the politically tricky issue a month before the 2004 presidential election." The article also asserted that U.S. State Department representatives had asked fellow members of the Arctic Council to delay drafting a policy of global warming in the Arctic until the following fall (U.S. Seeks Delay of Key Arctic Climate Report until after 2004 Election 2003).

Meanwhile, discussions about major messages in the ACIA were becoming more public, for example, in a congressional hearing on climate organized by the U.S. Senate Committee on Commerce, Science and Transportation (U.S. Senate 2004a). Informal consultations about how to get the process moving again resulted in a new start in March 2004. The deadline for an agreement on a policy text was the scheduled release of the ACIA at a ministerial meeting in Reykjavik, Iceland in November 2004. A lack of agreement would signal that the Arctic Council was unable to provide a policy response to the mounting scientific evidence of Arctic climate change, which would not bode well for the future policy role of this relatively new regional regime. The stalled policy process caused mounting frustration among the Permanent Participants.

In September, the international chair of the ICC, Shiela Watt-Cloutier, testified before the U.S. Senate Committee on Commerce, Science and Transportation. After descriptions of Inuit hunters falling through the "depleting and unpredictable sea-ice," she asked committee members to look closely at the role of the U.S. State Department. Referring to the ACIA, she said that "the Department is minimizing and undermining the effectiveness of this assessment process by refusing to allow policy recommendations to be published in a stand alone form just like the assessment itself" (U.S. Senate 2004b). A group of senators wrote to Secretary of State Colin Powell pointing out that "in order to fulfil our responsibilities to the American people, it is critical that we, as policy makers, have access to the latest scientific information and associated policy recommendations" (Eilperin 2004). Pressure mounted further when the ACIA report and an early draft of the policy document were leaked to the *Washington Post*, which ran an article about the U.S. resistance to policy recommendations (Eilperin 2004). The scientific findings also became public in a front-page article in the *New York Times* (Revkin 2004).

Just a few days before the Arctic Council ministerial meeting in Reykjavik in November 2004, the Senior Arctic Officials agreed on a text. The text became part of the Reykjavik Declaration and the Senior Arctic Officials' report to the ministers and was extracted to a separate document, which was presented publicly at the ministerial meeting (Arctic Council 2004). While the document had less substance than many participants and observers had hoped, it showed a consensus about the importance of climate change for the Arctic and that key actors did not want to put the future of the Arctic Council at stake. Final negotiations of the ACIA policy document were closed to outside observers and not documented in writing, but several participants have shared their analysis in personal interviews.[3] The following picture emerged.

The United States came under pressure, as it was quite alone in its resistance to a forceful policy statement. All other countries and the Permanent Participants stressed the importance of the Arctic Council and wanted to highlight observed climatic changes across the region, including their global implications, in part to influence global political negotiations. Only with the mounting publicity and the pressing deadline of the ministerial meeting did the United States agree to a policy document that recognized the significance of Arctic climate change. The other Arctic countries could have walked away from the negotiations, but decision making by consensus was of value to all key actors. Even if Canada followed the Nordic states in wanting a text that at least did not move international policy positions on climate change backward, countries acted to protect the integrity of the Arctic Council as a meeting place between Arctic states and indigenous peoples. Canada has a fossil-fuel-dependent economy and it faces significant challenges in slowing and reducing domestic GHG emissions (Stoett, this volume), but it is also a country with a relatively large indigenous Arctic population, with complicated debates about indigenous rights (Archer and Scrivener 2000; Broderstad and Dahl 2004). The federal government thus has a strong political interest in good relations with indigenous peoples.

The negotiations brought out two fundamental conflicts. One conflict involved the relationship among the scientific report, the overview, and the policy document within the assessment process. A key question was to what extent policymakers were bound by scientific statements that they had not negotiated. The United States repeatedly referred to the IPCC as the appropriate model, where the summary for policymakers is negotiated by government representatives word by word after the completion of the scientific document. Some of the toughest negotiations concerned the language referring to the findings of the ACIA scientific report. The final wording remained unresolved until the very last late-night negotiation, before the United States agreed. The final report includes the phrases, "Welcome with appreciation the scientific work," "Note with concern the impact throughout the region," and "Recognize that the Arctic climate is a critical component of the global climate system with worldwide implications" (Arctic Council 2004). However, it did not make any new policy commitments.

The lack of connection between science and policy can in part be explained by the way the ACIA process was set, lacking a boundary organization that could have better facilitated a science-policy interface (Cash, Clark, Alcock, et al. 2003; Guston 2001). In the ACIA process, science and policy issues were formally separated between scientists on the one hand and high-level policy representatives of the Arctic states and Permanent Participants on the other. The purpose of this division appears to have been to protect the perceived scientific integrity of the assessment. For example, the scientific overview document was the responsibility of the ACIA scientific steering committee, which was different from earlier summaries of AMAP's scientific assessments that were negotiated in the AMAP working group. However, this division between science and policy had the opposite effect on some policy audiences. Especially the U.S. State Department actively questioned the scientific integrity because it had not been available for government scrutiny (Nilsson 2007).

The other fundamental conflict in the ACIA process was the relationship between the regional Arctic arena and the global climate policy arena. Some countries brought in climate negotiators from the start and wanted to use the Arctic Council as a way of advancing international climate change policy that had not been possible under the UN Framework Convention on Climate Change and the Kyoto Protocol. The U.S. officials rejected this move. They argued that the Arctic Council, because of its regional scope, was not an appropriate forum for formulating any kind of climate change policy and further said that climate change policy issues should be addressed in global political negotiations. United States officials were not alone in viewing the UN Framework Convention on Climate Change as the proper arena for political negotiations, but only they actively challenged the idea of using the Arctic Council as a vehicle for climate change policymaking (Nilsson 2007).

When the process for developing the policy document became unclear and moved to informal talks among Senior Arctic Officials, indigenous peoples groups lost some of their influence as Permanent Participants. In response, the ICC went public with its criticism of the U.S. State Department. The ICC's allies in this game were climate scientists who did not want the assessment silenced and U.S. senators who also wanted to put climate change on the domestic political agenda. The U.S. Senate provided the indigenous peoples with an alternative public political platform through congressional hearings, in particular, in the Senate Commerce Committee under Chairman John McCain. The ICC strategy played directly into the negotiations for the policy document, and the internal U.S. politics strengthened the negotiating position for the countries that wanted a strong policy statement. There was thus a close interaction between the international and internal U.S. policy dynamics.

The dynamics involving the Permanent Participants can be described as a dual governance structure where indigenous political structures operate parallel with public agencies in the same geographical area (Broderstad and Dahl 2004). As many indigenous peoples' organizations have become transnational actors (Heininen 2004), the idea of dual governance has moved to the international level. However, the ACIA policy process illustrates that their status as equal partners to states is sometimes challenged, and the picture that emerges in this study is somewhat different from other descriptions of the Permanent Participants having "de facto equality" in the Arctic Council (Young and Einarsson 2004, 20). At the same time, their role as both national and transnational actors created opportunities: indigenous peoples' organizations publicly revealed dynamics of closed negotiations and exploited national U.S. political venues to increase pressure in the ACIA negotiations. Indigenous groups also collaborated with other countries that were not able to exploit domestic U.S. political dynamics in the same way.

The ACIA and Regional Climate Change Action

An important foundation for increased North American attention given to Arctic climate change issues, especially after 2004, was ACIA's synthesis of scientific and indigenous observations of climate change and its impacts. This synthesis not only provided a scientifically credible description of what was happening earlier in the region but also presented a human face to climate change by illustrating how climate change was affecting local communities. As such, ACIA contributed to a regional shift in the framing of Arctic climate change, away from costs of reducing GHG emissions that dominate many domestic debates and toward an increased focus on human and environmental consequences of climate change.

This shift in focus was facilitated by several factors. One factor relates to issues of scale, where focusing on a limited region made it possible to carry out a much more in-depth analysis of the consequences of climate change than in the IPCC assessments of climate change impacts. However, earlier U.S. national and subregional climate change impact assessments largely did not affect domestic debate, so other important factors enhancing the importance of the ACIA were also at play. One such factor was the fact that ACIA was a large-scale scientific assessment supported by the Arctic Council where established norms and structures affected both scientific and political cooperation. Scientifically, the international context of the ACIA created credibility and international visibility that was more similar to the IPCC than a national assessment. In addition, the assessment included indigenous knowledge, which made the ACIA credible and legitimate to a new set of important Arctic actors: indigenous peoples' organizations.

Even if the inclusion of indigenous knowledge was partly driven by indigenous peoples' organizations and the ACIA chair, the Arctic Council provided a platform from which Arctic indigenous people gained access to a combined scientific and policy process where they could voice their priorities. In that sense, the Arctic Council allowed new knowledge claims into the climate change debate, shaping the knowledge basis of the ACIA. Scientists also used the ACIA to speak about climate change with both the press and policymakers. Not only were scientists better informed about Arctic climate change issues as a result of the ACIA, but the launching of an accessible popular science report in press conferences throughout the Arctic also created a media demand for their knowledge.

The ACIA's scientific foundation also provided the ICC and other Arctic indigenous organizations with a platform from which to bring forth their views on climate change as a human rights issue. The framing of peoples and cultures at risk from climate change was prominent in an ICC petition to the Inter-American Commission on Human Rights (IAHCR). This petition, filed in 2005, detailed the ICC's claims that Inuit human rights were violated by the United States as the world's largest emitter of GHGs (Koivurova 2007). The IAHCR rejected the petition in 2006 but later invited the ICC and others to further discuss the connection between climate change and human rights (Center for International Environmental Law 2008). In addition, the Alaskan city of Kivalina and the Native Village of Kivalina together filed a lawsuit against nine oil companies, fourteen power companies, and one coal company in U.S. federal court in 2008 (Alaska Town Sues over Global Warming 2008). The plaintiffs argued that GHG emissions by these companies contribute to global warming and coastal erosion, threatening the existence of the Iñupiat community, which is located on a narrow barrier reef.

Even if some international environmental law scholars argue that victims of climate change are unlikely to find justice through international legal procedures (Koivurova 2007), the cases brought in the IAHCR and U.S. federal court illustrate that the actions and rights of Arctic indigenous peoples are moving to the forefront of North American climate change debate and politics. These legal efforts also connect to an increased emphasis on indigenous peoples' rights within the United Nations, which included the passing of the Declaration on the Rights of Indigenous Peoples by the United Nations General Assembly in September 2007 (Koivurova and Heinämäki 2006). However, there remain legal and political tensions between Arctic national governments and indigenous groups, including in the area of climate change (Bankes 2004; Broderstad and Dahl 2004; Selin and Selin 2008).

With increased emphasis on human rights, issues relating to the impacts of climate change come to the fore, including concerns about who will lose the most from a change in status quo. The structure and norms of the Arctic Council provided indigenous peoples' organizations with a platform that was much more adapted to their needs and desires than the Intergovernmental Panel on Climate Change or the UN Framework Convention on Climate Change, where state interests are more influential and where the knowledge base has been limited to climate change science. Indigenous peoples, however, are more likely to be able to participate in international norm-making in soft law agreements than in formally codified treaties, where states are the only legitimate actors under public international law (Koivurova and Heinämäki 2006).

The ACIA policy process also demonstrates that regional discussions of climate change are likely to be shaped by global policy dynamics. In this case, the Nordic countries used the process to push their agenda of increased control of GHG emission by using the new knowledge base to strengthen their argument about the need for action. The United States, on the other hand, tried to distance itself from this new scientific base in order to not reconsider its previous positions. In the case of ACIA, the global climate regime took prominence, and the resulting policy document did not go further than commitments already made in the global climate negotiations. The UNFCCC has cemented the framing of climate politics as a global issue with national governments as the legitimate actors, especially as it relates to mitigation. Regarding mitigation, the ACIA did not change this framing as it did not include any assessment of regional emissions of greenhouse gases. Based on the Arctic experience, there are thus clear limits to any prospects of using regional areas for overcoming deep differences at the global level.

Concluding Remarks

By bringing new actors into the core activity of knowledge production and framing of climate change, the ACIA brought new issues to the fore. In particular, the inclusion of indigenous peoples' knowledge helped feature current and potential future impact of climate change in ways that were vivid enough to be picked up by media. The impact of climate change is thus no longer only framed in terms of global averages. That might be the real power of a soft-law regional regime such as the Arctic Council, especially if the assessment process can muster scientific expertise, local knowledge providers, and policy communities in ways that make the results salient, credible, and legitimate to people both inside and outside the region.

Since the ACIA, the signs of Arctic climate change have become even more apparent. In the summer of 2007, the sea ice extent reached a record low since satellite monitoring began in 1979; that year the ice receded so much that the Northwest Passage opened up completely for the first time in human memory (National Snow and Ice Data Center 2007). Rapidly declining sea ice has created a renewed interest in the exploitation of natural resources and development of new shipping lanes. It has led to a renewed interest in territorial issues, not least between Canada and the United States. While Canada argues that much of the Northwest Passage is a part of internal water, the United States claims that the route is an international strait (Borgerson 2008; Paskal 2007). Security issues and economic consequences of Arctic climate change have also entered the U.S. foreign policy debates, in part fueled by Russia placing a flag on the bottom of the sea at the North Pole in the summer of 2007 (Borgerson 2008; Smith and Giles 2007).

After a period of increased international cooperation since the 1990s, it is becoming increasingly critical to analyze Arctic development through a more competitive security lens (Heininen 2008). Issues of economic and military power come to the fore, but also linkages to scientific endeavors and to environmental protection. For example, scientific data from investigations of the Arctic seabed may become useful for supporting claims for sovereignty of contested areas of the Arctic Ocean, while environmental concerns may become tools in the rhetorical battles that are likely to ensue in defining the future of the Arctic region (Smith and Giles 2007). In addition, the possible creation of a treaty for protecting the Arctic environment has gained increased attraction since ACIA. While the issue of an Arctic treaty is complex, it highlights the need to critically analyze the future of Arctic cooperation, including on climate change (Koivurova 2008; Koivurova, Keskitalo, and Bankes, 2009). The politics of the ACIA process may have been a mere bellwether for future Arctic politics.

Notes

1. Research for this chapter was carried out as a part of my doctoral dissertation work at the Department of Water and Environmental Studies, Linköping University. I thank my advisors at the Centre for Climate Science and Policy Research for their support and helpful comments.

2. The key findings are presented in the ACIA overview document (ACIA 2004):

- Arctic climate is now warming rapidly and much larger changes are projected.
- Arctic warming and its consequences have worldwide implications.
- Arctic vegetation zones are very likely to shift, causing wide-ranging impacts.
- Animal species' diversity, ranges, and distribution will change.
- Many coastal communities and facilities face increasing exposure to storms.
- Reduced sea ice is very likely to increase marine transport and access to resources.
- Thawing ground will disrupt transportation, buildings, and other infrastructure.
- Indigenous communities are facing major economic and cultural impacts.
- Elevated ultraviolet radiation levels will affect people, plants, and animals.
- Multiple influences interact to cause impacts to people and ecosystems.

3. Personal interviews were conducted with a promise of confidentiality. Interviewees included people with direct insights in the ACIA policy process from Canada, Denmark, Finland, Norway, Sweden, Iceland, the United States, the Inuit Circumpolar Conference, and the Saami Council.

References

ACIA. 2004. *Impacts of a Warming Arctic: Arctic Climate Impact Assessment*. Cambridge: Cambridge University Press.

ACIA. 2005. *Arctic Climate Impact Assessment*. Cambridge: Cambridge University Press.

AEPS. 1997. SAAO Report to the Ministers for the Fourth Ministerial Conference of the Arctic Environmental Protection Strategy (AEPS), June 12–13, Alta, Norway.

AHDR. 2004. *Arctic Human Development Report*. Akureyri, Iceland: Stefansson Arctic Institute.

AMAP. 1997. *Arctic Pollution Issues: A State of the Arctic Environment Report*. Oslo: Arctic Monitoring and Assessment Programme.

AMAP. 1998. *AMAP Assessment Report: Arctic Pollution Issues*. Oslo: Arctic Monitoring and Assessment Programme.

Alaska Town Sues over Global Warming. 2008. Associated Press, February 26.

Archer, Clive, and David Scrivener. 2000. International Co-Operation in the Arctic Environment. In *The Arctic. Environment, People, Policy*, edited by Mark Nuttall and Terry V. Callaghan, 601–619. Amsterdam: Harwood Academic Publishers.

Arctic Council. 2000. Notes from the Second Ministerial Meeting. Barrow, Alaska, October 12–13.

Arctic Council. 2004. Arctic Climate Impact Assessment. Policy Document. Issues by the Fourth Arctic Council Ministerial Meeting. Reykjavik, November 24.

Bankes, Nigel. 2004. Legal Systems. In *Arctic Human Development Report*, edited by AHDR, 101–118. Akureyri, Iceland: Stefansson Arctic Institute.

Bolin, Bert, Bo Döös, Jill Jäger, and Richard A. Warrick. 1986. *The Greenhouse Effect. Climate Change and Ecosystems*. Chichester, UK: John Wiley.

Borgerson, Scott G. 2008. Arctic Meltdown: The Economic and Security Implications of Global Warming. *Foreign Affairs* 87(2)(March/April): 63–77.

Broderstad, Else G., and Jens Dahl. 2004. Political Systems. In *Arctic Human Development Report*, edited by AHDR. Akureyri: Stefansson Arctic Institute.

Cash, David W., William C. Clark, Frank Alcock, Nancy M. Dickson, Noelle Eckley, David H. Guston, Jill Jäger, and Ronald B. Mitchell. 2003. Knowledge Systems for Sustainable Development. *Proceedings of the National Academy of Sciences (PNAS)* 100: 8086–8091.

The Center for International Environmental Law. 2008. The Inuit Case. http://www.ciel.org/Climate/Climate_Inuit.html.

Corell, Robert. 2004. Author's interview. March 24.

Corell, Robert. 2006. In a Global Warning: Scientists Say Global Warming Intensifies Storms, Raises Sea Levels. "60 Minutes," February 19.

Downie, David L., and Terry Fenge. 2003. *Northern Lights against POPs*. Montreal: McGill-Queen's University Press.

Eilperin, Juliet. 2004. U.S. Wants No Warming Proposal; Administration Aims to Prevent Arctic Council Suggestions. *Washington Post*, November 4.

Guston, David. H. 2001. Boundary Organizations in Environmental Policy and Science: An Introduction. *Science, Technology & Human Values* 26: 399–408.

Heininen, Lassi. 2004. Circumpolar International Relations and Geopolitics. In *Arctic Human Development Report*, edited by AHDR. Akureyri: Stefansson Arctic Institute.

Heininen, Lassi. 2008. *Overview of the Geopolitics of the Changing North*. Position paper l for the 5th Open Meeting of the Northern Research Forum (NRF), Anchorage, AK, September 24–27, 2008.

Hoel, Alf H. 2006. Climate Change. In *International Cooperation and Arctic Governance: Regime Effectiveness and Northern Region Building*, edited by Olav Schram Stokke and Geir Hønneland. London: Rutledge.

IPCC. 1997. *The Regional Impacts of Climate Change. An Assessment of Vulnerability. Summary for Policymakers. Special Report of Working Group II.* Geneva: IPCC.

IPCC. 2004. *16 Years of Scientific Assessment in Support of the Climate Convention.* Geneva: IPCC.

IPCC. 2007. *Climate Change 2007. Climate Change Impacts, Adaptation and Vulnerability. Working Group II Contribution to the Intergovernmental Panel on Climate Change Fourth Assessment Report.* Cambridge: Cambridge University Press.

Koivurova, Timo. 2007. International Legal Avenues to Addressing the Plights of Victims of Climate Change: Problems and Prospects. *Journal of Environmental Law and Litigation* 22(2): 267–299.

Koivurova, Timo. 2008. Alternatives for an Arctic Treaty—Evaluation and a New Proposal. *Review of European Community and International Environmental Law* 17(1): 14–26.

Koivurova, Timo, Carina Keskitalo, and Nigel Bankes, eds. 2009. *Climate Governance in the Arctic.* New York: Springer.

Koivurova, Timo, and Leena Heinämäki. 2006. The Participation of Indigenous Peoples in International Norm-Making in the Arctic. *Polar Record* 42(221): 101–109.

Mitchell, Ronald B., William C. Clark, David W. Cash, and Nancy M. Dickson, eds. 2006. *Global Environmental Assessments: Information and Influence.* Cambridge, MA: MIT Press.

National Snow and Ice Data Center. 2007. Arctic Sea Ice Shatters All Previous Lows. Press Release, October 1.

Nilsson, Annika E. 2007. *A Changing Arctic Climate: Science and Policy in the Arctic Climate Impact Assessment.* Ph.D. dissertation, Department of Water and Environmental Studies, Linköping University. Linköping Electronic Press. http://urn.kb.se/resolve?urn=urn:nbn:se:liu:diva-8517.

Nuttall, Mark. 2000. Indigenous Peoples: Self-Determination and the Arctic Environment. In *The Arctic. Environment, People, Policy*, edited by Mark Nuttall and Terry V. Callaghan, 377–409. Amsterdam: Harwood Academic Publishers.

Paskal, Cleo. 2007. How Climate Change Is Pushing the Boundaries of Security and Foreign Policy. Chatham House Briefing Paper. Energy, Environment and Development Programme EEDP CC BP 07/01. http://www.chathamhouse.org.uk/research/eedp/papers/view/-/id/499/.

Revkin, Andrew C. 2004. Big Arctic Peril Seen in Warming. *New York Times*, October 30.

Schram Stokke, Olav. 2006. International Institutions and Arctic Governance. In *International Cooperation and Arctic Governance: Regime Effectiveness and Northern Region Building*, edited by Olav Schram Stokke and Geir Hønneland, 330–354. London: Rutledge.

Selin, Henrik, and Noelle Eckley Selin. 2008. Indigenous Peoples in International Environmental Cooperation: Arctic Management of Hazardous Substances. *Review of European Community and International Environmental Law* 17(1): 72–83.

Smith, Mark A., and Keir Giles. 2007. *Russia and the Arctic. The "Last Dash North."* Defence Academy of the United Kingdom. Advanced Research and Assessment Group. Russian Series 07/26. http://www.da.mod.uk/colleges/arag/document-listings/russian/07(26)MAS-KG.pdf.

U.S. Seeks Delay of Key Arctic Climate Report until after 2004 Election. 2003. *Inside EPA* 24 (45): November 7.

U.S. Senate. 2004a. Committee on Commerce, Science and Transportation, 108th Congress, March 3.

U.S. Senate. 2004b. Committee on Commerce, Science and Transportation, 108th Congress, September 15.

Warriors and Heroes. 2005. *Rolling Stone*, Nov. http://3;www.rollongstone.com/politics/story/8742145/warriors_heroes.

Watt-Cloutier, Shiela, Terry Fenge, and Paul Crowley. 2006. Responding to Global Climate Change: The View of the Inuit Circumpolar Conference on the Arctic Climate Impact Assessment. In *2° Is Too Much! Evidence and Implications of Dangerous Climate Change in the Arctic*, edited by L. Rosentrater. Oslo: WWF—Worldwide Fund for Nature.

Young, Oran R. 2002. *The Institutional Dimensions of Environmental Change. Fit, Interplay, and Scale*. Cambridge: Cambridge University Press.

Young, Oran R. 2004. Institutions and the Growth of Knowledge: Evidence From International Environmental Regimes. *International Environmental Agreements: Politics, Law and Economics* 4(2): 215–228.

Young, Oran, and Niels Einarsson. 2004. Introduction. In *Arctic Human Development Report*, edited by AHDR. Akureyri: Stefansson Arctic Institute.

IV

Climate Action among Firms, Campuses, and Individuals

11

Business Strategies and Climate Change

Charles A. Jones and David L. Levy

Introduction

Corporations are central players in carbon governance through their roles as greenhouse gas (GHG) emitters, investors, innovators, technical experts, manufacturers, lobbyists, and marketers (Levy and Newell 2005). Businesses not only emit GHGs from their own operations, they also purchase energy-intense inputs and sell products that generate substantial emissions over their lifetime. Businesses transmit practices, technologies, and standards to their suppliers and customers, influencing GHG emissions along their supply chains. Business is also a political actor, influencing governmental policy and developing private codes and initiatives. Governments and NGOs have recognized that large firms possess organizational, technological, and financial resources needed to address climate change. This acknowledgment of corporate potential has occurred, not entirely coincidentally, in a period of growing concern with the limitations of the Kyoto Protocol (Najam, Christopoulou, and Moomaw 2006), and more broadly, in a response to a "governance deficit" at the international level (Haas 2004).

During the 1990s, much of the energy of North American business, particularly in sectors related to fossil fuels, was directed toward preventing any caps on GHG emissions. Indeed, industry groups such as the Global Climate Coalition and the Climate Council played a major role in the U.S. withdrawal from Kyoto (Levy 2005). Since then, many businesses have adopted a more constructive stance that acknowledges the reality of climate change and their responsibility for addressing the issue (Margolick and Russell 2004). In this respect, climate change is increasingly portrayed as a business opportunity rather than a burden (Lash and Wellington 2007). A 2006 report from Ceres—a leading coalition of investors, firms, and environmental organizations working collectively on climate change—typifies this optimistic view:

Companies at the vanguard no longer question how much it will cost to reduce greenhouse gas emissions, but how much money they can make doing it. Financial markets are starting to reward companies that are moving ahead on climate change, while those lagging behind are being assigned more risk. (Cogan 2006, 1)

This new approach toward GHG mitigation is reflected in high-profile corporate initiatives, such as GE's Ecoimagination and Wal-Mart's plan to cut GHGs from stores and transportation. Some sectors, such as agriculture and insurance, face risks from the physical impacts of climate change, including rising sea levels and more frequent and intense storms (Haufler, this volume). Civil society organizations such as the Investor Network on Climate Risk and the Climate Group have played an important role recently in highlighting the risks and opportunities facing various sectors and encourage companies to assess and manage these risks rather than ignore them (The Climate Group 2007a, 2007b; INCR 2008). A more proactive stance could provide companies with some protection against litigation and damage to their reputation (Wellington and Sauer 2005), as well as more influence in shaping new regulations and governance systems.

Meanwhile, local government and voluntary initiatives have emerged in response to the perceived lack of guidance from national and international authorities. In the United States, Canada, and Mexico, subnational authorities are formulating a multitude of policies (Farrell and Hanemann, this volume; Gore and Robinson, this volume; Rabe, this volume; Selin and VanDeveer, this volume). Recent agreements include the Regional Greenhouse Gas Initiative (RGGI) and the Western Climate Action Initiative; both are centered on cap-and-trade mechanisms for reducing GHG emissions. The prospect of mandatory cap-and-trade systems, standards for power generation, and subsidies for renewable energy are driving more active corporate climate strategies. Business journals and consultants proffer advice on carbon management systems that entail assessing risks, conducting emissions inventories, setting targets, and assigning responsibilities (Hoffman 2006).

The Pew Center on Global Climate Change and the Climate Group, two organizations promoting business action on climate change, have documented climate change actions taken by numerous companies as well as the related financial and environmental benefits (The Climate Group 2007a; Margolick and Russell 2004). These initiatives cannot be dismissed as mere public relations exercises, as they entail organizational changes and investments in the development of low-emission technologies and products. Yet, despite this beehive of corporate activity, global GHG emissions in 2005 were 28 percent higher than in 1990, and show no sign of declining (EIA 2006). The growth in North American emissions during 2000–2005 was slower than from 1990–2000, but still substantial (UNFCCC 2008). Emissions from manufacturing and construction have declined, reflecting new process tech-

Table 11.1
GHG emissions by country and sector (megatons CO_2 equivalent)

		1990	2000	2005	Percent change 1990–2005
European Union	Total	4258	4135	4193	−1.5
	Energy	1165	1122	1200	3.0
	Mfg & Const.	619	559	555	−10.3
	Transp.	700	841	880	25.7
United States	Total	5529	6391	6432	16.3
	Energy	1818	2293	2392	31.6
	Mfg & Const.	864	882	847	−2.0
	Transp.	1463	1812	1906	30.3
Canada	Total	596	721	747	25.3
	Energy	147	199	202	37.4
	Mfg & Const.	63	64	63	0.0
	Transp.	149	183	198	32.9
		1990	2000	2002	1990–2002
Mexico	Total	425	558	549	29.2
	Energy	105	157	153	45.7
	Mfg & Const.	57	54	51	−10.5
	Transp.	89	113	114	28.1

Source: UNFCCC 2008.

nologies, but emissions from transportation and energy continue to rise (see table 11.1).

The continuously contradictory political activity of much North American business toward climate policy initiatives, however, may appear puzzling. For example, the three major U.S. automobile manufacturers are members of the U.S. Climate Action Partnership, which advocates for national mandatory emission controls, while simultaneously fighting California's efforts to regulate automobile carbon emissions. To explore this paradox, we examine several dimensions of corporate responses to climate change. We argue that business is generally willing to undertake measures consistent with the emerging weak and fragmented system of global GHG governance. Indeed, business has played a substantial role in shaping this system. North American firms are undertaking a range of voluntary measures and are increasingly willing to accommodate a national mandatory carbon-trading system.

In this respect, interests of North American businesses and policymakers appear to be converging on regulatory systems setting relatively low carbon prices that do not threaten the core business models of politically and economically important industrial sectors and firms. Much North American business continues to oppose, however, more stringent subnational initiatives that target particular sectors and are believed to constitute more immediate economic threats. While low-emission energy technologies are attracting increasing attention from venture capital, most businesses have not yet incorporated climate change into their strategic planning for core products and markets. Instead, climate change strategy is oriented toward organizational preparation, corporate branding, carbon accounting systems, and modest efficiency measures.

The chapter begins with a brief summary of the history of business action toward climate change. The next section examines current corporate responses to climate change. This section begins with an analysis of reports and databases that survey corporate initiatives, followed by a discussion of three sets of corporate responses: business investment in clean energy and low-emission technologies; measures taken toward carbon accounting, reporting, and trading; and political action and membership in associations or alliances active on the climate change issue. The final section assesses the many complex and sometimes contradictory corporate responses to the climate change challenge.

A History of Business Response to Climate Change

Climate change presents a profound strategic challenge to business. Despite the considerable attention given to economic opportunities, the primary issue facing many sectors is the regulatory risk of higher costs for fuels and other inputs, and lower demand for energy-intense products (Wellington and Sauer 2005). Measures to control the emissions of GHGs most directly threaten sectors that produce and depend on fossil fuels, such as oil and automobiles, and energy-intensive industries including cement, paper, and aluminum. Companies also face considerable competitive risk, as changes in prices, technologies, and demand patterns disrupt traditional business models and make existing competencies obsolete. Investing in new technologies is a treacherous business, however.

It is therefore not surprising that a wide range of sectors responded aggressively to the prospect of regulation of GHG emissions. During the 1990s, U.S.-based companies were particularly active in challenging climate science, pointing to the potentially high economic costs of GHG controls and lobbying government at various levels. Businesses formed a strong issue-specific organization, the Global Climate Coalition, to coordinate lobbying and public relations strategies (Gelbspan 1997; Leggett 2000; Levy and Egan 2003). Canadian energy firms engaged in political action similar to their U.S. counterparts (Smith and Macdonald 2000). In contrast, state-owned Petroleos Mexicanos (Pemex) adopted a more cooperative climate strategy similar to European oil companies BP and Shell (Pulver, this volume).

These divergent strategies defy simple explanation, but studies of the oil and automobile industries identify institutional environments as important determinants of strategic responses (Levy and Kolk 2002; Levy and Rothenberg 2002; Pulver 2007; Rowlands 2000). Expectations concerning markets, technologies, and regulation vary with corporate histories, headquarters location, and membership in industry organizations. Senior managers of European companies believed that climate change was a serious problem and that regulation of emissions was inevitable, but they were optimistic about the prospects for new technologies. Key Mexican managers within Pemex were also convinced by these beliefs. American and Canadian companies, by contrast, tended to be more skeptical concerning the science, more pessimistic regarding the market potential of new technologies, and more confident of their political capacity to block regulation.

By 2000, key firms on both sides of the Atlantic began to converge toward a more accommodative position that acknowledged a need to curtail GHGs. Companies began to invest substantial amounts in low-emission technologies and to undertake a variety of voluntary schemes to inventory, manage, and trade carbon emissions. One source of convergence is the participation of senior managers in global networks, which tend to induce similar expectations and norms concerning appropriate responses (Levy 2005). Competitive dynamics and interdependence also create convergent pressures (Chen and Miller 1994). American auto companies were initially skeptical of hybrid vehicles, for example, but soon followed Toyota with plans of their own.

The shift in the position of American industry can also be linked to the evolution of new organizations supportive of a proactive industry role. Efforts by the GCC and other industry groups to challenge climate science sometimes produced a damaging backlash (Gelbspan 1997; Hamilton 1998). The growth of organizations committed to a climate compromise further undermined the GCC's claim to be the voice of industry on climate. The Pew Center and other groups provide not only a channel of policy influence for member companies, but also a vehicle for legitimizing the new position in the business community. These realignments have been stabilized by the growth of the win-win rhetoric of ecological modernization (Hajer 1995; Porter and van der Linde 1995), which puts its faith in technology, entrepreneurship, voluntary partnerships, and flexible market-based measures (Casten 1998; Hart 1997; Romm 1999).

The win-win concept is reinforced by widespread case study evidence that emission reductions can generate significant cost savings and open new market opportunities (The Climate Group 2007a). Environmentally oriented business associations, such as the Business Council for Sustainable Energy and the World Business Council for Sustainable Development, have adopted this language. Influential environmental NGOs in the United States, especially the World Resources Institute and Environmental Defense (Dudek 1996), have initiated partnerships with business to pursue profitable opportunities for emission reductions. Governmental agencies find win-win rhetoric attractive for reducing conflict in policymaking, as exemplified in the pitch made by the U.S. EPA for its Climate Leaders program (EPA 2007).

Current Business Responses to Climate Change

Growing corporate expectations of GHG regulation, pressure from civil society, and optimism regarding market opportunities are driving corporate responses along a number of dimensions. Simultaneously, continued uncertainty regarding the nature and timing of regulation, future carbon prices, and the impact on existing markets combine to make business cautious. This section examines several dimensions of the multifaceted business response.

Surveys of Business Initiatives

A growing number of surveys of corporate climate change actions have been conducted since the early 2000s. Four reports are analyzed here in some detail: by Ceres (Cogan 2006), by the Climate Group (2007a), by McKinsey (2007), and by Deloitte (2006). These reports have different criteria for inclusion and evaluation, but together they provide a reasonable indicator of corporate responses. The lack of standardized reporting, however, makes sectoral and geographic analysis difficult. We also examined data from the Pew Center's Business Environmental Leadership Council (BELC), which comprises forty-two large companies (thirty-six from North America) who have committed to supporting action on climate change. The Pew Center web site lists company profiles and emission reduction activities, but does not provide summaries or analysis.

The most recent report by the London-based Climate Group was based on a survey and responses to the 2006 Carbon Disclosure Project to describe emission reduction achievements of 137 organizations (84 corporations and 53 city and local governments from 20 countries) with "the most impressive results." The data are largely unverified and based on self-reporting. Some companies report cutting GHG emissions by more than 25 percent, though clearly these are best performers rather than representative cases. The geographic profile of the corporations is approximately 40 percent European, 40 percent North American, and 20 percent Japanese. Twenty-seven corporations reported emission reductions combined with cost savings, with an average emission reduction of 18 percent. The three most frequently reported mitigation measures were energy efficiency, waste management, and use of renewable energy. Dow Chemicals, for example, claims to have saved $4 billion between 1994 and 2005 from reduced energy use, while DuPont reported $3 billon saved between 1990 and 2005. A high proportion of companies also report development of management systems for carbon and engaging in carbon offset activity.

The McKinsey survey of over 2,000 executives highlights a gap between high levels of corporate attention and limited action. It suggests that the core business case for action is weaker than claimed by the more selective reports from the Pew Center and the Climate Group. The survey indicates that 82 percent of executives expect some form of climate regulation in their own countries in the next five years, while 60 percent considered climate change to be strategically important. Notably, 70 percent of executives see climate change as important in corporate reputation and brand management, but "relatively few companies, however, appear to be translating the importance they place on climate change into corporate action" (McKinsey 2007, 2). Among CEOs, 44 percent report that climate change is not a significant item on their agendas.

The ranking of drivers for action are also revealing: corporate reputation, customer preferences, and media attention are ranked first, second, and third. Drivers with low rankings include investment opportunity, competitive pressure, and physical threats to assets. Companies in the United States appear to be lagging behind their international counterparts. Only 51 percent of executives based in North America considered the role of climate change in corporate strategy to be very or somewhat important, the lowest proportion for any region; the corresponding figure for European executives was 65 percent. European executives were significantly more optimistic than their North American counterparts concerning the potential impact of climate change on profits over next five years.

Deloitte's survey of 80 large Canadian GHG emitters, primarily in the oil and gas, manufacturing, and power generation sectors, is in broad alignment with the McKinsey survey. Though 80 percent of firms ranked GHG emissions management as an issue of moderate to critical importance, half of the companies still do not include emission management in their overall risk management strategy. The survey found that 91 percent of respondents claimed to have the management capability to complete a GHG emissions inventory, and 84 percent had actually completed one. Nevertheless, only 46 percent said they had the capability to execute the purchase or sale of emission credits and only 40 percent had established internal emissions targets and schedules.

The Ceres survey of 100 of the largest firms in ten carbon-intense industries also found that U.S. firms lagged behind. The companies were scored with a 100-point

Table 11.2
Top ten firms in climate governance

Firm	Industry	Base	Score
BP	Oil and gas	UK	90
DuPont	Chemicals	U.S.	85
Royal Dutch Shell	Oil and gas	Netherlands	79
Alcan	Metals	Canada	77
Alcoa	Metals	U.S.	74
AEP	Electric power	U.S.	73
Cinergy	Electric power	U.S.	73
Statoil	Oil and gas	Norway	72
Bayer	Chemicals	Germany	71
Nippon Steel	Metals	Japan	67

Source: Cogan 2006.

checklist, with a mean of 48.5, based on a review of specific actions in certain governance areas including board oversight, management execution, public disclosure, emissions accounting, and strategic planning. Of those companies rated, all have significant operations in North America: seventy-two firms are based in the United States, two in Canada, nineteen in Europe, and seven in the Asia-Pacific region. The top-ten list includes only four companies from North America, five from Europe, and one from Japan (see table 11.2).

North American firms are thus somewhat underrepresented among the best performers, given their predominance in the group of companies rated. All the bottom twelve companies are moreover from the United States (see table 11.3). Ceres also found significant differences between industries. In general, chemicals, electric power, and automotive firms have the highest scores; air transport, food, coal, and oil the lowest; and industrial equipment, metals, and forest products are in the middle. The differences, however, between firms within industries are substantial: European oil and resource extraction companies, for example, fare much better than U.S. and Canadian ones. This suggests the existence of significant space for discretionary managerial action.

Business Investments in Clean Energy and Low-Emission Technologies

An important dimension of business responses to climate change is the rapid growth of renewable energy, energy efficiency, energy storage, and other low-emission technologies (see table 11.4). Global markets for fuel cells, biofuels, wind power, and solar power reached an estimated $77 billion in 2007 and are growing at annual

Table 11.3
Bottom twelve firms in climate governance

Firm	Industry	Base	Score
UAL	Airline	United States	3
Williams	Oil and gas	United States	3
ConAgra	Food	United States	4
Bunge	Food	United States	5
Foundation	Coal	United States	5
Southwest	Airline	United States	6
Murphy	Oil and gas	United States	6
Phelps Dodge	Metals	United States	6
Arch	Coal	United States	8
AMR	Airline	United States	9
PepsiCo	Food	United States	9
El Paso	Oil and gas	United States	9

Source: Cogan 2006.

Table 11.4
Revenue in global clean energy ($US billions)

Sector	2007	2017 (projected)
Biofuels	$25.4	$81.1
Wind power	$30.1	$83.4
Solar power	$20.3	$74
Fuel cells	$1.5	$16
Total	$77.3	$254.5

Source: Makower, Pernick, and Wilder 2008.

rates of approximately 30 to 40 percent (Makower, Pernick, and Wilder 2008). Worldwide markets for associated power control electronics, energy efficiency, materials, construction, and services are even larger, estimated at $115 billion in 2005, though growing more slowly (Makower, Pernick, and Wilder 2006). Consumer support for clean energy also points to important business opportunities. Surveys and hedonic analyses consistently indicate consumers are willing to pay a premium for renewable energy (Roe, Teisl, Levy, et al. 2001; Zarnikau 2003).

The size and growth of these markets, as well as a recognition that the regulatory drivers are likely to intensify, are drawing the attention of entrepreneurs as well as established firms. In the clean energy sector, U.S.-based companies attracted nearly $2.7 billion of venture capital in 2007, an increase of 71 percent since 2006. The share of clean energy in total U.S. venture capital investments was 3 percent in 2004, 6 percent in 2006, and 9.1 percent in 2007. The largest recipients are solar, biofuels, and energy efficiency technologies (Makower, Pernick, and Wilder 2008). Established firms in related industries are also investing in these sectors. In 2007, for example, GE reported wind revenues of $4.5 billion, and utilities Pacific Gas & Electric and Florida Power & Light announced multibillion dollar investments in large-scale solar thermal power (Makower, Pernick, and Wilder 2008). Investor interest has also led to the development of several stock indexes that track the clean-energy sector in North America and facilitate portfolio investments through mutual funds and exchange-traded funds.

Arrayed against these important opportunities are significant technological, institutional, and economic barriers to deployment of low-carbon technologies (Dias, Mattos, and Balestieri 2004; Goldemberg, Coelho, and Lucon 2004). The scale of changes needed is not matched by the technology investments thus far (Hoffert, Caldeira, Benford, et al. 2002; Pilke, Wigley, and Green 2008). Rapid growth is from a tiny base, and aside from wind, renewable energy is far from competitive with coal or gas-fired power. The U.S. Energy Information Agency predicts that less than 12 percent of total primary energy supply will be met by nonhydro renewables by 2030, most of that in the form of biomass (EIA 2008). Biofuels compete with food production, and cellulosic biofuels face significant technological and cost hurdles to commercialization. Efficiency gains in automobile transportation are largely offset by rising vehicle weight and miles traveled, and emissions from air travel are rising rapidly.

Clean energy markets present substantial market risks. Many technologies under development will prove to be dead ends, while new low-emission technologies often require radically new capabilities that threaten to undermine the position of incumbent companies and open industries to new entrants (Anderson and Tushman 1990; Christensen 1997). The most successful companies in solar photovoltaics, for example, have been Japanese electronics companies with expertise in silicon. Many of the small firms active in these areas remain in a precarious financial position, dependent on subsidies and new venture-capital investments.

Outside of the energy sector, business is still relatively complacent about climate change. In the insurance industry, for example, despite rising insured losses that many attribute to climate change, major North American firms are reluctant to take action on the issue due to a tradition of conservatism, their reliance on the federal government for disaster relief, and the lack of clear financial benefits from action (Haufler, this volume). Business in the agricultural sector has tended to be more concerned about the impact of carbon regulation on fuel prices than changing climatic patterns and extreme weather events. If anything, companies are pursuing adaptation rather than mitigation. Monsanto, for example, is investing in genetically modified seeds to cope with drought, which may become a more pressing problem in many parts of the world as the climate gets warmer and dryer.

Measures on Carbon Accounting, Reporting, and Trading

Carbon trading, in various forms, has emerged as the centerpiece of governmental policies and private initiatives to constrain carbon emissions (Aulisi, Farrell, Pershing, et al. 2005). In the absence of federal action, states and regions have been developing carbon trading systems, such as RGGI (Rabe, this volume; Selin and VanDeveer, this volume). In response, many firms are preparing for emissions trading by developing the capacity to inventory, report, and trade GHG emissions. Firms might be anticipating mandatory controls, attempting to shape future trading systems, establishing baselines to gain credit for early action, or hoping to gain a competitive advantage through early trading experience. Many larger companies need to develop an emissions management system for their European operations.

Firms are also participating in the Chicago Climate Exchange (CCX), which opened in October 2003. By 2008, CCX membership had grown to over seventy companies committed to reducing their emissions from North American operations. Due to the voluntary nature of the cap, carbon prices have been very low, around $3–5 per ton of carbon dioxide during 2007 and 2008, which is unlikely to induce significant emissions reductions. Moreover, the U.S. federal government sponsors voluntary industry programs. The EPA's Climate Leaders program enlists companies to set goals for emission reductions and to "strategically position themselves as climate change policy continues to unfold." The Department of Energy's Climate VISION (Voluntary Innovative Sector Initiatives: Opportunities Now) enlists trade groups to reduce their members' GHG intensity. Participants in these voluntary programs have not always met their commitments, however, and do not bear any consequences (Stephenson 2006).

A number of groups are exerting pressure on companies to track and report their emissions by asserting that carbon management and accounting provides a mechanism for managing and assessing climate-related business risks and opportunities (Lash and Wellington 2007). The Investor Network on Climate Risk and the Carbon Disclosure Project (CDP) attempt to leverage the influence of institutional investors to create demand for carbon accounting, with implications for asset valuations. In response, some of the largest investment banks, including Citigroup, JPMorgan Chase, and Morgan Stanley have issued restrictive guidelines for new coal investments that note that "investing in CO2–emitting fossil fuel generation entails uncertain financial, regulatory, and environmental liability risks" (Makower, Pernick, and Wilder 2008, 4). Plans to develop more than fifty new coal-fired plants in the United

States were delayed between 2006 and 2008 due to a combination of environmental and investor concerns (Makower, Pernick, and Wilder 2008).

The CDP was launched in 2000 as a London-based coordinating secretariat for institutional investors to gain insight into the climate risk profiles of the Financial Times 500 companies, though it now surveys a much larger and more international group of companies. By the end of 2007, the CDP comprised 385 signatory investors with more than $40 trillion in assets, including large investment firms such as Merrill Lynch and Goldman Sachs, and state pension funds. This represents very rapid growth from just thirty-five investors in 2003, with $4.5 trillion in assets. It should be noted that there are no costs or carbon commitments for signatory investors. More than 1,300 companies responded to the fifth survey by the Carbon Disclosure Project (CDP5) in 2007, reporting on various aspects of their carbon management (Innovest 2007).

Significantly, CDP5 indicates that the gap between corporate attention and action is beginning to close. In 2007, 76 percent of responding companies across all regions reported reduction targets with timelines, compared to 42 percent in the 2006 survey (Innovest 2007, 18). However, far fewer companies had begun to implement these programs, and Canada and the United States still lag behind Europe (no data are available for Mexico). This gap is particularly evident with respect to disclosure of GHG data and implementation of emission reduction programs with targets (see table 11.5). It should also be noted that 91 percent of companies in the London Stock Exchange "FTSE 100" index reported data, compared with only 54 percent of the U.S. Standard & Poor 500 companies and 34 percent of companies in the Canada 200. The North American response rate, however, improved significantly between CDP4 and CDP5. Table 11.5 summarizes responses by region.

The Investor Network on Climate Risk (INCR) is a smaller U.S.-based initiative of Ceres with about fifty signatories, representing approximately $1.75 trillion under management, including state treasurers and controllers, public pension funds, asset management firms, and venture capital funds. Some notable examples include the California Public Employees' Retirement System (CalPERS), the Pennsylvania State Treasurer, and Domini Social Investments LLC. As with the CDP, INCR encourages financial analysts, ratings agencies, and investment banks to address climate risks and opportunities. The INCR goes further, however, and secures commitments from the signatory investors. At the February 2008 summit, INCR launched an action plan with a goal of deploying $10 billion in additional investment in clean technologies over the next two years, and to aim for a 20 percent reduction over a three-year period in energy used in core real estate investment portfolios (INCR 2008).

The effort to enlist investors in the institutionalization of carbon accounting and management represents a sophisticated strategy on the part of environmental

Table 11.5
Carbon accounting, reporting, and trading

	Total responses	Firms see business risks	Firms see business opportunities	Disclosed GHG data	Senior management responsibility for climate change	Considered emissions trading opportunities	Implemented emission reduction programs with targets
Canada 200	86	85%	86%	66%	53%	27%	24%
S&P 500	269	81%	69%	65%	50%	36%	29%
FTSE 100	91	98%	82%	83%	53%	38%	41%
German 200	104	77%	80%	67%	38%	20%	35%
France 120	67	88%	84%	72%	34%	31%	43%

Source: Innovest 2007, 28.

groups. Outside the coal sector, however, it is difficult to gauge its success. Investors have been quick to sign up when it imposes no commitments, but more reluctant when they are called to play a more active role. The value of carbon reporting to investors is unclear; they certainly do not appear to be clamoring for this information. As with the Global Reporting Initiative, a broader program also launched by Ceres with a parallel logic to assess corporate social and environmental performance, carbon disclosure generates volumes of detailed information in a form that is difficult to compare, interpret, aggregate, and analyze (Brown, deJong, and Lessidrenska 2007).

Carbon accounting for purposes of emissions trading is a much more narrow project that does little to indicate the potential financial impact of climate risks. Carbon accounting is an exercise in commensuration, defined by Levin and Espeland (2002, 121) as "the transformation of qualitative relations into quantities on a common metric"; just as financial accounts reduce a firm's myriad activities to monetary terms, carbon accounting attempts to render complex organizational operations involving multiple gases and impacts in terms of a common, tradable currency. This commoditization of carbon is a political project, requiring a legal and bureaucratic infrastructure to define and measure carbon units, allocate and adjudicate property rights, and impose conditions for the transfer of credits across systems and jurisdictions.

The politics of carbon commensuration provide a degree of flexibility in reporting and exempt entirely certain regions and sectors. For example, emissions from military activities and international air travel and shipping are not counted under the Kyoto Protocol mechanisms. Moreover, the politics of carbon trading can produce systems with low prices, high transaction costs, and large-scale import of credits from uncapped countries and unverified sources, as with the Clean Development Mechanism and various retail carbon offset schemes (Haar and Haar 2006; Michaelowa and Jotzo 2005). These credits frequently do not even necessarily generate absolute GHG reductions, as they originate in projects that are compared to hypothetical "business as usual" cases of growing GHG emissions (Bumpus and Liverman 2008). Perversely, the value generated by these credits can also provide economic incentives for projects with net GHG emission increases that would not otherwise have been undertaken (Bradsher 2006).

Business Political Action

North American business generally has moved away from aggressive opposition to GHG controls toward a more accommodating position that acknowledges climate change as a serious issue and expresses a willingness to engage in a variety of carbon management measures. Nevertheless, there has been a resurgence of opposition to carbon regulation, particularly at the state level in the United States (Rabe, this volume). Some companies are simultaneously members of multiple organizations and initiatives with apparently conflicting agendas. About half of the organizations participating in the U.S. Department of Energy's Climate VISION, for example, are also members of CARE, which strongly supports coal power and opposes any emissions caps. This picture reflects the complex politics of the emerging climate compromise around a weak and flexible regime.

While the Business Council for Sustainable Energy serves as an industry association for the fledgling clean energy industry, the Pew Center has developed a more broad-based coalition of major firms around a more proactive position. The Pew Center helped launch the U.S. Climate Action Partnership (USCAP) in 2007 as a coalition of major businesses and environmental organizations advocating a national U.S. cap-and-trade system (rather than a patchwork of state and regional rules). They support mandatory limits but with modest reductions, credit for preregulatory action, and carbon price limits. Though USCAP calls for substantial long-term cuts in GHG emissions, it only calls for stabilization at 90 to 100 percent of current levels within ten years of policy enactment. The INCR has also joined the call for a mandatory national U.S. policy, with more drastic long-term cuts of 60 to 90 percent below 1990 levels by 2050.

Pockets of corporate resistance to U.S. emission controls remain. These include industry-funded groups such as the Coalition for Affordable and Reliable Energy, the American Council for Capital Formation, and the Center for Energy and Economic Development. The model legislation advocated by the American Legislative Exchange Council (Greenblatt 2003) and U.S. state ballot initiatives have attempted to limit state enactment of more aggressive climate change policy (Rabe and Mundo 2007). As in the 1990s, these organizations typically mount a multipronged attack: casting doubt on climate science, highlighting costs of emission limits, and opposing specific legislation.

The Competitive Enterprise Institute funded a series of advertisements in 2006 featuring the line: "Carbon dioxide: some call it pollution, we call it life" (Zabarenko 2006). Shortly after the release of the fourth assessment report of the Intergovernmental Panel on Climate Change in February 2007, the American Enterprise Institute offered a $10,000 incentive to scientists and economists to write papers challenging the IPCC findings. The American Enterprise Institute continues to receive significant funding for its climate change lobbying from ExxonMobil and many other companies in the energy sector.

Business organizations have mobilized to oppose local and regional initiatives, particularly those that target particular sectors. The Alliance of Automobile Manufacturers, which includes USCAP members General Motors and Ford as well as foreign companies with U.S. operations like Toyota, is vigorously contesting efforts by California and sixteen other states to exert direct regulatory control over vehicular carbon emissions (Hakim 2005). This industry pressure was widely seen as an important factor behind the EPA's ruling in December 2007 that California lacked authority to regulate vehicular carbon emissions (Broder and Barringer 2007). In April 2008, a coalition of states, cities, and environmental groups sued the EPA in a Federal Appeals Court in an attempt to force the agency to regulate GHG emissions from new cars and trucks. The suit built on a Supreme Court ruling in 2007 that the Clean Air Act gave the EPA the power and duty to regulate these emissions (Barringer 2008). Meanwhile, industry lobbyists are increasing their efforts to thwart these state-level initiatives (Stoffer 2008).

Corporate lobbying by the Associated Industries of Massachusetts was also implicated in former Governor Mitt Romney's decision to withdraw Massachusetts from RGGI in early 2006 (Selin and VanDeveer, this volume). The Associated Industries of Massachusetts, moreover, condemned Governor Romney's successor, Deval Patrick, for rejoining RGGI in 2007. It is evident that while North American business is slowly moving toward a "carbon compromise" based on a flexible trading system, it has not abandoned more aggressive lobbying and litigating when core interests are perceived to be at stake.

Concluding Remarks

While press releases and press coverage of private sector GHG reduction policies are increasingly common, overall GHG emissions continue to increase across North America. The coexistence of a beehive of corporate activity on climate change with few tangible outcomes presents an intriguing paradox. It might simply be too early to evaluate the impact of corporate efforts; some investments in innovation are unlikely to yield short-term gains, and emission reduction programs require the development of an institutional infrastructure for carbon management and trading. Nevertheless, this review suggests that business responses, especially in North America, are uneven and rather ineffective, at least in relation to the scale of action needed.

Outside the energy sector, corporate responses tend to be directed toward reputation management, organizational changes, and peripheral emission reduction programs rather than fundamental changes to business models, products, and technologies. Climate change is still seen as a corporate social responsibility concern rather than a core strategic challenge. These corporate responses can be understood in the context of a global GHG regime that is still fragmented, carries weak price signals, and is still largely voluntary outside Europe. This GHG regime is simply not yet up to the task of a radical restructuring of economic activity that could deliver emission reductions of 60 percent by 2050, as contemplated by the United Kingdom. Thus GHG emissions are caught on a treadmill; incremental improvements in efficiency and the growth of renewable energy are more than offset by economic growth, particularly in India and China.

Even the actions of climate leaders are somewhat limited and tentative. The operating GHG emission reductions achieved by BP and Shell, for example, are a tiny fraction of the emissions produced by the use of their products (The Climate Group 2005). The Alternative Energy Division of BP invests approximately $800 million a year, but this includes natural gas along with solar, wind, and hydrogen. This figure is still less than 4 percent of 2007 net profits and less than 0.3 percent of 2007 sales revenues. The Ecomagination campaign of GE amounts to seventeen products with sales of $10 billion within a diversified $150 billion revenue company, and R&D commitments of about 10 percent of the $14 billion GE invests in development. The products other than wind turbines mostly comprise incremental improvements to efficiency and production processes for existing products, as would be expected to occur in normal technological development.

While North American companies increasingly realize that climate change is a long-term issue to which they will need to develop market and technological responses, in the short term they face only modest political and economic incentives for action. The reliance on voluntary measures reflects a wider trend in environmental governance toward various forms of industry self-regulation (Cashore, Auld, and Newsom 2004). Ironically, it is largely the resistance of fossil-fuel-dependent countries and industries to more stringent regulation that has induced the fragmentation and flexibility of the current governance system (Levy and Egan 2003). North American efforts on climate governance are clearly gathering pace, but there is not yet a firm regulatory or economic incentive for firms to adopt radical changes in their strategies. This uncertainty presents an obstacle for corporate planning.

Furthermore, RGGI illustrates some of the problems of emerging efforts on carbon trading in North America (Selin and VanDeveer, this volume). The initiative only covers the power sector in its initial phase and has modest emission reduction goals. Initially RGGI includes a "safety valve" to allow for the import of relatively cheap external credits should the price of carbon exceed $8 a ton, effectively setting a price cap that is insufficient to drive substantial efficiency measures or a switch away from fossil fuels (Fischer and Newell 2003; Neuhoff 2005). Advocates for RGGI argue that once the trading infrastructure is in place, the cap can be ratcheted down when a political window of opportunity arises. It is unclear, however, how binding a cap will prove to be. Moreover, RGGI covers only a very small part of North America. Overall, proposed cap-and-trade systems are stimulating considerable corporate activities in North America, preparing the organizational infrastructure for emissions trading, but carbon prices are likely to be too low to induce fundamental market and technological changes.

In the absence of a significant price signal or other regulatory action, the basic economic and political forces that structure energy markets ensure the continued growth of fossil fuels for the foreseeable future. Coal remains the cheapest source of fuel for power generation, even with a modest carbon price in the range of $10–30. The oil industry maintains sufficient political influence to secure subsidies and favorable tax treatment. Supply limitations are beginning to constrain oil production, but at higher prices, vast reserves of oil shale and deeper ocean sources become viable (Stoett, this volume). Oil majors are also well diversified into natural gas, the demand for which is booming. Biofuels such as ethanol from corn can slowly be incorporated into existing infrastructure and business models, but they will supplement rather than substitute for oil as a liquid fuel.

North American companies appear to be hedging their bets by undertaking substantial organizational preparations and modest investments in new products and technologies, while acting to preserve the value of their technological and market assets in the medium term. Increasingly, business is recognizing the inevitability of carbon regulation, and moving to accept emissions trading as the heart of the emerging consensus around market-based instruments. Simultaneously, business is striving to shape this system so that it does not unduly disrupt existing markets. With state and regional policy initiatives threatening to impose more immediate and stringent caps on emissions, business is reverting to its oppositional stance of the 1990s (Levy and Newell 2005). A dramatic environmental "shock" or an unlikely assertion of political leadership might well be required to provide the necessary impetus for more radical change.

References

Anderson, Philip A., and Michael L. Tushman. 1990. Technological Discontinuities and Dominant Designs: A Cyclical Model of Technological Change. *Administrative Science Quarterly* 35(4): 604–633.

Aulisi, Andrew, Alex Farrell, Jonathan Pershing, and Stacy VanDeveer. 2005. *Greenhouse Gas Emissions Trading in U.S. States: Observations and Lessons from the OTC NO_x Budget Program*. Washington, DC: World Resources Institute.

Barringer, Felicity. 2008. Group Seeks E.P.A. Rules on Emissions from Vehicles. *New York Times*, April 3.

Bradsher, Keith. 2006. Outsize Profits, and Questions, in Effort to Cut Warming Gases. *New York Times*, December 21.

Broder, John M., and Felicity Barringer. 2007. E.P.A. Says 17 States Can't Set Emission Rules. *New York Times*, December 20.

Brown, Halina S., Martin deJong, and Teodorina Lessidrenska. 2007. *The Rise of the Global Reporting Initiative as a Case of Institutional Entrepreneurship*. CSR Initiative Working Paper, 36. Cambridge, MA: Kennedy School of Government, Harvard University.

Bumpus, Adam G., and Diana M. Liverman. 2008. Accumulation by Decarbonization and the Governance of Carbon Offsets. *Economic Geography* 84(2): 127–155.

Cashore, Benjamin, Graeme Auld, and Deanna Newsom. 2004. *Governing Through Markets*. New Haven, CT: Yale University Press.

Casten, Thomas R. 1998. *Turning Off the Heat*. Amherst, MA: Prometheus Books.

Chen, Ming-Jer, and Daniel Miller. 1994. Competitive Attack, Retaliation and Performance: An Expectancy-Valence Framework. *Strategic Management Journal* 15(2): 85–102.

Christensen, Clayton M. 1997. *The Innovator's Dilemma: When New Technologies Cause Great Firms to Fail*. Boston: Harvard Business School Press.

The Climate Group. 2005. *Carbon Down Profits Up*. 2d ed. London: The Climate Group.

The Climate Group. 2007a. *Carbon Down Profits Up*. 3d ed. London: The Climate Group.

The Climate Group. 2007b. *In the Black: The Growth of the Low-Carbon Economy*. London: The Climate Group.

Cogan, Douglas G. 2006. *Corporate Governance and Climate Change: Making the Connection*. Boston: Ceres.

Deloitte. 2006. *Forward Thinking: The Importance of Managing Greenhouse Gas Emissions, A Survey of Canadian Emitters*. Deloitte report. http://www.deloitte.com.

Dias, Rubens A., Cristiano R. Mattos, and José A. P. Balestieri. 2004. Energy Education: Breaking Up the Rational Energy Use Barriers. *Energy Policy* 32(11): 1339–1347.

Dudek, Daniel J. 1996. *Emission Budgets: Creating Rewards, Lowering Costs and Ensuring Results*. New York: Environmental Defense Fund.

Energy Information Administration. 2006. *Emissions of Greenhouse Gases in the United States 2005*. Washington, DC: Energy Information Administration.

Energy Information Administration. 2008. *Annual Energy Outlook 2008*. Washington, DC: Energy Information Administration.

Environmental Protection Agency. 2007. A Program Guide for Climate Leaders. Washington, DC: Environmental Protection Agency.

Fischer, Carolyn, and Richard G. Newell. 2003. *Environmental and Technology Policies for Climate Change and Renewable Energy*. Washington, DC: Resources for the Future.

Gelbspan, Ross. 1997. *The Heat Is On*. Reading, MA: Addison Wesley.

Goldemberg, José, Suani Teixeira Coelho, and Oswaldo Lucon. 2004. How Adequate Policies Can Push Renewables. *Energy Policy* 32(9): 1141–1146.

Greenblatt, Alan. 2003. What Makes ALEC Smart? *Governing* (October): 30–34.

Haar, Laura N., and Lawrence Haar. 2006. Policy-Making under Uncertainty: Commentary Upon the European Union Emissions Trading Scheme. *Energy Policy* 34(17): 2615–2629.

Haas, Peter M. 2004. Addressing the Global Governance Deficit. *Global Environmental Politics* 4(4): 1–15.

Hajer, Maarten A. 1995. *The Politics of Environmental Discourse: Ecological Modernization and the Policy Process*. Oxford: Clarendon Press.

Hakim, Danny. 2005. Battle Lines Set as New York Acts to Cut Emissions. *New York Times*, November 23.

Hamilton, Kirsty. 1998. *The Oil Industry and Climate Change*. Amsterdam: Greenpeace International.

Hart, Stuart L. 1997. Beyond Greening: Strategies for a Sustainable World. *Harvard Business Review* 75(1): 66–76.

Hoffert, Martin I., Ken Caldeira, Gregory Benford, David R. Criswell, Christopher Green, and Howard Herzog. 2002. Advanced Technology Paths to Global Climate Stability: Energy for a Greenhouse Planet. *Science* 298 (5595): 981–987.

Hoffman, Andrew J. 2006. *Getting Ahead of the Curve: Corporate Strategies That Address Climate Change*. Washington, DC: Pew Center on Global Climate Change.

Investor Network on Climate Risk (INCR). 2008. *Investor Network on Climate Risk Action Plan*. Boston: Ceres/INCR.

Innovest. 2007. *Carbon Disclosure Project Report 2007 Global FT500*. London: Carbon Disclosure Project.

Lash, Jonathan, and Fred Wellington. 2007. Competitive Advantage on a Warming Planet. *Harvard Business Review* 85(3): 94–102.

Leggett, Jeremy. 2000. *The Carbon War: Dispatches from the End of the Oil Century*. London: Penguin.

Levin, Peter, and Wendy Nelson Espeland. 2002. Pollution Futures: Commensuration, Commodification, and the Market for Air. In *Organizations, Policy, and the Natural Environment*, edited by Andrew J. Hoffman and Marc J. Ventresca. Stanford: Stanford University Press.

Levy, David L. 2005. Business and the Evolution of the Climate Regime. In *The Business of Global Environmental Governance*, edited by David L. Levy and Peter J. Newell. Cambridge, MA: MIT Press.

Levy, David L., and Ans Kolk. 2002. Strategic Responses to Global Climate Change: Conflicting Pressures on Multinationals in the Oil Industry. *Business and Politics* 4(3): 275–300.

Levy, David L., and Daniel Egan. 2003. A Neo-Gramscian Approach to Corporate Political Strategy: Conflict and Accommodation in the Climate Change Negotiations. *Journal of Management Studies* 40(4): 803–830.

Levy, David L., and Peter J. Newell, eds. 2005. *The Business of Global Environmental Governance*. Cambridge, MA: MIT Press.

Levy, David L., and Sandra Rothenberg. 2002. Heterogeneity and Change in Environmental Strategy: Technological and Political Responses to Climate Change in the Automobile Industry. In *Organizations, Policy and the Natural Environment: Institutional and Strategic Perspectives*, edited by Andrew J. Hoffman and Marc J. Ventresca. Stanford: Stanford University Press.

Makower, Joel, Ron Pernick, and Clint Wilder. 2006. *Clean Energy Trends 2006*. Clean Edge report. http:// www.cleanedge.com/reports/.

Makower, Joel, Ron Pernick, and Clint Wilder. 2008. *Clean Energy Trends 2008*. Clean Edge report. http://www.cleanedge.com/reports/.

Margolick, Michael, and Doug Russell. 2004. *Corporate Greenhouse Gas Reduction Targets*. Arlington, VA: Pew Center on Global Climate Change/Global Change Strategies International.

McGraw, Dan. 2008. High-Speed Solutions: The Idea of Passenger Rail Travel to Major Texas Cities Picks up Speed. *Fort Worth Weekly*, March 5.

McKinsey. 2007. How Companies Think about Climate Change: A McKinsey Global Survey. *McKinsey Quarterly*, December.

Michaelowa, Axel, and Frank Jotzo. 2005. Transaction Costs, Institutional Rigidities and the Size of the Clean Development Mechanism. *Energy Policy* 33(4): 511–523.

Najam, Adil, Ioli Christopoulou, and William R. Moomaw. 2006. The Emergent "System" of Global Environmental Governance. *Global Environmental Politics* 4(4): 23–35.

Neuhoff, Karsten. 2005. Large-Scale Deployment of Renewables for Electricity Generation. *Oxford Review of Economic Policy* 21(1): 88–110.

Pilke, Roger, Jr., Tom M. L. Wigley, and Christopher Green. 2008. Dangerous Assumptions. *Nature* 452(3): 531–532.

Porter, Michael E., and Claas van der Linde. 1995. Toward a New Conception of the Environmental-Competitiveness Issue. *Journal of Economic Perspectives* 9(4): 97–118.

Pulver, Simone. 2007. Importing Environmentalism: Explaining Petroleos Mexicanos' Cooperative Climate Policy. *Studies in Comparative International Development* 42(3–4): 233–255.

Rabe, Barry G., and Philip A. Mundo. 2007. Business Influence in State-Level Environmental Policy. In *Business and Environmental Policy*, edited by Sheldon Kamieniecki and Michael E Kraft. Cambridge, MA: MIT Press.

Roe, Brian, Mario F. Teisl, Alan Levy, and Matthew Russell. 2001. U.S. Consumers Willingness to Pay for Green Electricity. *Energy Policy* 29(11): 917–925.

Romm, Joseph R. 1999. *Cool Companies: How the Best Businesses Boost Profits and Productivity by Cutting Greenhouse Gas Emissions*. Washington, DC: Island Press.

Rowlands, Ian H. 2000. Beauty and the Beast? BP's and Exxon's Positions on Global Climate Change. *Environment and Planning* 18: 339–354.

Smith, Heather A., and Douglas Macdonald. 2000. Promises Made, Promises Broken: Questioning Canada's Commitments to Climate Change. *International Journal* 55: 107–124.

Stephenson, John B. 2006. *Federal Agencies Should Do More to Make Funding Reports Clearer and Encourage Progress on Two Voluntary Programs*. Washington, DC: Government Accountability Office.

Stoffer, Harry. 2008. Industry Targets State CO_2 Rules; Lobbyist Takes Automakers' Case to Statehouses across the Country. *Automotive News*, April 7.

UNFCCC. 2008. *Greenhouse Gas Inventory Data—Detailed Data by Party*. http://unfccc.int/di/DetailedByParty/Setup.do. Retrieved April 3, 2008.

Wellington, Fred, and Amanda Sauer. 2005. *Framing Climate Risk in Portfolio Management*. Washington, DC: World Resources Institute.

Zabarenko, Deborah. 2006. "Carbon Dioxide ... We Call It Life," TV Ads Say. Reuters, May 17.

Zarnikau, Jay. 2003. Consumer Demand for "Green Power" and Energy Efficiency. *Energy Policy* 31(15): 1661–1672.

12

Insurance and Reinsurance in a Changing Climate

Virginia Haufler

Introduction

Given the slow response of governments in North America and elsewhere to the challenge of climate change, it is sometimes argued that private sector actors should take independent action, both for the public good and to protect their own self-interest (Jones and Levy, this volume). One of the industries with potentially great leverage over this issue is the insurance sector, a fact only slowly acknowledged within both the industry and the policy community. Many effects of climate change will be felt most immediately by property owners and by those who finance and insure property. Insurers will suffer financial losses when severe weather or rising sea levels cause major property losses to their customers. As a result, they may have a direct stake in preventing global warming and not just responding after the fact. They could do this by providing incentives for their customers to limit emissions of greenhouse gases, investing their institutional funds in energy saving technologies and firms, and lobbying governments to address climate change.

This chapter examines the ways in which the insurance industry is responding to changing weather risks today and suggests how it may—and should—react in the future. The following section discusses the changing effects of natural disasters on the insurance industry, and the response by the North American insurance industry. It describes the decentralized nature of national regulatory institutions, and the integrated nature of the insurance market itself. In particular, the major reinsurers play a dominant role in transmitting information about the effects of climate change and about appropriate standards for insurance contracts. Direct insurers, reinsurers, and national regulators all interact with activists and customers in a multilayered system that influences responses to climate change.

The Insurance Industry and the Risks of Climate Change

The insurance industry in North America has been slow to recognize the risks it faces from climate change. The barriers to policy change and learning have been high. These obstacles have been slowly overcome by two trends: increasing weather-related losses, suffered particularly by international reinsurance companies; and activism by a coalition of investors and environmentalists, whose reporting and recommendations concerning climate issues and insurance have garnered increasing attention. Public authorities have played a relatively small direct role in promoting or facilitating changes, and there are few regulatory incentives for the insurance industry to change its behavior. Market incentives for the industry lead in divergent directions—either to strategies of financial risk-shifting or to strategies of fundamental change. Until recently, the insurance sector has chosen the former direction, but there is heightening pressure to choose the latter.

The insurance industry is directly impacted by severe weather, which may result in mounting losses to property and life. The predicted effects of climate change will cause rising sea levels, modifications in ocean circulation, and changes in marine ecosystems. All of these changes will have a significant effect on the North American continent, placing increased stress on coastal resources and threatening low-lying islands, including Manhattan. Some agriculturally productive regions in the plains of the United States and Canada may experience severe droughts, and the "bread-basket" of North America may become a parched wasteland. Coastal resort areas of Mexico may be affected by rising waters as the glaciers shrink from warmer weather patterns. Pressure on habitats is increasing, as we see in the spread of insects from the southern regions of North America to areas farther north. Diseases common to warmer southern areas, such as malaria, may also spread to the north (Stone 1992, 448).

All of these changes pose an increased risk of loss to property and commerce, affecting in turn those who insure against these losses. Just about every type of insurance may be affected. Obviously property insurance is key, because of weather-related damage, but health and life insurance may also be involved as diseases spread into new geographic areas. More specialized insurance may be affected as well. For instance, companies selling directors and officers liability insurance may have to take into account the impact of shareholder lawsuits brought for breach of fiduciary duty if the directors and officers do nothing to prepare for or prevent climate change losses. The insurance sector itself is one of the largest institutional investors in the world, and changes in climate and weather patterns will influence insurers' decisions on where to invest. The impact of climate change will not be all negative for insurers, however. There are also opportunities for more entrepreneurial insurers to devise new products that build upon climate mitigation policies and programs, such as insurance for risks involved in trading carbon credits via newly established exchange mechanisms.

An insurer traditionally has three primary methods of responding to "excessive" risk: by withdrawing from particular markets and putting limits on what it is willing to insure; by raising the prices for what it sells; and by shifting risk to others through financial, legal, organizational, and political practices. To one degree or another, however, these methods can all be problematic from the standpoint of the public interest.

First, insurers often withdraw from a particular geographic or product market, either temporarily or permanently, when losses are too high. This leaves those in a disaster no recourse except drawing on the public treasury. For instance, in coastal areas of the United States, individual states have established insurance "pools" of last resort for homeowners unable to obtain property insurance; these state-backed insurance plans were overwhelmed by new applications after the hurricanes of 2005 (Price of Sunshine 2006). When a family loses their house and has no insurance, they must dig into their own personal resources to rebuild. But when thousands of people lose their houses in a disaster, the government must step in to provide the resources to rebuild entire communities.

Second, insurers may raise their prices to recover from severe losses and rebuild their own reserves, but this will drive people away from purchasing insurance. Again, the gap in coverage may need to be filled by public spending or by the establishment of publicly funded and managed insurance programs. This is what has happened in most coastal areas of the United States, where insurers have raised rates significantly in anticipation of high losses. Those who contemplate buying expensive insurance may find home ownership more difficult to afford and mortgages may be more difficult to obtain. Thus, withdrawal from markets and raising prices off-load risk onto governments.

Third, insurers may pursue financial, legal, and organizational changes to displace risk onto other private entities, including their customers. The insurers seek to limit their financial risks by defining their liabilities vary narrowly. For instance, past history in the environmental field demonstrates that insurers will change contract terms, interpret them in a restricted fashion, and litigate in response to losses that exceed their expectations. They aim to limit their potential exposure, defining what they cover very narrowly in an effort to avoid paying claims. Toxic exposure cases in the United States, including asbestos and mold, have led insurers to exclude these risks from new contracts, reject claims from old cases, and litigate extensively.

In the face of potential losses from the impact of global warming, however, the insurance sector could instead redesign insurance contracts to provide incentives to customers to reduce their impact on the climate and protect their property from the effects of climate change. This fourth option is only beginning to be considered seriously. Insurers could, for instance, charge a lower price to customers who reduce their carbon emissions. In addition, insurers have the option of using their influence as institutional investors to direct financial resources to investments that reduce carbon emissions or expand alternative energy sources. This more progressive stance has been strongly promoted by climate change activists, but it has been resisted by the U.S. insurance sector, which would rather leave these issues to the federal government. This contrasts with the experience of major international (mostly European) reinsurers, who have actively partnered with international organizations and nongovernmental organizations to consider the role of insurance on climate change issues.

Despite three decades of concern about the impact of climate change on the insurance industry, however, it has been estimated that less than one company in a hundred has examined climate change issues to any significant degree (Mills 2007). One analysis described the problem facing insurers as a dual gamble: first, a gamble on the frequency, severity, and consequences of natural catastrophes; and second, a gamble about whether or not climate change dramatically overturns our current understanding of weather-related risks (Crichton 2005).

The Insurance Industry: International, Regional, and National Connections

Insurance, in its most fundamental form, is a mechanism for transferring financial risk. The insurance firm obtains payment in the form of premiums, and the customer receives in return a promise that the insurer will provide a payment when a specified risk occurs. Economic historians identify the development of institutions that manage risk as a key facilitator in the expansion and development of the modern economy (North 1990). The insurance industry is composed of many different types, or "lines," of insurance, including life, health, property and casualty, auto, and various specialized forms of insurance such as political risk or directors and officers liability. The industry includes the direct insurers (companies and agents), who sell to customers; brokers, who bring customers and insurers together; adjusters, who evaluate and administer claims; and reinsurers, who provide insurance to the direct insurers. In addition, there are many different service providers, from rating agencies to specialist insurance providers.

The insurance industry is one of the largest industries in the world, writing about $3,426 billion in premiums, of which $1,452 billion is in non–life insurance business (Sigma 2006). The industry is highly organized, with well-developed industry associations, extensive interindustry contracting, and explicit norms regarding acceptable practices. Insurers are highly dependent on maintaining a good reputation with customers and government regulators, since they offer a service that requires customers to pay in advance for a service to be provided in the future (if at all). The different participants in the insurance sector are entwined with each other via contracts, associations, and a common background, increasingly reinforced by the consolidation, and mergers and acquisitions, across previous sectoral and national boundaries. These service industries tend to be close to the customer base, and the health of that base is considered crucial to the maintenance of a modern economy. But this is also an extremely competitive market that can be affected by extreme swings in the business cycle.

The main hubs of insurance activity are in London and New York. Big reinsurers are based in Germany and Switzerland, and one of the biggest direct insurers is in Tokyo. In North America, the big insurers are based in the northeastern United States. Insurance is regulated at the state and local levels, though the industry has experienced national consolidation and international mergers and acquisitions. The Canadian market is similar to the market in the United States, though certain types of insurance are provided directly by provincial governments, and there is more extensive federal regulation. The major insurance companies include a few Canadian companies, such as Cooperators General, and a number of international insurers, such as ING, State Farm, and Lloyd's. Most Canadian insurance is sold through brokers, often by local branches of international firms (Keller and Amodeo 2001).

The Mexican insurance market is dominated by five companies prominent in auto, life, and property insurance; all have experienced significant foreign investment in recent years. The top tier of companies includes both Mexican national insurance firms and branches of international firms such as ING, Zurich, and AIG. The Mexican government's move to privatize state-owned insurance in recent years has attracted foreign investment. As in many developing countries, the insurance market is underdeveloped, and most risks are not insured and simply become losses (Kreimer, Arnold, Barham, et al. 1999). In general, the major reinsurance companies, primarily based in Europe, operate internationally, with companies such as Munich Re active in all three markets. Local and regional markets are intertwined, so that severe losses from storms in the Gulf of Mexico region, for example, can reduce the amount of insurance available in the Great Plains of Canada.

Historically, the main natural disaster risks for Canada have included droughts, wind storms, floods, fires, and earthquakes. In recent years, statistics for weather-related disasters show a sharp increase. For Mexico, the main risks have been wind storms, earthquakes, floods, and drought. The United States tends to face a higher degree of flood risk and suffers damages from a regular hurricane season (Souter 1991). The types of insurance coverage offered in these markets reflect these differences. Given the size and regularity of natural disasters in the U.S. market, it dominates discussion of the links between weather and insurance. The major reinsurers are more at risk in the U.S. market than in the other North American markets. Therefore, most of the interesting changes with regard to insurance and climate change primarily reflect the relationships among U.S. direct insurers, U.S. regulators, and international reinsurers.

Reacting to Severe Weather

The insurance industry began to consider manmade climate change as a threat following a series of weather-related disasters in the 1980s and 1990s, such as hurricanes and floods. According to Munich Re, in the 1960s there were twenty-seven major natural disasters with $76.7 billion in economic losses. In the decade of the 1990s, there were ninety-one major disasters valued at $514.5 billion (Munich Re 2004). Property-casualty insurers worldwide experienced record-breaking losses, and the loss trend since then has been continually upward. In comparison to the 1960s, the 1980s had 3.1 times more overall economic losses from major natural disasters; 4.8 times more insured losses; and 5.0 times as many major catastrophes (Leggett 1993, 3). As table 12.1 shows, the United States alone has suffered major losses due to natural disasters in the first decade of the twenty-first century.

Table 12.1
The ten most costly catastrophes in the United States: Insured losses (not total losses) (in $US millions)

Rank	Date	Catastrophe	Amount of loss at time of event	Equivalent amount of loss in 2007
1	Aug. 2005	Hurricane Katrina	41,100	43,625
2	Aug. 1992	Hurricane Andrew	15,500	22,902
3	Sept. 2001	World Trade Center/Pentagon attack	18,800	22,006
4	Jan. 1994	Northridge, CA earthquake	12,500	17,485
5	Oct. 2005	Hurricane Wilma	10,200	10,933
6	Aug. 2004	Hurricane Charley	7,475	8,203
7	Sept. 2004	Hurricane Ivan	7,110	7,803
8	Sept. 1989	Hurricane Hugo	4,195	7,013
9	Sept. 2005	Hurricane Rita	5,627	5,973
10	Sept. 2004	Hurricane Frances	4,595	5,043

Source: ISO's Property Claims Services Unit, Insurance Information Institute.

Reports of unusually severe natural disasters and their potentially dire effects on insurance profitability and solvency began to appear in business journals in the 1980s. At the World Insurance Congress in July 1991, a representative of Continental Corporation noted that 1989 and 1990 were both record-breaking years for catastrophic losses; she mentioned the possibility this might be related to global warming but did not take a definitive stance (Souter 1991). In 1992, the Munich Re corporation assessed losses from more than 500 natural catastrophes worldwide and noted this was a hundred more than in the previous year. Swiss Re did an analysis demonstrating the increasing size and frequency of catastrophes (Gordes 1997). Insurers became increasingly reluctant to provide insurance coverage in areas subject to these natural disasters, including many island states. These early 1990s reports stimulated discussion within the insurance industry of the relationships among extreme weather, climate change risks, and insurance.

Jeremy Leggett, of Greenpeace International, was one of the first to make the link between insurance losses and global warming. In 1992, Leggett began to urge the insurance industry to take action against global warming, making numerous presentations at industry conferences (McIwaine 1992). In an effort to mobilize insurers, he published a widely noticed article citing previous insurance studies and linking their results to climate change (Leggett 1993). Leggett argued that the standard industry response—raising premium rates and deductibles and restricting the terms and conditions for insurance policies—was a short-sighted solution to a major problem. He believed the long-term health of the industry depended on reducing greenhouse gas emissions to prevent, not accommodate, climate change. At this time, Greenpeace was looking for a business group that could be organized to oppose the fossil fuel interests, which adamantly resisted multilateral negotiations over limiting carbon emissions (Gordes 1997; Sabar 1994b). Leggett cited numerous statements by insurers that indicated a growing concern that climate change was implicated in their current losses or could potentially become a severe problem in the future (Gordes 1997; Leggett 1993).

Munich Re, the largest reinsurance company in the world, called on governments in 1994 to stabilize greenhouse gas emissions and keep their Rio commitments (Abbott 1994). A year later, just prior to the Berlin IPCC conference, Munich Re reported on further natural disasters, linked them to possible global warming, and called for a reduction in carbon emissions. Gerhard Berz of Munich Re stated, "There is no longer any doubt to us that a warming of the atmosphere and oceans is causing an increased likelihood of storms, tidal waves, hailstorms, floods and other extreme events" (Thiel 1995). In what amounted to a call to action, in 1995 H. R. Kaufman, general manager of Swiss Re, stated, "There is a significant body of scientific evidence indicating that last year's record insured loss from natural catastrophes was not a random occurrence.... Failure to act would leave the industry and its policyholders vulnerable to truly disastrous consequences" (Betting on Global Warming 1995, 23). During the Berlin conference, representatives of Munich Re, Swiss Re, and Lloyd's of London lobbied for emission reductions, in the hope this would decrease the probability of a rise in the number of natural disasters.

The major Norwegian insurer, Uni Storebrand, began lobbying other companies in Switzerland, Germany, and Britain to organize more actively on climate change issues and participate in international negotiations. Major insurance companies Uni Storebrand, General Accident and National Provident in the U.K., and Gerling in Germany, formed a corporate environmental alliance, drawing up a letter of intent linked to a United Nations Environmental Programme (UNEP) statement. The UNEP program director at the time worked closely with the industry and cosponsored a "Statement of Environmental Commitment," in which the signatories promised to incorporate environmental considerations into their risk management and to adopt industry best practices in this regard (Jagers and Stripple 2003; UNEP 1996). They would regularly make public reports of their environmental actions and would realign their asset management to include environmental considerations (Kirk 1995). By November 1996, sixty-two insurers from around the world had signed on to this statement.

In 1996, UNEP sponsored a conference on the insurance industry and the environment in London—the First International Conference of the UNEP Insurance Initiative: Implementing Environmental Commitment in the Insurance Industry. Nearly a hundred insurance companies from around the world participated in the conference, which focused on ways the industry could implement its commitment to incorporate environmental considerations into "best practices." Participants focused on eight areas: the handling of claims for losses; managing insurers' assets; designing insurance products; preventing losses; managing physical assets; mobilizing the company; environmental reporting; and lobbying (UNEP 1996). This eventually became one element in UNEP's overall strategy to organize the financial sector as a whole on environmental issues. The UNEP's work with the insurance industry is part of its larger Financial Initiative, begun in 1991, in which it obtains environmental commitments from banks, investment houses, and the wider financial community. The UNEP Financial Initiative—UNEP FI—was expanded in 1997 to include an Insurance Industry Initiative, or UNEP III. The majority of signatories are European firms (see table 12.2).

A UNEP Insurance Industry Initiative position paper on Climate Change from 1996 clearly pointed out the potential effects of climate change. It discussed not only the losses that might be suffered by property insurers but also warned that life insurers and pension funds may be affected by climatological effects on human

Table 12.2
UNEP FI signatories by region 2006

Region	Percentage
Europe	56%
Asia Pacific	27%
North America	10%
Latin America	3%
Africa	3%
Middle East	1%

Source: http://www.unepfi.org/signatories.

health. In addition, it said long-term investors such as the insurance industry might be affected by major changes in economic activity. The report argued that market forces alone would not make this shift efficiently or effectively, and it concluded that the precautionary principle must be the basis for decision-making (UNEP 1996). The insurers participating in UNEP III threw their support behind the Framework Convention on Climate Change. They urged countries to achieve early and substantial reductions in carbon emissions and argued for increased participation by nongovernmental organizations, including business, in any negotiations.

Over the next few years, during the struggle to ratify the Kyoto Protocol, the UNEP III met a number of times. A decade later, its membership had increased and the group's commitments had also expanded. The group focused primarily on the Kyoto process, and members were expected to lobby their governments to ratify the Kyoto Protocol as a counterweight to industries that had mobilized against it. They argued that the insurance industry could not take a "wait and see" attitude, but must adopt the precautionary principle, that is, they were urged to act ahead of irreparable change (Insurers Can Kick-Start Kyoto, 2000 Jagers and Stripple 2003). By 2008, the UNEP FI Insurance Working Group consisted of seventeen companies, including big insurers and reinsurers such as AIG, Swiss Re, and Munich Re. No Mexican insurer participated, and only one Canadian cooperative insurance company joined.

Outside of the UNEP III, insurers were beginning to reassess their risks and respond with new policies by the late 1990s. One of the more innovative mechanisms they developed was designed to shift some of the risk of natural weather disasters to broader capital markets through "securitizing" the risk in the form of catastrophe bonds or insurance-linked securities (Jagers, Paterson, and Stripple 2004). European and U.S. insurers also invested in a new Risk Prediction Initiative, to assess whether past history regarding the relative risks of different events, including natural disasters, would be an accurate predictor of future risks. European reinsurers, particularly

Swiss Re and Munich Re, had already established research and statistical services on disasters in the 1970s. Given the perception that natural disasters had increased in frequency and severity, and given the debate over climate change, by the late 1990s American insurers were beginning to be willing to invest millions in developing more technical capacity and knowledge on this subject (Insurers Can Kick-Start Kyoto, 2000).

Since the late 1990s, the weather helped propel the global warming agenda. In both 1995 and 1996, weather-related losses broke all previous records. For example, in just a matter of hours, Hurricane Andrew—a category 5 hurricane that did not even make landfall in the most developed areas of the state—caused $16 billion of insured losses and wiped out the premiums collected over the previous twenty years (Gordes 1997). In 2004, global losses linked to weather totaled $145 billion, with insurers covering $45 billion. In 2005, weather-related losses topped $200 billion and insured losses were around $70 billion (UNEP 2005). The next year losses were unusually low, but they spiked up again in 2007, reaching $75 billion—an increase of 50 percent over 2006. The number of natural disasters globally hit a high of 950 in 2007, the highest figure recorded by Munich Re since it began keeping records. According to one report, global weather-related losses have been trending upward, outstripping increases in population, inflation, and non-weather-related events (Berkeley Lab 2006).

A decade after the first efforts to involve insurers in climate change politics, the European insurers became open advocates for taking account of climate risks in their business. In 2005, the Association of British Insurers (ABI) produced a report arguing that climate change could increase the financial costs of extreme weather around the world. The report stated that "even quite small increases in the intensity of major storms (hurricanes, typhoons, windstorms), as predicted by the latest climate change science, could increase damage costs by at least two-thirds by the end of the century. The most extreme storms could become even more destructive, making insurance markets more volatile, as the cost of capital required to cover such events increases" (Association of British Insurers 2005, 3). Swiss Re announced in 2006 that it would partner with RNK Capital LLC to sell insurance for Kyoto-related risk in carbon credit transactions. It has established a specialist unit to address climate change mitigation and to take advantage of emerging market opportunities (Canadian Underwriter Daily News 2006). Swiss Re also lobbies governments to combat climate change. It participates in the Carbon Disclosure Project and is a member of both the International Emissions Trading Association and the UNEP Finance Initiative, including the Insurance Industry Initiative. As Chris Walker, managing director of the Greenhouse Gas Risk Solutions unit of Swiss Re, said recently:

It is the nature of our business to identify risks in the long term, and I see strong communication of those risks as an obligation for a reinsurer. If you start talking about an issue for a number of years it creates a groundswell of interest and awareness. If we can do this with climate change then it will be good for our clients and good for Swiss Re. (Climate Group 2006)

In North America, the U.S. industry remained outside this mobilization, despite the efforts of Greenpeace to enlist them in the cause. Neither Canadian nor U.S. insurers signed the environmental pledge cosponsored by UNEP at the time it was put forth. The U.S.-based insurers in particular viewed UNEP as a European initiative and UNEP showed little interest in working with American insurers. The U.S. insurers suffered losses similar to those of the European reinsurers, but they viewed the problem as simply a series of events that reduced their financial reserves and undermined their financial health, not as a part of some larger concern. The legal and political system in the United States is such that insurers often pay more for catastrophes than in other jurisdictions. As one British insurance lawyer put it, "Experience shows that if a catastrophe happens in the United States, you can expect to pay up to 30 times more damage claims than you would elsewhere in the world" (Souter 1991, 31). The link the American insurers made was not between global warming and disasters, but between overdevelopment and litigation that increased the costliness of disasters, requiring government intervention.

Until recently, only a few American insurers mentioned global climate change as a threat to their business. Frank Nutter, of the Reinsurance Association of America, has been the primary liaison between the U.S. industry and Greenpeace, and initially expressed doubts about the link between climate change and insurance loss (Sabar 1994a). In February 1995, the U.S. Insurance Institute for Property Loss Reduction, the Reinsurance Association of America, the Office of Vice-President Gore, and Timothy Wirth, Undersecretary for Global Affairs, sponsored a meeting at the White House on climate change attended by a number of American insurers at which they agreed to review the link between environmental change and recent losses. Presentations were made by climate scientists and European insurers with little effect (Gordes 1997). In the mid-1990s, insurers and reinsurers in Bermuda, the United States, and Europe established the Atlantic Global Change Institute (AGCI) to conduct research on climate risks that affect business. It focuses on making available to insurers the latest scientific advances in predicting climate patterns (Atlantic Global Change Institute 1996).

Ten years later, in 2006, the influential U.S. National Association of Insurance Commissioners finally set up a task force to study climate change, a belated effort to consider the risks to insurers (Hsu 2006, A10). The task force was set up in part because Midwestern politicians of both political parties experienced unusual droughts and severe weather that raised questions in their minds about the cause of so much bad weather. One task force member remarked, "I'm a financial guy, not an activist ... I don't know if we're prepared to be another Netherlands. But it does seem that we are too often in the position of cleaning up after the elephants run by" (Jackson 2006). The Canadian Institute for Catastrophic Loss Reduction, funded by insurers, has directly addressed climate change issues, and they have become more prominent in climate change debates (Canadian Underwriter Daily News 2005).

The U.S. industry response to severe weather patterns has been a traditional one: lobbying the U.S. government. Industry lobbyists argue that public authorities must establish a federal disaster fund as a safety net for the industry, on the grounds that major catastrophes threaten the solvency of insurers and their solvency is crucial to the economic health of the nation (Gordes 1997). The Canadian property and casualty industry–funded Institute for Catastrophic Loss Reduction (ICLR) began lobbying for action on climate change in 2005, with a call to the prime minister to develop a strategy on climate change (Canadian Underwriter Daily News 2005). So far, neither U.S. nor European insurers have opted to change their premium prices based explicitly on climate risk assessments (Awful Weather We're Having 2004). There is still a great deal of uncertainty regarding models of weather patterns and how the distribution and impact of changes will affect insured property and lives. The string of hurricanes in 2005 is now being incorporated into the most recent risk prediction models, however, and may lead to higher premiums in future years (Higher Insurance Premiums: Blame Climate Change 2006).

Lloyd's, the London-based insurance market, issued a major report in 2006 arguing that the industry has to reevaluate its models for underwriting, investing, and pricing its products or it could face financial stresses and even collapse (Price of Sunshine 2006, 79). In January of 2007, the chairman of Lloyd's said it was "taking steps to address climate change by investing in new scientific research in the United States and U.K. Lloyd's also insures new green technology, including a third of insurance for waste-to-energy recycling plants and a quarter of the world's wind farms" (Lloyd's Insurance International 2007). Lloyd's, the Association of British Insurers, other market participants and the nonprofit Prince of Wales Business Leader's Forum have since developed a set of guidelines, the Climatewise principles, for how insurers should integrate climate change into their business. These principles are somewhat vague but they nevertheless represent a big step forward. The Climatewise principles are: to lead in risk analysis; to inform public policymaking; to support climate awareness among customers; to incorporate climate change into investment strategies; to reduce the environmental impact of business; and to report and be accountable.

While the European insurers have developed climate prediction models and new measures of carbon emissions in order to benchmark progress and valuate firms (Jagers, Paterson, and Stripple 2004), the U.S. insurance industry is doubtful that state regulators will allow them to raise their prices based on models of the future, as opposed to traditional pricing based on historical data. Predictive models are viewed by regulators with suspicion, and historical data is perceived to be impartial and fair in determining underwriting results (Mills, Lecomte, and Peara 2001). In 2005, Massachusetts was the only state that allowed changes based on future estimates, as insurance is regulated at the state level in the United States (Mills, Roth, and Lecomte 2005; News Briefs: Climate Changes Insurance 2006). The slow response in the U.S. insurance sector, however, has prodded other actors to respond. There has been a long-standing interest on the part of environmental advocacy groups in persuading the financial sector to use its leverage over other firms to provide incentives to adopt more sustainable practices.

The Ceres coalition of environmentally responsible investors and environmental interests has been one of the most active groups prodding the insurance sector to account for climate change in its business practices. In 2005, Ceres sponsored a prominent report on climate change and insurance and launched a Northeast and Canada Climate Program (Mills, Roth, and Lecomte 2005). It also organized twenty institutional investors, with $800 billion in assets, to ask thirty publicly held insurance companies to create risk analyses of climate change and report these to the public by August 2006. These investors include state treasurers from California, Connecticut, Kentucky, Maryland, New York, North Carolina, Oregon, and Vermont; two of the largest public pension funds; the New York City comptroller; the Illinois State Board of Investment; and others. They are all members of the Investor Network on Climate Risk (INCR), a nonprofit group that is working to influence climate change policies (Awful Weather We're Having 2004; Higher Insurance Premiums: Blame Climate Change 2006; Institutional Investors to Insurance Industry: Act Now on Climate Change 2005). This pressure from Ceres is credited with pushing AIG to take a publicly proactive stance on climate issues.

In August 2006, Ceres produced a second report describing new insurance activities, such as providing customers with credits for "green" buildings and providing incentives for investing in renewable energy. Surprisingly, over half of the activities were reported by U.S. companies. For instance, Fireman's Fund Insurance has new "green" coverage for commercial building owners, while Marsh and AIG are both providing carbon emission credit guarantees and other products that will be useful for the European carbon trading market. A follow-up report one year later highlighted significant progress in the insurance sector, identifying twice as many climate-related products and services in the industry as the year before (Mills 2007). In March 2007, Ceres and other groups issued a Climate Policy Call to Action, including leading insurers and major noninsurance companies, which called on the U.S. federal government to adopt stringent policies to reduce greenhouse gases (Mills 2007, 36).

Shifting Risk or Preventing Climate Change?

One notable feature of the insurance sector response to climate change issues is the significant variation between European and American insurance cultures. While both have sought to shift risks to other entities, in Europe the insurance industry has been more proactive in directly addressing climate change as an issue of concern. This can be explained by two major factors: first, the structure of the industry itself; and second, the differing institutional environments. Reinsurance is a major sector of insurance activity in Europe, with Swiss, German, and British firms very active in reinsuring business around the world. They have generally borne the losses from hurricanes and floods in North America, and therefore are in some sense more materially affected by potential climate change than North American direct insurers, who off-load their risks to reinsurers and the government. Despite the handwringing over the high cost of recent disasters, the last few years have been some of the most profitable for U.S. insurers, in part because the losses were borne by overseas firms or reinsurers, who buffer their risk with returns on their investments in global capital markets (Hsu 2006, A10).

The general institutional environment is also quite different on each side of the Atlantic. The institutional environment influences how businesses respond to change (Jones and Levy, this volume). In the insurance case, European governments have been more proactive on climate change issues, and the U.S. government has gone in the opposite direction. With European public leadership, the private sector is more willing to take a stance on climate change issues, and it has the option of working in partnership with public authorities. With little or no national government leadership on this issue in North America, the insurers are unwilling to acknowledge the need for change. In general, business accepts a larger role in social issues in Europe because of a history of corporatism, social welfare, and high expectations from society.

At a more technical level regarding industry practices, European insurers are more willing to invest in the kind of research data and model development that generate new knowledge, understanding, and underwriting standards than U.S. firms. It was only in the mid-1990s that American firms agreed to establish the Risk Prediction Institute to invest in research capacity. Firms in the United States are limited to some degree by intensely competitive and shareholder-driven markets that punish firms immediately for performing below expectations. This circumstance is combined with a legal and regulatory framework that limits U.S. firms' ability to use new techniques and pricing structures. However, some observers have noted that even the more progressive European insurers have not acted strongly to mitigate climate change, despite their rhetoric. Additionally, there has been no systematic research on the potential effects of climate change on the industry as a whole (Mills, Lecomte, and Peara 2001; Mills, Roth, and Lecomte 2005).

This state of affairs contrasts with the European industry, which regularly reports on environmental issues and the possible impact of climate change (Mills, Lecomte, and Peara 2001). Swiss Re has been issuing reports on trends in natural disasters for decades. In Germany, the Allianz Group established a "Climate Core Group" to study the issues, and it is working with the government on how to respond (Mills, Lecomte, and Peara 2001). European firms tend to have in-house scientific research capabilities that American firms do not, providing a voice within the corporate organization for future-oriented planning (Gordes 1997). However, a few U.S. industry research centers have produced reports in the last few years that note the possibility that climate change could be a factor in insurance losses. Most North American insurers appeared to believe that the research on this topic is not conclusive enough to warrant active efforts to reduce carbon emissions; therefore, they simply recommended continued research instead of action (Gordes 1997). A statement from Wallace Hanson, president of the Property Loss Research Bureau, reflects the common attitude:

> The industry mindset is: Is this part of the normal cycle? Or, as Greenpeace suggests, is it something that society is bringing on itself and will get worse? This is the fence companies are sitting on. I feel that fossil fuels may be the cause, but I'm afraid of throwing a whole lot of resources at it and finding out it's something completely different. (Gordes 1997, 11)

Another reason for the conservative position of American insurers may be the liability system in the United States. In the past few decades, insurers have been forced by the court system to pay for environmental cleanup beyond what they thought they had contracted for originally. Long after the relationship between the insurance company and the customer has been ended, the insurer may still be held liable for pollution and environmental damage. This may encourage insurers to simply withdraw from markets, where possible, instead of dealing with liability in cases of property damage from climate change.

Unlike their European counterparts, American insurers simply do not perceive the possibility of financial opportunities from climate change action. European insurers perceive good financial prospects for investing in emissions trading, renewable energy, climate friendly technologies, and new insurance products that help customers manage environmental risks. Also, European insurers plan to become directly involved in carbon-trading markets by providing incentives for industry to adopt more environmentally friendly technologies through the terms of their insurance contracts (Allianz Group 2006). Swiss Re, as mentioned above, is already planning to insure carbon transactions. It has established a specialist unit within the company called Greenhouse Gas Risk Solutions, which focuses on mitigating and managing risk and pursuing opportunities. It seeks first mover advantages in creating an investment fund for energy efficiency and renewable energy in Europe (The Climate Group 2006).

In Europe, insurers are more sensitive about their reputations, which are more easily affected by public perceptions regarding their responsiveness on the environment (Dlugolecki 2004). The United States has to date lacked such a robust green market, which would provide incentives for reform, although this may be changing. This does not mean that U.S. insurers are completely unaware of possibilities for profit—they just have been very slow to recognize it. For example, AIG, working closely with Ceres, recently announced it planned to get involved in mitigating and profiting from climate change by providing products for European carbon-trading markets. Marsh and McLennan, a major insurance broker, is positioning itself as a consultant on climate risk, including the threat of increasing lawsuits (Lavelle 2006).

Another reason for the difference in response is due to the paradoxical role of the U.S. public authorities. The U.S. government has a large role in insulating insurers from particular kinds of risk, with extensive government programs for both flood and crop insurance (Mills, Lecomte, and Peara 2001). At the same time, the government has done almost nothing on climate change mitigation. Government action in the former case, and inaction on the other, directs insurance industry attention away from this issue. There is an assumption by insurers that the U.S. government will pick up the slack if the private sector does not provide insurance, as well as an awareness that any action by insurers on global warming issues probably would not elicit support from the government (Mills, Lecomte, and Peara 2001). The U.S. regulatory system discourages the use of new predictive models, and the tax system provides disincentives for the industry to build up reserves for future disasters. There is also strict regulation of the insurance sector, and any attempts to raise prices or withdraw from the market generate regulatory scrutiny (Mills, Lecomte, and Peara 2001).

Concluding Remarks: Choice and Consequences

Both public and private actors are increasing their involvement in climate change governance. In order for real and effective policy change to occur, there must be a combination of market incentives and regulatory action. The insurance industry has an opportunity today to influence the direction of markets and the behavior of customers in ways that slow down or even prevent disastrous levels of global warming. Insurers, by their actions, may regulate their customers' behavior and contribute to climate change governance (Haufler 1997, 1999).

There are three main options for insurers in the face of these debates over climate change, its potential effects, and the definition of their own interests: do nothing, withdraw from business affected by climate change, or use insurance as a leverage to prevent climate change (Leggett 1993). These three options are not mutually exclusive, and some firms are attempting to pursue more than one at the same time. Many insurers, particularly in North America, are adopting the first option: do nothing in the hope that the most dire predictions are simply wrong and recent natural disasters are a fluke and not a trend. These firms, despite the pressure from reinsurers and from increasing dissemination of knowledge about the risks of climate change, define their interest in terms of immediate short-term calculations of profit and loss. They are reluctant to give up a market that still remains profitable for many. If natural disasters continue to increase in number and severity due to climate change, then the ultimate risk will be placed on governments, since the insurers will experience extreme losses, go bankrupt, or finally withdraw entirely from particular markets. United States and Canadian insurers are not alone in this attitude; in fact, some point to it as a particular problem in developing countries, where insurance markets are not yet well developed (Cheung 1995).

Many insurers are counting on the government to provide the funding to recover from disasters and to supplement the private market with public insurance funds. We have certainly seen this recently with the reaction to losses from Hurricanes Katrina and Rita and the millions of dollars being spent on recovery in the affected states. State governments in the United States are also looking to the federal government to establish new disaster insurance funds. In Mexico, the destruction caused by flooding and high winds is either covered by the government or borne entirely by individuals and businesses. At the global level, we see the same dynamic at play. The Alliance of Small Island States, which will be the first to feel the effects of rising sea levels, has proposed that governments establish a global insurance institution to fund the costs of climate change in their countries. This would be a public insurance project, not the kind of private market activity discussed in this chapter.

A second option is to explicitly accept climate change and its effects and assume that it is an unstoppable force. But the goal would be to make sure there are sufficient financial resources for the insurance industry to remain solvent and to prevent harm to other financial actors such as banks and institutional investors. Private markets would do what they do best—signal what adjustments others should make through the price and availability of insurance (Stone 1992). This might provide a smooth transition to a world less dependent on fossil fuels, but the pace of change may instead lead to extreme volatility in prices and availability, which is what we see now in the North American market.

Under this option, private insurers would need to consider climate risks more directly in determining where and what to insure and how much to charge. Many areas, particularly coastal ones in the United States and Mexico, would no longer be insured by them at all. Coinsurance, perhaps through insurance pools, would become more common. Other financial sectors could take up some of the risk, for instance through developing new products such as catastrophe futures and weather derivatives to hedge against very high risks and losses. Thus, risk would be transferred from those experiencing losses and placed on the capital markets instead of on insurers (Jagers, Paterson, and Stripple 2004, 5). But it is through the terms of insurance contracts and the types of insurance they sell that insurers have a degree of leverage over customers. For instance, North American insurers could consider imposing higher premiums on companies that do not have environmental management systems, which is an option being considered by European insurers. Ceres has a project that is exploring whether shareholder lawsuits can be brought against corporate directors who can be accused of putting their companies' assets at risk by not addressing climate change. This liability could be used to influence investment decisions. The insurers that supply directors and officers liability insurance could be hard hit by this, and they may require their customers to implement new environmental policies to reduce the risk.

A third option would be to actively work to prevent climate change from occurring, instead of simply redistributing the losses. This entails two strategies: using industry leverage over customers, and using its influence over policymakers. In the first instance, insurers could create new financial products and services that help companies reduce their carbon emissions (such as through risk consulting and carbon credit finance); facilitate investment in renewable energy technologies (such as through carbon credit insurance and structured finance); and provide incentives for companies to improve their governance and performance on climate change (such as through directors and officers liability policies that provide additional protection to companies that have taken steps to reduce their emissions). This would be a more optimistic strategy, at least for insurers, in that it would entail developing new market opportunities. Insurers would do what they do best—package risks and sell financial coverage for losses. They would look upon climate change as a profitable opportunity, not just as a source of disastrous losses.

Alternatively, however, insurers could engage in political activism through UNEP III and other international and national venues. This strategy relies on the government not as a source of deep pockets to pay for losses in a disaster, but as a regulatory institution to force change on industry as a whole. The originators of the UNEP insurance environmental alliance have established expertise and begun developing both the normative and technical requirements of a proactive stance. As a result of their expertise and activism, they helped establish a new international effort to develop new norms regarding the role of insurers in climate change debates. In Canada, insurers have explicitly called upon the government to provide a larger framework of sustainability for addressing climate change risk. The Harper administration, however, turned its back on Canadian commitments under the Kyoto Protocol (Stoett, this volume), which made it less open to the insurers' demands. And of course, the Bush administration rejected the Kyoto Protocol and direct action by governments to mitigate climate change.

It is clear that the insurance industry is beginning to adapt, but it is also evident that it is not as yet making any profound changes in its business methods. A few companies are pursuing progressive strategies that attach environmental conditionality to the products they sell. There is also increasing discussion among insurers about the need to prevent climate change and to engage in political action, mostly notably through the activities of Ceres and its partners. There are moreover a number of reasons to think that insurers will make more significant changes in the coming years. The extreme weather of the last few years may have opened a window of opportunity. Governments, especially in Europe, are beginning to adopt policies that facilitate industry strategies premised on reducing greenhouse gas emissions. In addition, there has been increased activism directed at, and coming from, the financial sector as a whole. This includes the projects of Ceres; the larger Carbon Disclosure Project; and such financial sector initiatives as the Equator Principles, which regulate project finance on social and environmental values.

When the insurance sector as a whole becomes more completely committed to mitigating and profiting from climate change risk, they will inevitably have a profound influence on the shape of the economy. Individuals and firms will face potentially higher costs for doing nothing about climate change. This will provide a powerful incentive to change behavior. While it will take significant political action by governments to make a real difference in reducing greenhouse gas emissions, the power of the market and the private actors that trade in it may have an equally significant impact.

References

Abbott, Alison. 1994. Insurance Company to Back Out of Some Climate-Linked Risks. *Nature* 372(6503): 212–213.

Allianz Group. 2006. Faber: We Simply Cannot Ignore Climate Change. Munich: Allianz Group.

Association of British Insurers. 2006. Financial Risks of Climate Change. London: Association of British Insurers.

Awful Weather We're Having. 2004. *Economist*, October 10, 74.

Berkeley Lab. 2005. Berkeley Lab Scientist Sees Risk to Insurance Industry from Climate Change. *Research News*, August 11.

Betting on Global Warming. 1995. *Environment* 37(1): 23–27.

Canadian Underwriter Daily News. 2005. February 15. http://www.canadianunderwriter.ca.

Canadian Underwriter Daily News. 2006. June 13. http://www.canadianunderwriter.ca.

Cheung, Michele. 1995. Risky Business. *Business Mexico* 5 (November 11): 19–21.

The Climate Group. 2006. Swiss RE—Corporate, Reinsurance. Swissre.com.

Crichton, David. 2005. Insurance and Climate Change. Paper Presented at the Conference on Climate Change, Extreme Events, and Coastal Cities: Houston and London. Houston, Texas, February 9.

Dlugolecki, Andrew. 2004. A Changing Climate for Insurance: A Summary Report for Chief Executives and Policymakers. London: Association of British Insurers.

Gordes, Joel. 1997. Climate Change and the Insurance Industry: Uncertainty among the Risk Community. CT: Environmental Energy Solutions.

Haufler, Virginia. 1997. *Dangerous Commerce: Insurance and the Management of International Risk*. Ithaca, NY: Cornell University Press.

Haufler, Virginia. 1999. Self-Regulation and Business Norms: Political Risk, Political Activism. In A. Claire Cutler, Virginia Haufler, and Tony Porter, eds., *Private Authority in International Affairs*. Albany, NY: SUNY Press.

Higher Insurance Premiums: Blame Climate Change. 2006. *Consumer Reports*, February.

Hsu, Spencer. 2006. Insurers Retreat from Coasts. *Washington Post*, April 30.

Institutional Investors to Insurance Industry: Act Now on Climate Change. 2005. GreenBiz.com, December 5.

Insurers Can Kick-Start Kyoto. 2000 *Environmental Finance*, May.

Jackson, Derrick Z. 2006. Insurance Industry Feels the Heat of Global Warming. *Boston Globe*, March 15.

Jagers, Sverker, and Johannes Stripple. 2003. Climate Governance beyond the State. *Global Governance* 9(3): 385–410.

Jagers, Sverker, Matthew Peterson, and Johannes Stripple. 2004. Privatizing Governance, Practicing Triage: Securitization of Insurance Risks and the Politics of Global Warming. In *The Business of Global Environmental Governance*, edited by David Levy and Peter Newell. Cambridge: Cambridge University Press.

Keller, George, and Frank Amodeo. 2001. The Canadian Insurance Market. IRMI.com

Kirk, Don. 1995. Insurers Pledge to Reduce Environmental Risk. *Business Insurance* 29(49): 63.

Kreimer, Alcira, Margaret Arnold, Christopher Barham, Paul Freeman, Roy Gilbert, Frederick Krimgold, Rodney Lester, John D. Pollner, and Tom Vogt. 1999. Managing Disaster Risk in Mexico: Market Incentives for Mitigation Investment. In *Disaster Risk Management Series*. Washington, DC: World Bank.

Lavelle, Marianne. 2006. Insurers May Cash In on Climate Change. *U.S. News and World Report*, June 5, 46–47.

Leggett, Jeremy. 1993. Climate Change and the Insurance Industry: Solidarity among the Risk Community. Amsterdam: Greenpeace International, May 24.

Lloyd's Insurance International. 2007. Press Release, January.

McIwaine, Kate. 1992. Australian Insurers Advised to Heat up Participation in Global Warming Debate. *Business Insurance*, November 30.

Mills, Evan. 2007. From Risk to Opportunity: 2007 Insurers Response to Climate Change. *Ceres Report*, November.

Mills, Evan, Eugene Lecomte, and Andrew Peara. 2001. *U.S. Insurance Industry Perspectives on Global Climate Change*. Berkeley, CA: Berkeley National Laboratory.

Mills, Evan, Richard J. Roth, Jr., and Eugene Lecomte. 2005. *Availability and Affordability of Insurance under Climate Change: A Growing Challenge for the U.S.* Ceres Report, September 2005.

Munich Re. 2004. Annual Review: Natural Catastrophes 2003. In *TOPICS Geo*. http://munichre.com.

News Briefs: Climate Changes Insurance. 2006. *Environmental Science and Technology* 40(3): 633–663.

North, Douglass. 1990. *Institutions, Institutional Change and Economic Performance*. Cambridge: Cambridge University Press.

Price of Sunshine, The. 2006. *Economist*, June 10, 76–79.

Sabar, Ariel. 1994a. Eco-Underwriting: Greenpeace Woos the Insurance Biz. *Mother Jones* 19(4): 17.

Sabar, Ariel. 1994b. Greenhouse 101. *Whole Earth Review* 19(4): 17.

SIGMA. 2006. World Insurance in 2005. *sigma*, 2006, 1. http://swissre.com.

Souter, Gavin. 1991. Response Plan Eases Recovery from Disasters. *Business Insurance*, July 15, 30.

Stone, Christopher. 1992. Beyond Rio: "Insuring" against Global Warming. *American Journal of International Law* 86(3): 445–489.

Thiel, Stefan, and Bill Powell. 1995. While the Earth Burns. *Newsweek*, April 10, 44.

UNEP. 1996. First UNEP Conference on Insurance and the Environment. http://www.unepfi.org.

UNEP. 2005. 2005 Breaks a String of Disastrous Weather Records. Press Release, Montreal, December 6, 2005.

13

Campus Climate Action

Dovev Levine

Introduction

According to Yale University President Richard Levin, "Universities have begun to take the lead, along with enlightened corporations as well as municipal and provincial governments, in setting standards for carbon emissions that are substantially more restrictive than those adopted by national governments" (Levin, 2008). By cutting emissions 17 percent[1] while experiencing significant physical growth, Yale demonstrates that meaningful climate actions are achievable and cost-effective (Blum 2008). This is just one example of many campus climate actions developing across North America. Universities can play significant roles in addressing contemporary climate change and energy challenges for several reasons: they are often the size of small cities and have budgets comparable to larger firms; they educate millions of students each year; they are often the first in their respective jurisdictions to engage in a variety of climate actions; and they play host to a great deal of climate activism, research, and policymaking. In these respects, campus climate action comprises a significant portion of developing multilevel climate governance in North America.

In the absence of federal leadership on climate change, several states, provinces, municipalities, and civil society actors have made ambitious commitments to reduce GHG emissions (Rabe 2004, 2008; Rappaport and Creighton 2007). A multitude of local governments and private sector actors are working on creative solutions for addressing the threat of global warming (Gore and Robinson, this volume; Jones and Levy, this volume). As a growing number of actors engage in efforts to address climate change, new opportunities to tackle the climate challenge emerge through public-private partnerships, bottom-up pressures, and transnational networks. Decisions affecting GHG emissions and strategies for adapting to the impacts of climate change are made by a myriad of state and nonstate actors working at all levels, from the global to the local, and often connected through transnational networks (Betsill 2005). These actions send an ever-stronger signal to municipal, state, provincial, and federal policymakers that more aggressive climate action is possible, cost-effective, and politically supported (Moser and Dilling 2007, 397).

Though it is unlikely that bottom-up initiatives can supplant the need for more concerted national and multilateral climate policymaking, these initiatives demonstrate concern and can also help in providing new solutions and pushing national governments to act more forcefully (Speth and Haas 2006, 123). Campus action adds to this flurry of expanding local-level climate action in North America. A multitude of campus actors engage in action within their campuses and local communities, manifesting many functions of local government and private firms, even as they seek to engage state, national, and international debates and policymaking. As such, campus climate action may be seen as increasingly important within the multiscale processes of climate governance. Multilevel environmental governance perspectives would therefore be remiss if they did not explore the role of campus climate action.

This chapter examines campus climate action in North America. The first section explores initiatives developing across campuses, with a focus on curriculum, operations, research, and education. This is followed by a discussion of the major drivers behind the expansion of campus climate action, including saving money, staying abreast of future regulations, and gaining public relations benefits. The next section examines four ways in which college and university climate action is important beyond the campus level: by leveraging collective buying power, modeling cost-effective actions, demonstrating the feasibility of GHG reductions, and providing citizen education. The subsequent section discusses important limits and constraints on colleges and universities as they continue developing climate actions. The chapter ends with concluding remarks on the value of campus climate action in the context of North America multilevel governance.

What's Happening on Campus?

The University of New Hampshire categorizes its campus climate actions into four general themes: Curriculum, Operations, Research, and Education (CORE).[2] This CORE framework is used here to organize a brief overview of current trends of North American campus climate actions.

Curriculum: What Is Taught

Colleges and universities are increasingly incorporating climate-based curricula into program offerings in political science, public policy, environmental studies, engineering, business, and other departments. Drivers for this development include an increasing demand for workers in climate-related fields and student demand for such curricula. Schools are also beginning to use their climate offerings as a recruitment tool.

Anticipating more call for new technical and design skills related to climate, colleges and universities across the nation are offering degree programs in the technology and policy fields. For example, the Oregon Institute of Technology has developed the country's first four-year undergraduate degree program in renewable-energy systems. The program trains fifty students and graduated its first class at the end of the 2007–2008 academic year (Schneider 2008). In 2007, Arizona State University launched the world's first degree-granting school devoted exclusively to sustainability. Their new School of Sustainability, which launched its initial class of several dozen students in September 2007, offers bachelor's, master's, and doctorate degrees in sustainability. Dozens of other schools are integrating new sustainability classes into curricula, with many combining educational tracks across disciplines, such as a business major that may require classes in environmental science or an economics or engineering major that includes the study of atmospheric science (Nicholson 2007).

This increase in curricula focused on climate change is driven largely by student interest as growing numbers consider careers in alternative energy fields. For example, the number of undergraduates at the University of California, Berkeley, enrolled in introductory energy courses has almost tripled, and a new graduate class in solar photovoltaics attracted seventy students, making it one of the largest courses in the College of Engineering. One faculty member noted, "Over the last two years, demand for energy courses is off the charts" (Anderson 2007). There is furthermore a notable increase in climate-based curriculum offerings in many MBA programs. Along with customary classes on subjects such as finance, accounting, and marketing, more than half of U.S. business schools require a course in environmental sustainability or corporate responsibility, up from 34 percent in 2001 (Knight 2007). Schools are using their climate-focused curricula in recruitment efforts. Along with rankings in *U.S. News and World Report*, MBA programs now also tout their placement in the Beyond Grey Pinstripes Index, which assesses efforts of business schools to teach principles of environmental stewardship (Steptoe 2007).

Similar developments are evident in Canada. For example, environmental studies programs are drawing increasing interest from Ontario high-school students when they apply to universities: "We are seeing huge demand," said Nancy Weiner, associate registrar at the University of Waterloo, where first-year applications for environmental studies are up 62 percent from last year (Church 2008). Applications to environmental programs—including those in science and engineering, as well as programs in sustainability, resource management, environmental policy, and even green tourism—jumped dramatically for the 2007–2008 term, even doubling at some schools. At the University of British Columbia, inquiries about the environmental science program doubled between 2006 and 2007, while applications to the natural resource conservation program increased by 33 percent during the same timeframe (Agrell 2007). Overall, the Province of Ontario has experienced a 50 percent increase in students picking environmental programs for a major (Church 2008).

A spotlight on climate change also has broad pedagogical benefits within curricula. The public debate over national and international climate goals in the popular press is invaluable. A focus on climate change in the media encourages faculty, students, and administrators to consider the larger context while they pose challenges to public officials and society at large. A key question is: How can we reduce our contribution to the problem? From a faculty perspective, the international scope and the long-term intergenerational impacts predicted for climate change expand the range of issues that are relevant. Climate change issues offer opportunities to connect global and local issues and institutions, examine the roles of science and technology in societies and communities, and inspire research to address interrelated social and ecological challenges.

Operations: Meeting Energy Needs

Consistent with the heightened national interest in energy and climate change issues, as well as concerns about energy price volatility, college and university administrators and staff are paying more attention to energy use and its costs. Like many municipalities and firms, colleges and universities have dramatically stepped up their energy efficiency efforts, seeking to install more energy efficient lighting, heat and air conditioning, and laboratory equipment. Colleges and universities are also establishing purchasing policies for vehicles and a host of office and other types of equipment to encourage energy efficient purchasing. Also, more and more colleges and universities are pursuing the use of renewable energy, either by purchasing green power at a premium from an electric utility or other supplier or by building renewable energy generation facilities on or near their campuses (see tables 13.1 and 13.2).

In these ways, many colleges and universities are positioning themselves as pioneers in the North American shift to renewable energy production and consumption. Innovative and exemplary actions can be found across the continent. In 2006, more than 200 U.S. colleges and universities were either purchasing renewable energy or producing renewable energy onsite for a portion of their energy needs (National Wildlife Foundation 2008). Seventeen of these institutions are buying all of their electricity from green sources, like wind energy, solar energy, and small-scale hydropower generators (Monastersky 2007). Furthermore, ninety-one campuses have their own solar panels, thirty-three have wind turbines, and at least forty institutions have adopted green building policies (Newman and Fernandez 2007). The recent collective action by a consortium of thirty-four Pennsylvania-based institutions purchasing 92.2 million kilowatt hours of wind energy constitutes the largest

Table 13.1
U.S. higher education institutions where all electricity is from green sources (as of October 2007)

Institution	State
Brainbridge Graduate Institute	Washington
Colby College	Maine
College of the Atlantic	Maine
Concordia U. at Austin	Texas
Connecticut College	Connecticut
Evergreen State College	Washington
Lander U.	South Carolina
New York U	New York
Paul Smith's College of Arts and Sciences	New York
St. Mary's College of Maryland	Maryland
Saint Xavier U. (Ill.)	Illinois
Southern New Hampshire U.	New Hampshire
Unity College	Maine
U. of California at Saint Cruz	California
U. of Central Oklahoma	Oklahoma
Warren Wilson College	North Carolina
Western Washington U.	Washington

Source: U.S. Environmental Protection Agency.

Table 13.2
U.S. higher education institutions that buy the most green electricity (as of October 2007)

Institution	Annual usage of green power in kilowatt-hours	Share of total electricity use
New York U.	118,616,000	100%
U. of Pennsylvania	112,000,000	29%
Pennsylvania State U.	83,600,000	20%
California State U. System	78,333,573	11%
Duke University	54,075,000	31%
U. of California at Santa Cruz	50,000,000	100%
Texas A&M U. System	43,350,000	15%
City University of New York	41,400,000	10%
Northwestern University	40,000,000	20%
Western Washington U.	38,008,000	100%

Source: U.S. Environmental Protection Agency.

nongovernmental aggregated commitment to wind power in the United States (U.S. Department of Energy 2006).

Higher education is also a leader in the construction of energy-efficient combined heat and power systems, commonly known as cogeneration systems. By early 2008 more than 140 campuses in the United States and Canada had their own cogeneration systems.[3] This is similar to the activity of states such as California, which sources about 17 percent of its electricity from cogeneration (Global Power 2007). The University of New Hampshire, in another example, has invested $45 million into deriving much of its energy needs from landfill gas used to power its new cogeneration facility. Given that energy prices charged by university vendors have grown rapidly over the previous five years, University of New Hampshire officials believe the facility will recoup the investment after only about ten years, because it significantly reduces the need to purchase commercial natural gas. It is estimated that powering the cogeneration facility with landfill gas will reduce 2008 GHG emissions by 67 percent below 2005 levels.[4]

In addition, the environmental magazine *Grist* named the University of British Columbia one of the top fifteen universities in the world when it comes to taking action to reduce GHG emissions and becoming more energy efficient. Through its many GHG emissions reduction projects, the University of British Columbia has succeeded where most provinces and territories and the federal government have failed, by both meeting and exceeding Canada's GHG reduction target under the Kyoto Protocol (6 percent below 1990 levels by 2012). Furthermore, the $2.6 million saved annually through a host of energy efficiency projects will enable the university to pay off loans it took out to undertake actions to reduce GHG emissions, as well as to fund its office of sustainability (Mason 2007).

Research: Fostering Technological and Policy Advances

Over the last decade, many colleges and universities have sought to assess and/or enhance their climate and energy-related faculty and research programs, driven in large part by increased research funding and emerging societal needs. Leading colleges and universities play host to a substantial fraction of North American climate research and a large number of researchers from these institutions are involved with the Intergovernmental Panel on Climate Change. Faculty members engage policy debates through research and assessment of possible GHG taxation and cap-and-trade schemes through research projects that cross social, natural, and physical science disciplines. Advances in research on solar power, wind power, and biomass fuels are growing exponentially. Furthermore, much of the research on increasing energy efficiency takes place on North American campuses.

Anecdotal evidence indicates that a growing number of colleges and universities are setting up centers where engineering students work on alternative fuels, business students calculate the economics of those fuels, and political science majors examine policy-related issues as they apply to both industrialized and developing countries. However, it is difficult to exactly quantify the growth of these research centers, which include an array of sustainability centers, environmental institutes, and global warming initiatives. Many of these centers do not stand alone but are located inside an existing school (Deutsch 2007a). But so far it is evident that these colleges and universities have the research capacity to steer scientific advancements in climate -related fields, and they have become increasingly involved in related issues of economic development and technological innovation (Sa, Geiger, and Hallacher 2008).

Engagement: How Colleges and Universities Interact with Others

Engagement refers to the level with which colleges and universities interact with other organizations and local communities. Universities belong to a number of climate-related associations, and their membership is increasing. These associations can produce significant outcomes. A prominent example is the Association for the Advancement of Sustainability in Higher Education (AASHE), which coordinates the American College and University Presidents Climate Commitment.[5] The AASHE also shares directories of campus sustainability contacts and promotes opportunities for networking and attending conferences (Scherer 2006). Other university-based associations include the Ivy Plus Sustainability Working Group, founded to help the growing cadre of campus sustainability coordinators compare strategies and solutions (Blum 2008), and the Pennsylvania Environmental Resources Consortium, a coalition of forty colleges and universities.

Canadian schools also form networks. For example, Concordia, Université du Québec à Montréal, Université de Montréal, and McGill are working together as part of Sustainable Campuses, a national program of the Sierra Youth Coalition to challenge universities and colleges to adopt sustainable policies and practices. The program began in 1998, involves sixty-six Canadian campuses, and is linked to the Campus Climate Challenge, a similar North America–wide program (Cooper 2007). Because the vast majority of Canadian funding to universities flows through provincial governments, there are few private universities. Interuniversity meetings of upper-echelon administration at the vice-president level are relatively frequent, creating opportunities for the communication of effective climate change strategies. The climate change discourse and approach on Canadian campuses also more commonly include social, economic, and traditional environmental considerations in climate and sustainability goals than those in the United States (Yale 2007).

North American colleges and universities also engage many private- and public-sector actors. A recent initiative under the Clinton Foundation allows colleges to enter into a purchasing consortium for energy-efficient products. Under this partnership, five of the world's largest banks—ABN AMRO, Citi, Deutsche Bank, JPMorgan Chase, and UBS—offer $1 billion each in financing for energy-efficiency retrofits so that colleges can avoid capital spending or increases in monthly operating expenses. Eight energy services companies are signed on to perform such projects. This Clinton Foundation initiative was created under the rationale that by ramping up the number of projects, these companies can cut down on marketing costs and charge less for their services. Companies also agreed to make systemic retrofits and not to advise colleges to make only quick fixes with immediate paybacks (Powers 2007). In the government sector, the program manager for the EPA's green-power partnership says, "Colleges and universities have been some of the earliest and strongest participants in the program" in green-energy purchases (Monastersky 2007). Furthermore, the collective purchase of wind power by thirty-four institutions under the umbrella of the Pennsylvania Environmental Resources Consortium constitutes the largest nongovernmental aggregated commitment to wind power in the United States.

Climate-based student activist groups also figure significantly into intercampus activity. For example, student groups at 570 schools signed up in 2007 for the Campus Climate Challenge, a campaign sponsored by thirty environmental groups (Green 2007). These engagement activities have had significantly productive outcomes. Campus Climate Challenge partners have secured more than 200 campus carbon reduction policies, including 28 renewable energy purchase agreements, 22 green fee project funds, and 16 sustainable transportation policies. In 2007, the student-led Step It Up Campaign organized events—involving students and nonstudents—in 1,400 communities in fifty states. During events, ranging from routine rallies to highly creative actions involving people dressed as polar bears, participants demanded an 80 percent reduction in U.S. GHG emissions by 2050. Step It Up's day of action on November 3, 2007, was even larger, with seventy-one members of Congress attending and all the major Democratic candidates for president endorsing the campaign's goals (Namikas 2008).

Driving Campus Climate Action

Campus leadership in climate initiatives emerges from many different sources, including administrators, faculty, and students. Having a permanent sustainability coordinator position is increasingly an indicator that a college or university takes these issues seriously (Rappaport and Creighton 2007, 74). These positions have in common the mandate to bridge the university's operational initiatives to its teaching and research—to make the nuts and bolts count toward the big teachable ideas, and ultimately lead to incorporating sustainable and climate friendly ideals throughout campus operations. In short, the sustainability coordinator plays "traffic cop, diplomat, and facilitator," (Blum 2008) institutionalizing a host of environmental and social concerns within university administrations. Such jobs are becoming fixtures on many campuses, with approximately 100 full-time and 150 part-time campus sustainability managers across the United States by 2008 (Xiong 2008). Administration-driven action is also seen in the growing number of college and university presidents committing to making their campuses carbon-neutral under the American College and University Presidents Climate Commitment (ACUPCC 2007).

Faculty from a host of academic disciplines and departments lead much climate action on campuses. For example, the University of New Hampshire's Climate Change Task Force, a presidential-level group, consists of faculty from a range of disciplines, including Political Science and Earth Sciences, who meet with members of the university's Office of Sustainability Programs and other administrators to provide guidance on climate actions, including building guidelines for new campus structures and energy and transportation issues. Committee members shape and help run energy- and climate-related programs on campus, and faculty members and the students they supervise often provide in-house assessment and analysis in support of climate and energy-related ideas and initiatives.

Much campus climate action is also fueled by students who, for example, often push actors and organizations throughout the higher education sector to purchase wind energy. Students organize referendums on student fees for renewable energy purchases, with students often voting strongly in favor of measures to purchase clean energy (Putnam and Phillips 2006). At the University of California, Santa Cruz, for example, money raised from a $3 per quarter student fee increase led to the purchase of 50 million Kilowatt hours of renewable energy certificates. The purchase, combined with the university's existing electrical contract for 5 million Kilowatt hours of renewable power, means the campus offset 100 percent of its electrical consumption (University of California, Santa Cruz 2006). Students also engage energy conservation initiatives, assess university policies and practices, and push administrators to set and achieve more aggressive GHG reduction goals.

Many climate initiatives in areas of curriculum, operations, research, and education are, of course, driven by a belief that climate change poses a significant risk and that colleges and universities should be at the forefront of mitigation and adaptation efforts. There are, however, also other—more practical—reasons for university leaders and administrators to engage in climate-related action, including opportunities to save money, stay abreast of future regulations, and gain public relations benefits. Across North America, colleges and universities are increasing their electricity use (Rappaport and Creighton 2007, 230). Combined with increasing energy rates, many colleges and universities are reporting doubling and tripling of their energy costs over the past decade (Putnam and Phillips 2006). While most renewable energy efforts require significant amounts of up-front capital, most colleges and universities base their decisions on long-term planning (Creighton 1998). In ten years, savings from energy efficiency projects at the University of California, Santa Barbara have totaled over $36 million.

Pursuing enhanced energy efficiency, increasing the use of renewable energy sources, and lowering GHG emissions is not only fiscally and socially responsible but also helps colleges and universities stay ahead of the regulatory curve. For public universities in particular, state and provincial policies are increasingly mandating greater energy efficiency and, at times, increased use of renewable energy. California led the way in August 2006 with a law requiring all public and private companies and agencies in the state to reduce global warming emissions to 1990 levels by 2020—or 25 percent from 2006 levels. A bill passed in summer 2007 in North Carolina requires all new or renovated state-owned buildings—including those in higher education institutions—to be 20 to 30 percent more energy- and water-efficient than previous minimum standards. Light emitting diode (LED) exit lights and low-flow bathroom fixtures are now required. Similarly, Ohio public universities are required to reduce their energy intensity 20 percent below 2004 levels by 2014 (National Wildlife Foundation 2008).

Public relations benefits are accrued by campus climate action, as well. Campus energy conservation programs can assist with marketing efforts at no additional expense. For example, media coverage of energy and climate initiatives on campus can help to attract students and faculty to campus; garner positive attention from local, state, and federal officials; increase alumni donations and other fundraising opportunities; and help to "brand" a university. For example, when Colorado State University offered students living in residence halls the opportunity to purchase 100 percent of their electrical needs from wind power, the news was reported in more than twenty-five newspapers nationwide. Similarly, media coverage of environmental, sustainability, and energy-related majors and other initiatives, including those at the University of New Hampshire, the University of Washington, Oberlin, Stanford, and the State University of New York, garner regional and national attention (Rimer 2008; Schneider 2008). Such publicity is a public relations boon and can make colleges and universities more desirable to prospective students and their parents (Morris 2005).

Impacts beyond Campus

North American federal development of effective climate policies has been undermined by a combination of a lack of political will and a failure to understand and quantify the costs and benefits of preventive or remedial action (McCormick 2005, 83). Through their collective influence, climate actions by colleges and universities have the potential to shape developments that extend beyond the borders of campuses in at least four ways: leveraging buying power; modeling cost-effective actions; demonstrating feasibility of GHG reductions; and citizen education.

Collective Buying Power

Many colleges and universities are like towns or small cities in their size and financial influence, and they have enough purchasing power to help shape demand for products and services that produce fewer GHG emissions than the conventional alternatives. Whether this comes in the form of signing a long-term contract for the campus to be powered by wind power or to purchase electric and hybrid vehicles for the university's fleet, the cumulative impact of these decisions can make a real difference (Rappaport and Creighton 2007, 312). Collectively, U.S. colleges and universities spend over $360 billion annually and hold roughly the same amount in endowment investments. Including staff and administrators, the total population today on U.S. college and university campuses is around 20 million individuals. Add the hundreds of thousands of business suppliers and countless other commercial and nonprofit entities that interact with colleges and universities, and the economic clout of North American higher education is sizable (National Wildlife Foundation 2008).

Colleges and universities spend an estimated $2 billion per year on energy, including purchases of close to 1.1 billion kilowatt hours of green electricity, enough to power 87,000 average homes for a year, according to the U.S. Environmental Protection Agency (Monastersky 2007; Rappaport 2008). In fact, in 2005, the higher education sector was the largest purchaser of wind energy in the United States (Calhoun and Cortese 2005). For example, the University of Pennsylvania signed a ten–year contract to purchase 29 percent of its total energy needs from wind-generated resources. This infusion of capital to the wind generator subsequently enabled construction of a twelve-turbine, 20 megawatt wind farm in Pennsylvania (Sustainability Endowments Institute 2007). Commitments from colleges and universities across the continent help to establish consistent and reliable revenue streams for renewable energy investors and producers, helping to support investments in lower-carbon energy production and helping to make green power more affordable and competitive.

The collective buying power of colleges and universities thereby contributes to renewable energy investments and generation expansions. In this way, colleges and universities help lower the cost of new low-emissions technology across local, national, and international markets. Similar dynamics are visible also outside the energy sector as companies that serve campus communities respond to other types of climate–related demands. For example, the Sodexho Alliance, which handles food and cleaning contracts for 900 schools, in recent years has seen a 20 percent annual increase in the number of accounts demanding food that is local, organic, and grown sustainably. Increased offering of such food in many cases reduces the overall carbon footprint of the food served on campus (Green 2007). In addition, purchasing policies for more energy-efficient office and laboratory equipment and green building serves to expand markets for these products and services.

Modeling for Others

Illustratively, a recent report by the American College and University Presidents Climate Commitment program (ACUPCC 2007) was entitled "Climate Leadership for America: Progress and Opportunities in Addressing the Defining Issue of Our Time." As this title suggests, campus climate initiatives actively seek to experiment with policies to address climate change and to demonstrate their success to broader local, national, and international communities.

Another way that college and university climate actions are important beyond campus is through modeling of cost-effective initiatives that may be used by other sectors (Energy Action 2005). Many authors conclude that North American climate change policy is conditioned by the *perception* that economic costs of GHG reductions are exorbitant (Luterbacher and Weib 2001, 78). The Bush administration, for example, consistently opposed the Kyoto Protocol in part because it "would cause serious harm to the U.S. economy" (Roberts 2007, 4). There is, consequently, a need to demonstrate and communicate cost-effective climate actions using concrete facts and statistics (such as quantified data on GHG emissions reductions and cost savings), which may persuade others to take action. Demonstrating cost-effective achievement of GHG reduction targets, these actions contribute to bottom-up political and economic pressures on local and national policymakers as separate social groups form coalitions for change (Moser and Dilling 2007, 506; Selin and VanDeveer 2007).

Across North America, campus-based clean energy activities are leading to net savings. Elsewhere, savings from energy efficiency projects at the University of California, Santa Barbara have totaled over $36 million (National Wildlife Foundation 2008). One of the first state-funded green buildings in Maryland is going up at St. Mary's College (Harley 2007). A Harvard University cogeneration plant was the first power project to win approval under new Massachusetts restrictions on GHGs (Global Power Report 2007). At the national level, over 500 college presidents have signed the American College and University Presidents Climate Commitment, pledging to assess their GHG emissions and develop a strategy for reducing them, or buying offsets, with the goal of becoming carbon neutral (Barringer 2007). This commitment makes higher education the first sector of society to commit to becoming climate neutral (Brooks 2007).

One key variable accounting for policy change is the degree of domestic pressure (Speth and Haas 2006, 123). By sending a signal to municipal, state, provincial, federal, and international policymakers that more aggressive climate action is both possible and cost-effective, campus actions arguably contribute to making expanded public- and private-sector action more likely (Moser and Dilling 2007, 397; Speth and Haas 2006, 123). Whatever misconceptions there may be about cost barriers to launching different kinds of initiatives to reduce GHG emissions, the net-positive experiences of many schools are helping to prove those assumptions wrong (NWF 2008). Such demonstrations offer concrete evidence for climate policy advocates to use in political debates across public, private, and civil society sectors at multiple governance levels (Selin and VanDeveer 2007).

Setting and Achieving GHG Reduction Goals

Roughly 1,000 North American colleges and universities have enrollments of 5,000 or more students. With some of those institutions having weekday populations, including all faculty and staff, of 60,000 or more, many colleges and universities are the largest energy users in their regions (National Wildlife Foundation 2008). The total population on U.S. campuses is around 20 million individuals, accounting for 5 percent of all U.S. commercial-building-sector emissions (NWF 2008). The higher-education sector represents about 3 percent of the U.S. gross domestic product and 2 percent of the workforce. Thus, successful campus-based reduction actions are significant in a North American context. While there is no evidence suggesting that the entire higher-education sector has reduced its GHG emissions, a host of examples across the continent demonstrate that colleges and universities can continue to grow even as they reduce their GHG emissions.

Many colleges and universities are leading the way in setting and achieving GHG emissions reduction goals. Some have either met or exceeded the GHG emissions reduction target under the Kyoto Protocol that was rejected by the Bush administration (a 7 percent reduction below 1990 levels by 2012). For example, Yale University has cut its GHG emissions by 17 percent, with projects underway expected to cut another 17 percent by 2009. These cuts are achieved at a cost of less than 1 percent of the annual operating budget (Blum 2008). Such actions might be limited in the actual amount of GHG emissions that are reduced, but they send a symbolic message. These early actions are a way to get a first commitment upon which bigger commitments can be built. Small-scale actions and successes may slowly change the political climate, which in turn enables larger policy and political changes (Moser and Dilling 2007, 506).

Even if colleges and universities cannot solve the climate problem on their own, they are gathering important experience with policy options (McKinstry 2004). In other words, campus reduction targets and achievements are important not only as a beneficial end in themselves but also because they have the potential to create a ripple effect throughout North American societies. Colleges' and universities' role as laboratories for invention means that many campus innovations carry with them an academic imprimatur that may encourage other public- and private-sector actors to copy their work (Namikas 2008). Such expanded efforts are critical: as Yale President Richard Levin puts it, "We're showing it can be done, but our carbon savings are miniscule compared to what needs to happen. The answer has to come from governments" (Blum 2008).

Educating Citizens

Shifting the North American economy to renewable energy requires training a new generation of professionals. According to a 2006 study, a national economy committed to meeting the 20 percent renewable portfolio standard could create 164,000 new jobs in the renewable energy industry by 2020—significantly more jobs than the fossil fuel sector is expected to generate. Another study on the manufacturing implications of renewables asserts that 43,000 firms nationwide could create more than 850,000 new jobs if all components needed for the wind, solar, geothermal, and biomass industries were manufactured in the United States. This national development would represent nearly $160 billion of manufacturing investment. With 4,200 U.S. colleges and universities enrolling 17 million students, the higher-education sector is meeting these training needs with new programs in clean energy technology (NWF 2008).

Universities are also meeting the need for educating future voters on policy options for North American climate action. Through courses in environmental studies, political science, economics, and other areas, some analysts suggest that by teaching individuals about the likely ecological consequences of their actions (including their political activities and consumption patterns), such curricula may result in higher reductions of GHG emissions than those resulting from government-imposed tax schemes (Luterbacher and Weib 2001, 116). Importantly, universities also use informal education to teach climate change. This is critical to broader North American climate policy development, as students learn by observing what they see around them. Arguably, the more they see climate awareness in campus operations, the more likely they are to be engaged in such efforts in their lives, both during and after their college years. For example, Oberlin College uses a comprehensive system to monitor and display electricity consumption in dormitories, with the goal of providing real-time feedback that allows students to better conserve environmental resources.[6]

As campuses increasingly engage in climate actions such as onsite wind turbines and solar turbines, it is conceivable that they will become normal and expected parts of the landscape, and subsequently resistance to such efforts will wane over time (Selin and VanDeveer 2007). Over the long term, says Anthony Cortese of Second Nature, educational institutions will have their biggest impact on the climate through their students. "We have 100 percent of the educational footprint because we train all the future K–12 teachers and we train all the leaders in every sector of society," he says (Monastersky 2007).

Limits and Constraints

This chapter has focused attention on the climate action leaders among colleges and universities and on what impact such leadership may have on larger political and social institutions. However, it is important to not overstate climate action on campuses; relatively few colleges and universities have taken serious, concrete steps to move toward a cleaner energy future. Of the approximately 4,200 colleges and universities in the United States, only 200—less than 5 percent—purchase or produce clean energy, and most of these have implemented small-scale clean-energy projects. Furthermore, of the 504 signatories to the American University and College Presidents Climate Commitment, less than 20 percent have committed to purchasing or producing 15 percent of their energy from renewable resources within one year of signing the commitment.[7] There are several explanations for these limitations to campus action, including external obstacles, competing priorities, lack of incentives for energy efficiency, failure to complete GHG inventories, and lack of uniform approaches to collection of emissions data.

Colleges and universities also face political constraints. For example, a project by the University of Massachusetts to construct a cogeneration plant was delayed for more than twenty years because the local power utility, which did not want to lose the university as a customer, used its influence in the state legislature to oppose the plant (Skolfeld 2006). It is conceivable that similar scenarios play out across the United States and Canada, impeding campus climate actions and the influences such actions can have on public- and private-sector decision making. For example, by discouraging energy efficiency, these influences act against efficiency-gaining measures, such as California's decoupling of utility income from the amount of electricity sold, which encourage utility investment in energy efficiency; "California really represents what the rest of the country could do if it paid a bit more attention to energy efficiency," says Greg Kats, managing principal at Capital E, an energy and clean-technology advisory firm. "California is the best argument we have about how to very cost-effectively both reduce energy consumption and cut greenhouse gases. And they've made money doing it" (Mufson 2007).

Limitations on college and university climate action also include competing priorities. A majority of campus decisions are shaped by marketlike behavior in which short-term bottom-line economics concerns outweigh long-term climate-friendly decision making. This is driven in large part by the competition for student enrollments, which sometimes leads to ever-more lavish offerings to students, many of which are highly energy-consuming, including extravagant residence halls, student centers, and even water parks (Winter 2003). An additional obstacle is that only a minority of colleges and universities have inventoried their climate-altering gas emissions using a systematic approach (Rappaport and Creighton 2007, 23). Since data are not being gathered and reported consistently, quantitative assessment of trends is premature (Rappaport 2008). The development of a uniform emissions accounting system could bring about participation from more colleges and universities, as it would provide a baseline for assessing the amount of capital and investment needed to cut GHG emissions from different operations.

Campus climate actions are constrained by the frequent absence of federal, state, and provincial policies that support and reinforce campus greening efforts, including policies that promote significant improvements in energy efficiency across the economy, funding for research and development of technologies that reduce GHG emissions and create jobs, support for demonstration projects, and making capital available to less affluent institutions to improve energy efficiency (Rappaport 2008). This all adds up to concern about whether university administrations are both committed to and capable of making the long-term investments in staff, infrastructure, behavioral changes, and renewable energy required to achieve significant GHG reductions. Indeed, the doubts are widespread enough that many institutions with otherwise strong environmental credentials have not signed the climate neutrality pledge (Monastersky 2007).

Furthermore, there are worries that few colleges and universities will be able to reduce their emissions enough to achieve carbon neutrality in the short term. Most colleges may be able to fulfill near-term carbon neutrality pledges only by purchasing offsets, in other words, by paying someone else to reduce emissions on their behalf (Rappaport 2008). Amid all the carbon commitments, many college officials harbor skepticism about the current plans to zero-out emissions. Concerns include the fact that institutions will define carbon neutrality for themselves, deciding, for example, whether to consider emissions from student vehicles as part of their carbon footprint. Also, no institution has set a firm deadline for achieving total carbon neutrality (Deutsch 2007b). Some campus leaders are openly skeptical: "I think the idea of carbon neutrality is basically a myth," says David W. Oxtoby, president of Pomona College, who signed the commitment but worries that colleges cannot erase their climate impact without taking unreasonable steps (Monastersky 2007).

Concluding Remarks

In the absence of national leadership on climate change, many North American colleges and universities have made ambitious commitments to cut their GHG reductions and developed a host of related programs and initiatives. These activities may have resulted in rather limited GHG emission reductions to date, but they send an ever-stronger signal to municipal, state, provincial, and federal policymakers that more aggressive climate action is possible, cost-effective, and politically supported. These small-scale actions are critical for the spread of technological and policy innovation. They also contribute to bottom-up political and economic pressures that begin to take off when heretofore separate social organizations and groups form coalitions and thereby leverage their respective strengths for greater impact. Thus, small-scale actions and successes can slowly change the political climate, which in turn enables larger policy and political changes. Through their collective influence, colleges and universities contribute to these changes every day as multilevel climate governance is expanded all over North America.

Notes

1. Yale's 2007 greenhouse gas emissions are 17 percent below 2005 levels.

2. The CORE framework is used at the University of New Hampshire to incorporate sustainability principles into all facets of campus activity. Special thanks goes to Tom Kelly, director of the UNH Office of Sustainability, for providing information on this framework. For more information, see http://www.sustainableunh.unh.edu/coreframework.html.

3. For list of campuses, see http://www.energy.rochester.edu/us/list.htm.

4. Author's interview with Brett Pasinella, University of New Hampshire's Office of Sustainability, June 2008.

5. For more information, see the American College and University Presidents Climate Commitment at http://www.acupcc.org.

6. For more information, see http://www.oberlin.edu/dormenergy/dorme.htm.

7. For a listing of the ACUPCC signatories that have committed to 15 percent renewable energy purchases or production, see http://www.aashe.org.

References

Agrell, Siri. Welcome to Green U. *The Globe and Mail*, June 6.

American College and University Presidents Climate Commitment (ACUPCC). 2007. *Climate Leadership for America: Progress and Opportunities in Addressing the Defining Issue of Our Time*. 2007 Annual Report. http://www.presidentsclimatecommitment.org/reporting/documents/ACUPCC_AR2007.pdf. Accessed April 2, 2008.

Anderson, Leonard. 2007. More U.S. College Students Studying Clean Energy. Reuters, March 28.

Barringer, Felicity. 2007. For New Center, Harvard Agrees to Emissions Cut. *New York Times*, September 18.

Betsill, Michele M. 2005. Global Climate Change Policy. In *The Global Environment: Institutions, Law, and Policy*, edited by Regina S. Axelrod, David L. Downie, and Norman J. Vig. Washington, DC: CQ Press.

Blum, Andrew. 2008. Carbon Neutral U. *Metropolis Magazine*, February 20.

Brooks, Courtney. 2007. A Commitment to the Environment. *Boston Globe*, May 26.

Calhoun, Terry, and Anthony D. Cortese. 2005. *We Rise to Play a Greater Part: Students, Faculty, Staff, and Community Converge in Search of Leadership from the Top*. Ann Arbor, MI: Society for College and University Planning.

Church, Elizabeth. 2008. Green Studies in Big Demand at Ontario's Universities. *The Globe and Mail*, January 19.

Cooper, Russ. 2007. Greener Education. *Canadian Geographic* 127(5): 22.

Creighton, Sarah Hammond. 1998. *Greening the Ivory Tower: Improving the Environmental Track Record of Universities, Colleges, and Other Institutions*. Cambridge, MA: MIT Press.

Deutsch, Claudia H. 2007a. A Threat So Big, Academics Try Collaboration. *New York Times*, December 25.

Deutsch, Claudia H. 2007b. U.S. Colleges Are Making Conservation a Priority. *International Herald Tribune*, June 14.

Energy Action and Apollo Alliance. 2005. *New Energy for Campuses Report*. Energy Action and Apollo Alliance. http://www.campusactivism.org/server-new/uploads/new_energy_for_campuses_report.pdf. Accessed April 1, 2006.

Geiselman, Bruce. 2007. It's a Gas; N.H. University to Launch Major LFG Project. *Waste News*, August 20.

Global Power Report. 2007. Harvard Cogeneration Project Is First in Mass. to Come under New GHG Monitoring Program. *Global Power Report*, September 20.

Green, Heather. 2007. The Greening of America's Campuses. *Business Week*, April 9.

Hartley, Megan. 2007. Students Add Conservation to Curriculum. *Washington Post*, March 11.

Knight, Rebecca. 2007. Global Warming Has Become a Hot Topic on the Curriculum. *Financial Times*, January 29.

Levin, Richard. Leading by Example: From Sustainable Campuses to a Sustainable World. Climate Lecture Series. University of Copenhagen. January 21, 2008.

Mason, Gary. 2007. B.C.'s School of Greener Learning. *Globe and Mail*, August 25.

McCormick, John. 2005. The Role of Environmental NGOs in International Regimes. In *The Global Environment: Institutions, Law, and Policy*, edited by Regina S. Axelrod, David L. Downie, and Norman J. Vig. Washington, DC: CQ Press.

Monastersky, Richard. 2007. Colleges Strain to Reach Climate-Friendly Future. *Chronicle of Higher Education*, December 14.

Morris, John P. 2005. The Hidden Economics of Campus Sustainability. *Facilities Manager* (May/June).

Moser, Suzanne C., and Lisa Dilling. 2007. Toward the Social Tipping Point: Creating a Climate For Change. In *Creating a Climate for Change: Communicating Climate Change and Facilitating Social Change*, edited by Suzanne C. Moser and Lisa Dilling. Cambridge: Cambridge University Press.

Mufson, Steven. 2007. In Energy Conservation, Calif. Sees Light. *Washington Post*, February 7.

Namikas, Kiki. 2008. Boiling Point. *Earth Island Journal* 22(4): 59–61.

National Wildlife Foundation. 2008. *Higher Education in a Warming World: The Business Case for Climate Leadership on Campus*. Reston, VA: National Wildlife Foundation.

Newman, Julie, and Lisa Fernandez, eds. 2007. *Strategies for Institutionalizing Sustainability in Higher Education*. Report on the Northeast Campus Sustainability Consortium 3rd Annual Conference and International Symposium, November 2–4.

Nicholson, Tom. 2007. For Sustainability and Clean-Energy Majors, Global Warming Is Hot. *Engineering News-Record* 44(259): 15.

Putnam, Andrea, and Michael Phillips. 2006. *The Business Case for Renewable Energy: A Guide for Colleges and Universities*. Alexandria, Virginia: APPA. Alexandria, Virginia: National Association of College and University Business Officers. Ann Arbor, MI: Society for College and University Planning.

Rabe, Barry G. 2004. *Statehouse and Greenhouse: The Emerging Politics of American Climate Change Policy*. Washington, DC: Brookings Institution Press.

Rabe, Barry G. 2008. States on Steroids: The Intergovernmental Odyssey of American Climate Policy. *Review of Policy Research* 25(2): 105–128.

Rappaport, Ann. 2008. Campus Greening beyond the Headlines. *Environment* 50(1): 33–38.

Rappaport, Ann, and Sarah Hammond Creighton. 2007. *Degrees That Matter: Climate Change and the University*. Cambridge, MA: MIT Press.

Raustiala, Kal. 2001. Nonstate Actors in the Global Climate Regime. In *International Relations and Global Climate Change*, edited by Urs Luterbacher and Detlef F. Sprinz. Cambridge, MA: MIT Press.

Rimer, Sara. 2008. How Green Is the College? Time the Showers. *New York Times*, May 26.

Roberts, J. Timmons., and Bredley C. Parks. 2007. *A Climate of Injustice: Global Inequality, North-South Politics, and Climate Policy*. Cambridge, MA: MIT Press.

Sa, Creso M., Roger L. Geiger, and Paul M. Hallacher. 2008. Universities and State Policy Formation: Rationalizing a Nanotechnology Strategy in Pennsylvania. *Review of Policy Research* 25(1): 3–19.

Scherer, Ron. 2006. "Sustainability" Gains Status on U.S. Campuses. *Christian Science Monitor*, December 19.

Schneider, Keith. 2008. Majoring in Renewable Energy. *New York Times*, March 26.

Selin, Henrik, and Stacy D. VanDeveer. 2007. Political Science and Prediction: What's Next for U.S. Climate Change Policy? *Review of Policy Research* 24(1): 1–27.

Skolfield, Karen. 2006. Beyond the Bluster. *UMass Magazine Online*, Spring 2006. http://umassmag.com/Spring_2006/Beyond_the_Bluster_1070.html. Accessed July 7, 2006.

Speth, James Gustav, and Peter M. Haas 2006. *Global Environmental Governance*. Washington, DC: Island Press.

Sprinz, Detlef, and Martin Weib. 2001. Domestic Politics and Global Climate Policy. In *International Relations and Global Climate Change*, edited by Urs Luterbacher and Detlef F. Sprinz. Cambridge, MA: MIT Press.

Steptoe, Sonja. 2007. Getting Schools to Think and Act Green. *Time Magazine*, August 10.

Sustainability Endowments Institute. 2007. *College Sustainability Report Card 2007*. Sustainability Endowments Institute. http://www.greenreportcard.org/. Accessed April 1, 2008.

University of California, Santa Cruz. 2006. EPA Ranks UC Santa Cruz the Sixth Largest 'Green Power' Purchaser among Campuses. Press Release, January 30.

U.S. Department of Energy. 2006. *Colleges and Universities*. Washington, DC: U.S. Department of Energy.

Winter, Greg. 2003. Jacuzzi U.? A Battle of Perks to Lure Students. *New York Times*, October 5.

Xiong, Chao. 2008. Sustainability Is Everyone's Job, But Some Get Paid for It. *Star Tribune*, January 21.

Young, Abby. 2007. Forming Networks, Enabling Leaders, Financing Action: The Cities for Climate Protection Campaign. In *Creating a Climate for Change: Communicating Climate Change and Facilitating Social Change*, edited by Suzanne C. Moser and Lisa Dilling. Cambridge: Cambridge University Press.

14

Communicating Climate Change and Motivating Civic Action: Renewing, Activating, and Building Democracies

Susanne C. Moser

Introduction

Governments are critical for bringing about the transition to a more sustainable human interaction with the environment as they set priorities and policies and may model new behavior. Yet, civil society is also indispensable in bringing about change. It is no small challenge to communicate effectively in order to engage civil society in this task. Many argue that the federal governments of North America are failing to provide needed leadership on climate change. In the absence of committed top-level leadership, bottom-up pressure is building to force policy changes at the federal level. This volume provides convincing evidence of growing action on climate change at various levels and in different sectors of North America (Farrell and Hanemann, this volume; Gore and Robinson, this volume; Rabe, this volume; Selin and VanDeveer, this volume). At the same time, a social movement for climate protection is beginning to emerge (Moser 2007b).

Broad sections of U.S. and Canadian societies, however, are not yet fully on board regarding the need for comprehensive climate change action and meaningful behavioral changes (Rabe, this volume; Stoett, this volume). In Mexico, civic mobilization around climate change has been barely evident at all in the early years of the twenty-first century (Pulver, this volume). This chapter examines civic mobilization around climate change primarily in the United States, and to a lesser extent in Canada and Mexico, in relation to climate governance efforts in public and private sectors from local to international levels. It focuses on how greater civic engagement on climate change can be fostered. Civil society can play at least two critical roles in climate change governance: (1) it can mobilize to push for policy changes at any level of government, and (2) it can enact behavioral changes consistent with needed mitigation and adaptation strategies.

If North American societies are to engage in these two types of civic responsibilities, however, communicators of climate change must go beyond merely conveying climate change knowledge and more effectively encourage and enable individuals to take part in the societal transformation necessary to address climate change successfully (Moser and Dilling 2004, 2007a). Climate communicators have not yet fully taken on this challenge, but climate change presents an opportunity to renew U.S. society and democracy with greater civic engagement, build enduring democratic institutions in Mexico, and activate civic engagement more fully in Canada. The next section explores and compares public opinions about climate change across North America, demonstrating that deeper civic engagement has not yet been achieved in any of the three countries.

Public Opinions about Global Warming in the United States, Canada, and Mexico

Expressing concern about climate change and general support for government and industry actions, while indicative and encouraging, is very different from actively engaging in civic action. Understanding public opinions and attitudes is important, however, to increase civic engagement. In the United States, according to a 2007 national poll, 72 percent of Americans are convinced that global warming is not just a problem of the future, but happening now, and a majority (57 percent) think it is largely caused by human activities (Leiserowitz 2007). Other polls have found that Americans feel more confident than ever in their understanding of the basics of global warming, yet only 22 percent say they understand the issue "very well." Questions testing their knowledge on the science or politics reveal that individual knowledge is superficial at best (Nisbet and Myers 2007).

After several years of expanding media coverage, concern about climate change among Americans was slightly lower in 2008 than it was in 2007: 37 percent in 2008 said they personally worried a great deal about global warming, compared with 41 percent in 2007. Public concern in both these years was also only a few percentage points higher than in 1989, with climate change ranking ninth among twelve local and global environmental problems (Carroll 2007; Jones 2008). Importantly, American opinions differ markedly along party lines. In 2007, Democrats (75 percent) exhibited greater personal worry about global warming than Independents (59 percent) and Republicans (34 percent). Democrats are also more convinced than Republicans that climate change will affect them and their children, and they are more likely than Republicans to view the scientific debate over human-induced climate change as settled (ABC News/TIME/Stanford University 2006; Brewer 2005a, 2005b Carroll 2007; Saad 2006, 2007).

Nonetheless, between two-thirds and four-fifths of Americans support a wide range of policy measures to reduce greenhouse gas emissions (PIPA 2005). The American public specifically supports measures that mandate emission reductions by industry and automakers, and it generally favors incentives and rebates over increased taxes. Americans, however, are resistant to any regulations that would directly affect their personal choices by increasing costs, restricting options, or demanding personal behavior changes (GfK Roper Public Affairs and Media 2007). As repeated U.S. election campaigns show, global warming and related energy-policy matters have not by any significant measure entered electoral debate or featured strongly in American voting decisions. A 2007 survey indicated that differences on energy and climate would not deeply affect Americans' choices among candidates in the 2008 presidential election (Leiserowitz 2007). Thus, there is a gap between concern about global warming and personal behavior.

A 2007 multinational poll found that equal percentages of Americans and Canadians (89 percent in both countries) had heard either "a great deal" or "at least some" about global warming, while only 73 percent of Mexicans shared this same level of familiarity, with nearly another quarter of Mexican respondents having heard "not so much" (BBC World Service 2007). The poll also found that while a majority of Americans (71 percent) thought human activity constitutes a "significant cause of global warming," more Canadians (77 percent) and even more Mexicans (94 percent) believed so. In 2007, 77 percent of Canadians were convinced that global warming was real, and 70 percent believed that the science behind human-induced climate change was "true" (Angus Reid Strategies 2007b). GlobeScan polls in 2000, 2003, and 2006 (asking similar questions in each survey year) found increasing numbers of Canadians and Americans viewing climate change as a "very serious problem," and overall higher, but slightly declining percentages of Mexicans agreeing (GlobeScan 2006; Leiserowitz 2008). Consistent with these judgments of seriousness, Mexicans (83 percent) are more likely than Canadians (72 percent) and Americans (59 percent) to state that it is "necessary to take major action very soon" to address global warming (BBC World Service 2007).

Climate change has risen in importance to Canadians over time. Canadians broadly support fulfilling Canada's Kyoto commitments (Stoett, this volume). While they favor incentive approaches to reducing emissions such as improving energy efficiency and conservation, they also widely reject policies that would restrict car travel (Angus Reid Strategies 2007a). However, Canadians vary in the level of action they take with respect to their global warming beliefs: surveyor-identified *skeptics* (23 percent) don't believe in global warming and are completely opposed to action; *agnostics* (16 percent), not having made up their mind on global warming one way or the other, tend to not act consciously in climate-friendly ways, nor do *converts* (22 percent), but they feel guilty about their lack of environmentally conscious behavior. *Believers* (22 percent) are far more environmentally conscious and behave accordingly, and *activists* (18 percent) act most environmentally conscious and fervently try to convert others to do the same (Angus Reid Strategies 2007c). This survey suggests that about six out of ten Canadians either doubt the need for action and/or do not act on their beliefs for action.

Similarly detailed polling data on Mexican attitudes and beliefs are not available (and cross-national comparisons of survey data from separate polls are inherently difficult due to differences in methodologies and the questions asked). The comparatively low level of economic development in Mexico, however, gives Mexicans a far smaller per capita carbon footprint than their North American neighbors. Nevertheless, the greater vulnerability of many Mexicans to developing and possible impacts of climate change compared to Americans and Canadians could be a reason for them to actively engage on the issue. However, as Pulver (this volume) describes, the Mexican public has yet to do so.

The next section lays out why climate change communication, if it is to have a constructive role in democratic governance, must accomplish more than just informing and alerting a citizenry to the reality of climate change (which it has largely accomplished in the United States and Canada) and explaining the science, its causes, and its implications (which it has done, achieving uneven results). Effective communication as a tool and force of democracy would be considered "effective" if it brought about civic engagement, both in the sense of political activism and behavioral change (Moser 2007a).

Communication and Civic Action

Communication plays an essential role in mobilizing and sustaining civic action. It expresses and supports the fundamental work of civic engagement in a democracy. Interestingly, *communication* and *community* share the same etymological root. "To communicate" derives from a Latin word that means "to impart," "to share," and "to make common"; in turn, the word "common" derives from the two roots *com* "together" and *munia* "public duties" (Harper 2001). This etymology links communication closely to the ideal of civic action. Practically, communication and community can also mutually foster each other, whereas unsuccessful communication can alienate individuals from acting in the public sphere and hence completely fail to be an instrument of citizenship. Thus, creatively designed and skillfully executed communication can be declared effective if it serves as a tool for building and sustaining the community that acts on a *res publica* (a matter of public interest) such as climate change, and in helping individuals create, and feel part of, a civic community.

The civic movement literature distinguishes "being a citizen," that is, in a narrow sense merely being an individual member of a city, country, or otherwise defined community, from "participating in civic action." The former may be quite divorced from public and political life, relegated to being a self-interested individual acting on his or her own needs and wants, consuming goods and services, and otherwise ensuring that these personal desires are met (through complaints, advocacy, volunteering, or voting). In this capacity, people can help reduce their energy use and reduce the use of those technologies that produce large amounts of greenhouse gas emissions. The role of communication in this case would be to foster individual behavioral changes. Individuals engaging in civic action, by contrast, use their actions to express commitment to, and awareness and support of, a larger common goal.

Civic action is public action by members of a community in response to a public matter of great concern. "'Public work' is work done by ordinary people that builds and sustains basic public goods and resources—what used to be called 'our commonwealth'" (Boyte and Kari 1996). In that sense, public work contributes to sustaining the moral fabric that is at the heart of "community." The role of communication in such "public work" focuses on how the issue of concern is framed; how its causes, implications, and solutions are explained; how dialogue occurs; and how it draws on and feeds social capital (Daniel, Schwier, and McCalla 2003). As a means to create common cause and understanding, communication makes connections across issues and thus helps build a public that is engaged on climate change.

Common Obstacles to Civic Engagement on Climate Change

For communication to achieve these objectives is no small order. It is difficult enough to communicate the unwieldy problem of global climate change to various lay publics; it is extraordinarily difficult to overcome the lethargy, habits of thought and action, and institutional arrangements that underlie current emissions-generating behaviors. In the United States, individuals frequently must overcome their widespread disenfranchisement from the political process. In Canada, where civic engagement may be considered higher than in the United States, if measured by voter turnout in elections, individuals may not see the need for other forms of civic engagement beyond voting, leaving climate change solutions to technical experts (Henderson 2008). In Mexico, the challenge of political engagement may be more fundamental: civic skills building, spreading of democratic norms, overcoming poverty, and fostering basic education.

In short, there are internal psychological and cognitive processes that may prevent an individual from engaging on this issue, as well as social, political, economic, and other structural external barriers to such engagement. These barriers are discussed further next.

Psychological-Cognitive Barriers

Processing of climate change information may undermine the motivation of individuals to engage on the issue if they have emotional responses to it that are demotivating. These responses could include: a sense of being powerless and overwhelmed; denial; numbing; feeling exempt from the threat; blame; wishful thinking or rationalization that the problem will be resolved by experts; displacement of attention on other problems; apathy; fatalism; or other forms of "capitulatory imagination" (Immerwahr 1999; Loeb 2004). These types of cognitive and emotional responses are particularly common in response to issues which are scary, ill-understood, difficult to control, overwhelming, and in which people are complicit, such as global climate change (Moser 2007c).

Common cognitive barriers to more active personal engagement on climate change include: not understanding the issue, the causes, the relevance of climate change impacts to one's life, or the possible solutions; misunderstanding, confusion, or disagreement with the actions, policies, or strategies proposed by advocates or policymakers to address climate change; an unattractive future vision painted in people's imagination (often one of doom); and lack of resonance with the framing and language in which climate change is being discussed (synthesized in Moser and Dilling 2007b).

Social Barriers

Individuals are embedded in social networks, hold social identities, engage in social interactions, and adhere in varying degrees to social norms that circumscribe appropriate or inappropriate behavior. If engaging in civic action on climate change portrays a particular social identity, produces a social stigma, or reflects social norms that are in conflict with people's desired identity and accepted norms, they are unlikely to engage in this particular type of civic action. If civic engagement takes "too much" time or resources, and is inconvenient or too demanding given other daily concerns and competing obligations, even people sympathetic to the cause may not get involved. Finally, individuals—who are deeply embedded in society through socialization, institutions, modern-day habitual activities, or the provision of basic needs—may not question or see alternatives to common emission-generating behaviors, and may resist calls for alternative behaviors (Tribbia 2007).

Political Barriers

Individuals may be generally disinterested in political matters, prefer to leave political activism to others, be genuinely unfamiliar with forms of civic engagement, and/or feel deeply disenfranchised from the political process. Some may hold a belief that government or industry or some other "other" will find a technological fix or policy solution (political transference). Others may not believe that existing institutions are failing in their responsibilities, thus seeing no need for activism, especially if it is inconvenient. A related response is blaming others for the problem and/or projecting responsibility for remedial action onto them. Still others, wedded to tradition and habit, may refuse to do anything different or new. Scientific uncertainty about the causes, urgency, or solutions of a problem can serve as a convenient rationale to hold on to the status quo (e.g., Klandermans and Oegema 1987; Leighley 1995; Macnaghten and Jacobs 1997).

Other Barriers

Even if the internal psycho-cognitive and external social and political barriers could be overcome, a person may still face structural barriers. These include the lack of a convenient or economically feasible alternative technology, existing laws and regulations, lack of public infrastructure, and existing political institutions and electoral processes that are heavily controlled by vested interests. Information channels and communication infrastructure may also hinder engagement, even in the information age. Heavy filters against the overabundance of information, declining newspaper readership, continued reliance on television as the main news source (especially for Americans), and increasing reliance on, and high selectivity among, internet news sources can limit depth of coverage and understanding of any issue (The Pew Research Center for the People and the Press 2004).

Moreover, the political economy of the media industry, with its increasing concentration of media ownership and consequent narrowing of the range of diversity of voices heard in mass communication channels, frequently does not offer individuals the breadth of views that may allow them to develop a well-informed opinion. More typically, people exist in rather homophilous environments: because of individuals' similarity in sociodemographic backgrounds, they tend to have access to similar kinds of information, issue framings, and so on, and therefore are more likely to communicate with each other but are rather isolated from other equally homophilous sections of society (Mcpherson, Smith-Lovin, and Cook 2001; Rogers 2003).

Communication Strategies to Mobilize Civic Action on Climate Change

Understanding major barriers to civic engagement can give communication efforts a clearer focus and infuse them with a longer "shelf life" than the average ten-second sound bite or ten-week outreach campaign. This section focuses on specific communication strategies that build on a recognition of these barriers to engagement. In particular, it highlights elements of communication strategies that focus both on elevating the motivation to engage and lowering the barriers to doing so.

Audience Choice

Best practice in communication begins with consciously and strategically selecting an audience and understanding that audience's mental models and level of understanding of climate change as well as its interests, values, and concerns. This deeper understanding helps communicators make connections to issues already of concern to a given audience and to frame climate change in a language that resonates.

The strategic selection of audiences reflects the fact that not all audiences have equal influence over particular changes that must be made to reduce emissions. Even individuals inclined to use non-CO_2-emitting energy sources, or more energy-efficient transportation or appliances, will be hindered in acting on their inclination if their preferred choices are not available. Rather than engaging in a massive social marketing campaign to change travel or purchasing behavior, the more strategic campaign would select those with direct influence over regional transportation planning, vehicle emission standards, or energy efficiency standards of appliances, while only secondarily fostering market demand for these alternatives. This simplistic example serves to illustrate the basic rule: those targeted for a particular type of civic engagement must be in the position to actually effect the desired change.

Framing Climate Change

Naming and framing an issue is one of the additional fundamental challenges for communicators, especially for an "invisible" global problem such as climate change. Frames are mental structures that influence perspectives. Expressed and suggested through language, images, gestures, and the messengers who use them, such frames shape our goals and plans as well as the way we act and what we think are good or bad outcomes of our actions (Lakoff 2004). Audience-specific communication thus means making global climate change "local" in more than the geographic sense. While people generally relate better to the things they can directly feel, experience, or see, making global warming "local" means reasonably connecting it with anything that is currently or persistently salient to them. Every chosen frame also entails a certain set of solutions. For example, climate change framed as an energy problem (a common frame used in Canada) primarily engages those providing energy, those setting energy policy, and those advocating for various energy choices.

Climate change as a moral problem, by contrast, brings in a different set of players, such as spokespeople claiming high moral standing and people of faith. While their practical solutions may still focus on energy questions, their motivation and possibly their deeper "solutions" might include moral commitment, renewal, or reorientation. This is an emerging framing among faith communities, such as factions of the Evangelical church in the United States, including the Interfaith Climate

Change Network, Web of Creation, and the Eco-Justice Programs of the National Council of Churches of Christ. A moral framing could also be a possibility in Mexico if the liberation theology wing of the Catholic Church were to take up climate change as a cause for mobilization (Norget 1997).

Such a framing would not necessarily resonate with business leaders, whose primary concern is with the bottom line, investments, markets, and competitiveness in domestic and international economies. Local and state governments, students, or low-income communities in less-developed Mexico or in the poorer U.S. and Canadian cities would have still different concerns, understandings, and values that effective communication must tap into (Agyeman, Doppelt, Lynn, et al. 2007). Different audiences need to be addressed in audience-specific ways that match frame, message content, and language with their specific information needs, preexisting knowledge, and concerns. Frames also are critical for sustaining civic engagement through challenging periods; crossing social divides, thereby aiding the forming of coalitions; and assisting in the deeper societal transformation ultimately needed to address this immense challenge (Moser 2007a).

The task of framing and reframing, as an issue evolves in public consciousness and the political process, involves identifying those frame(s) that promise to be most powerful to a particular group of social actors. Frames are strategic tools of social movements and countermovements precisely because of their power to mobilize some actors while disengaging or disregarding others. That is, they can be employed to either unite factions or split and create opposition between them (Goffman 1974; Lakoff 2004). The history of public debate of climate change in the United States (but more recently also in Europe and Australia), in which climate contrarians have deeply influenced the framing and discussion of the issue, attests to the power of framing and the power of access to media channels that promote these frames (McCright and Dunlap 2001, 2003).

In the early years of the twenty-first century, communication of climate change in the United States is witnessing an important transition, where the issue is no longer just framed as an "environmental" issue but also as a social, economic, technological, educational, security, and moral issue. For example, the Apollo Alliance, invoking the compelling national focus on putting the first man on the moon, envisions a future of clean energy; technological, economic, and moral leadership; and secure employment. Leaders in the environmental justice community who have taken up the climate issue tend to focus on fairness, health, safety, and well-being. Al Gore's film, *An Inconvenient Truth*, prominently framed global warming as a moral issue, and a behavior change campaign launched in early 2008 by his Alliance for Climate Protection evokes an optimistic, "can do" attitude, emphasizing the power in numbers and business opportunities in acting on climate change.

These alternative frames help individuals, organizations, and communities already active on other issues see how their work might be impacted by climate change. It also helps people not yet concerned with (or skeptical of) global warming find common cause and ground. In short, not every conversation must begin or end with climate. Instead, the climate change conversation can be entered through a myriad of doors.

Messenger Choice

To reach audiences heretofore unengaged, it is also important to select the messenger(s) carefully. In the United States, scientists, environmental NGOs, contrarians, and the media have dominated climate change communication in the past, resulting in high levels of problem awareness, but also in a perception of global warming as a largely technical/scientific, (still) uncertain, and controversial environmental issue. While one may argue that scientists must remain important communicators of climate change for the foreseeable future, there is an equally legitimate argument for bringing a greater diversity of people into the needed discourse, thereby reaching into sections of civic society yet to be engaged, crossing important social divides. To reach these goals, the choice of messenger is a critical strategic decision. Effective communication matches messengers with the message and the audience.

In the first match, it is critical to understand messengers as part of the framing: Former Director of the Central Intelligence Agency James Woolsey talking about the need to reduce oil consumption as a matter of national security (while also benefiting the climate) is an example of matching messenger with message content and frame (Warriors and Heroes 2005). Messengers also need to be credible to the audience being addressed. The CEOs of companies involved in the Pew Center for Global Climate Change's Business Environmental Leadership Council are more persuasive spokespeople to other business leaders because they are like them and understand the pressures and issues CEOs have to deal with on a daily basis. Such "people like us" (or PLUs) are important for an audience's personal comfort, identity, and group-internal norms and cohesion. Often, PLUs (especially if the audience knows and trusts them personally) have greater credibility and legitimacy than someone who does not know an audience's circumstances as well.

Beyond Information and Emotional Appeals to Create Urgency

To overcome the psychological and cognitive barriers to engagement, communicators must be critically aware of the role of information and emotions in behavior change. While a minimum amount of information is necessary to understand a problem and its causes, implications, and solutions, information and understanding by itself typically do not suffice to motivate behavioral change or civic engagement (Moser and Dilling 2007b; O'Connor, Bord and Fisher 1999). In some instances, simply learning more about an issue can lead individuals to lower their concern or sense of responsibility for it and seduce them to believe that they have actually "done something" (Kellstedt, Zahran and Vedlitz 2008; Rabkin and Gershon 2007). Thus, even the most well-intended information dissemination effort can become a dead end. Efforts such as the speaker trainings through Al Gore's Climate Project, or the 2008 nationwide educational effort Focus the Nation, are hugely ambitious, but these efforts are primarily focused on helping people to more effectively disseminate information and to educate a broader segment of the American population. Efforts such as these, if they are to contribute to a sea change in American thinking and action, must not end with these first-order goals but serve as springboards for deeper behavioral and political engagement.

Similarly, trying to get people to "care more" about an issue through appeals to fear or guilt can backfire and produce the opposite results (that is, denial, numbing, and disengagement) unless a series of conditions are met that actually enable people to translate their concern and fear into appropriate actions that reduce the danger (Moser 2007a). A communication strategy that does not very quickly tell people that there are feasible solutions with which to begin to address the problem, and what specific and appropriate actions individuals can take to help, is more likely to hinder than help the outreach and engagement effort. Moreover, because people feel manipulated and numbed by exposure to overtly guilt- or fear-evoking messages, emotional appeals are frequently not enough to break through disinterest, apathy, and information filters. Surprise and novelty are more promising.

Thus, rather than inundate audiences with more information or scary images of a gloom-and-doom future, it is critical now for communicators to constructively engage and support individuals and communities by creating a sense of feasibility, collectivity, and urgency arising from fact, experience, common sense, and a moral sense of responsibility. Such messages—several of which are illustrated in this volume's chapters—are only beginning to emerge in 2007–2008 and tend to include the following elements:

- Global warming is not a future problem but a present challenge (illustrated through already observed climatic changes and impacts on regional and local levels).
- A concerted collective effort is needed to address global warming, and many people, communities, and businesses are already involved.
- Any delay now makes later solutions more difficult and expensive (for example, illustrated with ecosystems that cannot adapt, or social systems close to a threshold beyond which they may not be able to adapt).

• Examples are available of people and communities who have taken first steps and actually saved energy and money, improved their quality of life, or enhanced their business operations or local economies (e.g., through less traffic congestion, cleaner air).
• We already have models (and metaphors) for acting responsibly and reasonably in our long-term interest without sacrificing terribly in the present (saving for retirement or college, insurance, etc.).

Scientific Confidence, Practical Solutions, and Hope

Looking over the past twenty years of research, what is remarkable is not how much remains uncertain, but how strong the scientific consensus on climate change has actually grown. At the same time, a public impression remains—fed by climate contrarians and common "balancing" media practices—that there still is scientific controversy over the basic notion of human-caused climate change (ABC News/TIME/Stanford University 2006; Leiserowitz 2007). Scientists themselves share in the responsibility for this situation, partly because of their common emphasis on remaining uncertainties, and partly because of their point-by-point engagement with climate contrarians. There is good reason to do so—misinformation should never be left standing unchallenged, and opportunities to educate the public should not be missed. But this pattern has left the proenvironmental and scientific side on the defensive, as it is far more powerful to dictate the frame than to respond to someone else's (Lakoff 2004).

To the extent communicators continue to focus on persuading the public of climate science, three tasks stand out to strengthen public resolve. First, scientists and educators must continue to convey the state of the science and how the confidence in scientific understanding of climate change has grown over time. Second, they must never overstate the scientific confidence with which aspects of climate change are known. But to retain credibility while conveying confidence, communicators should lead with what is most certain, and discuss remaining uncertainties in light of what is well understood.

Typically, people respond constructively to uncertainty (because they live with uncertainty all the time!) when they have some bearings that help them navigate unknown territory. In fact, it is an unsubstantiated claim that people need to have certainty in important matters before they can act (such as in decisions to go to war, invest in the stock market, or act on medical diagnoses). Finally, communicators should provide context for the evolving scientific understanding of climate change, that is, that it is the nature of science to always push back the frontiers of the unknown, and in the process, to stumble upon findings that require revisions of what was previously thought to be known.

It is, however, at least as important to communicate clearly established facts, the risks of not acting, and which solutions are already available (while even bigger answers and solutions are being developed). As the polls cited above suggest, most Americans, Canadians, and Mexicans are already convinced that climate change is real, serious, and already underway, even if this belief is not very solid or anchored in deep scientific understanding of the issue (ABC News/TIME/Stanford University Press 2006). Once people are engaged and realize the magnitude of challenge that climate change presents, however, they instinctively want to know what can be done, and what *they* can do. People want practical solutions. Those inclined to engage in civic action may be particularly predisposed to taking or supporting personal actions, but also to supporting larger political efforts.

The polls also suggest that most people do not know which solutions are most useful, available, feasible, or which are to be prioritized, and many cannot see their own role in tackling the problem. Moreover, polls show that distant policy solutions are preferred over personally enacted or felt ones. Thus, the communication challenge is to answer the question of what any one person can do, but also how such individual actions are part of a larger collective effort. Importantly, communication must counter the sense of futility and powerlessness individuals experience vis-à-vis this global problem through practical solutions, help and support from others, encouragement, and empowerment (DeYoung 2000; Gärling et al. 2003; Kaplan 2000).

At the same time, communicators must not mislead their audiences: larger policy and structural changes are also needed. An appeal to the deeply held value of "everyone doing their part" is maybe the most direct effort to counter the temptation of free-riding in this collective-action challenge. Tapping into people's desires for a better future, their social identities and aspirations, and cultural values that promote individual and collective action and engagement for the greater good (e.g., ingenuity, responsibility, stewardship, being a good team player, and leadership) can all increase people's motivation besides the more instrumental reasons (such as personal economic gain, competitiveness, legal compliance, and so on).

Finally, to counter overly pessimistic portrayals of the future and the pervasive sense of futility, individuals need a sense of real hope—not an overly optimistic, false promise of a future that is most likely unattainable, but constructive encouragement to work toward a future worth fighting for. No assumption is made here that any collection of individuals would want the same future or that they would be inspired by the same thing. Frequently, however, examples of what others are already doing successfully and constructive communal engagement in an action can generate hope for a better future—one in which individuals are part of active, engaged communities working toward livable, enjoyable, and fulfilling lives.

New Communication Forums

Mass communication channels bear the clear advantage of reaching large numbers of people, and fast. Getting media coverage is for many who communicate on climate change an unsurpassed measure of success, garnering visibility, attention, and the ability to reach policymakers, who claim to pay close attention to the news. In the increasingly fast-paced political economy of today's media landscape, where science reporting is declining, reporting staff is cut, and consequently the extent and depth of climate news coverage is threatened, it can be a jackpot experience to get a thirty-second clip on the evening news or a brief quote in the morning paper (Project for Excellence in Journalism 2006; Readership Institute 2002). Communicators cannot afford to abandon such communication channels and opportunities.

"Retail" communication does not benefit from the economy of scale of mass communication channels, nor does it have the same level of visibility. Yet, while mass media can help raise awareness and set social and political agendas (and thus connect to activities in other governance networks), they are never as persuasive and engaging as one-on-one conversations. Communication in smaller groups, through existing networks and forums, and where feasible in new groupings, will be a critical element to foster greater civic engagement. Such smaller forums offer the opportunity for communicators, and in fact all involved in the dialogue, to help individuals stay engaged on an easily overwhelming task, sort through complex issues, understand difficult trade-offs, and change ingrained habitual thoughts and behaviors. Thus, communicators would be well advised to identify such smaller forums to fully engage as many sources of social support as possible.

There is good reason why the most successful behavior change programs (e.g., Alcoholics Anonymous, Weight Watchers and various social marketing campaigns) are group-based. Typically, interpersonal and small-group dialogue can address people's needs much better than mass communication received in the privacy (and isolation) of one's living room. Neighborhood-based eco-teams, green-living projects on campuses, science cafés, and church-based discussion and support groups illustrate these insights, and many such examples exist already across North America. In such small settings, the power of social norms, accountability, identity, and personal ties is brought to bear on the barriers and resistance to change. They also allow individuals to be acknowledged and appreciated for their efforts, to unfurl the influence of role models, and to provide very immediate positive feedback on and social support for one's actions.

A Compelling Positive Vision

Finally, most news about climate change in the media, from scientists and environmental advocates, involves projections of frightening futures, possible doom for treasured environments and species, and mental images of disaster and havoc. Frequently used phrases like "climate chaos" or "climate crisis" are suggestive of this tendency. Scenarios of our global climate future are indeed very difficult to face, and consequently, many would rather not confront them. While empirical studies of U.S. or other audiences' emotional responses to the frequently scary images evoked in the news are scarce, anecdotal evidence suggests that many Americans actively avoid thinking about these possible climate futures because they are too frightening. An added challenge is that citizens alive today are unlikely to see greenhouse gas concentrations in the atmosphere return to preindustrial levels, or even to 2005 levels, even with a concerted global mitigation effort. This and the next generation may well become witness to a deteriorating climate for many regions of the world.

Communicators face the challenging task of giving a realistic assessment of the challenges ahead while trying to avoid doom-and-gloom or false hopes. The time lags built into our social and climate systems require that communicators think hard about what success would look like, and how to sustain civic engagement when positive feedback is not forthcoming from an unforgiving atmosphere. Defining a positive vision of a worthwhile future must therefore become a key focus of communication, outreach, and civic engagement efforts in coming years, including defining easily identifiable measures of progress (Olson 1995). Communicators must convey these indicators of forward achievement just as much—and maybe even more so—than what is wrong or not yet happening. Thus, pointing to positive examples and on-the-ground successes in other communities, states, and sectors (e.g., the stories emerging from many of the chapters in this book) will serve as the milestones that people need to hear. While it is unrealistic to expect that citizens will stay focused on climate (or any other issue) through the ups and downs of issue attention cycles, a vision of a compelling positive future will be essential as a compass through challenging times (Downs 1972).

Conclusion

In closing, I return from the challenge of linking communication with civic engagement to the opportunity embedded in climate change. If climate change communication can be improved to foster the kind of civic engagement that climate change demands, this tremendous challenge may well serve a much-needed democratic renewal in the United States, and perhaps more active forms of civic engagement in Canada, while helping to build a more democratic society in Mexico.

Burgeoning levels of activity at lower levels of government and in civil society have characterized America's response to climate change in the last years of the twentieth and the first few years of the twenty-first century. Local and state governments, pioneering businesses, religious communities, campuses across the country, traditional environmental and social advocacy groups, and a range of newly created groups have emerged as "grassroots leaders" on climate change. Even if and when they succeed in building sufficient political pressure on federal leaders to force nationwide policy changes, their role in societal response to climate change is not complete.

What the already-existing civic engagement illustrates is that countless leverage points exist to initiate and realize social change (from the bottom-up, top-down, and across sectors). Smaller changes plow the ground for bigger ones while spreading an important symbolic message to those who are not yet engaged. It is the typical pattern of pioneers and early adopters to create the conditions for a majority of actors eventually to adopt some innovative practice or technology (Rogers 2003). Given the long-term nature of climate change, civic engagement as a reflection of a community's or society's social capital will be essential in dealing with the impacts of climate change and addressing not just mitigation but eventually also adaptation needs.

Effective communication is an essential tool in mobilizing, linking, and uniting people for civic engagement, which is essential for the governance of climate change at all levels of society. The tasks of attaining deeper understanding of climate change, persuading people of its urgency, constructively and respectfully debating the value choices that underlie societal responses, envisioning a positive future, and practically supporting individuals and groups in actually changing behavior and policies, point to an important shift needed in future communication efforts. Mass media and the internet have been and will remain critical in creating and maintaining networks of information distribution, and in serving as an alert- and rapid-response system.

While existing communication efforts can be improved, more dialogic forms of communication are also needed. Such dialogs serve not only to exchange information and increase knowledge of climate science but also to develop common visions for a better future, address value differences, and form or revitalize social bonds to support the necessary behavioral and social changes. It is this much-needed face-to-face communication that stirs the hope that communication could play an essential role in forming trustful social bonds, building and maintaining social capital, facilitating civic engagement on climate change, and ultimately rejuvenating the democratic political process in the United States.

Canadians' level of active engagement with climate change is quite similar to that of Americans: low, safe, small-scale, neighborhood-based activities (e.g., social marketing campaigns to change energy consumption behavior), regionally based research, visioning and policymaking efforts, and a relatively active and well-networked, but much smaller, NGO community working toward sustainable energy policies. By contrast, in Mexico, civic engagement to date has been focused on more local environmental issues. If the climate change issue were to engage civil society in Mexico, it might play an important role in the process of building democratic virtues and forms of governance (Norget 1997). Given the very different level of development, energy consumption, vulnerability, and response capacity, the tone, foci, and language with which to engage Mexicans on climate change are likely to be very different from those used by its northern neighbors.

References

ABC News/TIME/Stanford University. 2006. Global Warming Poll 3/14/06: Intensity Spikes in Concern on Warming; Many See a Change in Weather Patterns. http://abcnews.go.com/images/Politics/1009a1GlobalWarming.pdf.

Agyeman, Julian, Bob Doppelt, Kathy Lynn, and Halida Hatic. 2007. The Climate-Justice Link: Communicating Risk with Low-Income and Minority Audiences. In *Creating a Climate for Change: Communicating Climate Change and Facilitating Social Change*, edited by Susanne C. Moser and Lisa Dilling. Cambridge: Cambridge University Press.

Angus Reid Strategies. 2007a. Angus Reid Climate Change Survey: Commit to Kyoto, but Don't Curb Car Travel, Say Canadians. http://www.angusreidstrategies.com/uploads/pages/pdfs/2007.04.03%20Enviro%20Policy%20Release.pdf.

Angus Reid Strategies. 2007b. Angus Reid Climate Change Survey: Global Warming a Reality and a Threat, Canadians Say. http://www.angusreidstrategies.com/uploads/pages/pdfs/2007.03.21%20Enviro%20Press%20Release.pdf.

Angus Reid Strategies. 2007c. Angus Reid Climate Change Survey: Rich and Educated Less Likely to Act Green, Today or Tomorrow. http://www.angusreidstrategies.com/index.cfm?fuseaction=news&newsid=36&page=27.

BBC World Service. 2007. All Countries Need to Take Major Steps on Climate Change: Global Poll. Survey conducted by GlobeScan and the Program on International Policy Attitudes (PIPA) at the University of Maryland. http://www.bbc.co.uk/pressoffice/pressreleases/stories/2007/09_september/25/climate.shtml.

Boyte, Harry C., and Nancy N. Kari. 1996. *Building America: The Democratic Promise of Public Work*. Philadelphia: Temple University Press.

Brewer, Thomas L. 2005a. U.S. Public Opinion on Climate Change Issues: Implications for Consensus-Building and Policymaking. *Climate Policy* 5(1): 2–18.

Brewer, Thomas L. 2005b. U.S. Public Opinion on Climate Change Issues: Update for 2005. *Climate Policy* 5(4): 359–376.

Carroll, Joseph. 2007. Polluted Drinking Water Is Public's Top Environmental Concern: Concern about Global Warming Inching Up. Gallup Poll, conducted March 11–14.

Daniel, Ben, Richard A. Schwier, and Gordon McCalla. 2003. Social Capital in Virtual Learning and Distributed Communities of Practice. *Canadian Journal of Learning and Technology* 29(3). http://www.cjlt.ca/index.php/cjlt/article/view/85/79.

DeYoung, Raymond. 2000. Expanding and Evaluating Motives for Environmentally Responsible Behavior (ERB). *Journal of Social Issues* 56 (3): 509–526.

Downs, Anthony. 1972. Up and Down with Ecology: The Issue-Attention Cycle. *Public Interest* 28: 38–50.

Gärling, Tommy, Satoshi Fujii, Anita Gärling, and Cecilia Jakobsson 2003. Moderating Effects of Social Value Orientation on Determinants of Proenvironmental Behavior Intention. *Journal of Environmental Psychology* 23(1): 1–9.

GfK Roper Public Affairs and Media, Yale School of Forestry and Environmental Studies. 2007. The GfK Roper Yale Survey on Environmental Issues. Fall 2007: American Support for Local Action on Global Warming. New Haven, CT: Yale University. http://environment.yale.edu/pubs/Roper-Yale-Environment-Poll/.

GlobeScan. 2006. GlobeScan Poll: Global Views on Climate Change. Washington, DC: World Public Opinion, Program on International Policy Attitudes. http://www.worldpublicopinion.org/pipa/pdf/apr06/ClimateChange_Apr06_quaire.pdf.

Goffman, Erving. 1974. *Frame Analysis: An Essay on the Organization of Experience*. New York: Harper & Row.

Harper, Douglas. 2001. *Online Etymology Dictionary*. http://www.etymonline.com/.

Henderson, Bill. 2008. Climate Change in Canada: Ottawa Slumbers as Premier Campbell Cancels the Olympics. *Counter Currents*, March 14. http://www.countercurrents.org/henderson140308.htm.

Immerwahr, John. 1999. Waiting for a Signal: Public Attitudes toward Global Warming, the Environment and Geophysical Research. Paper presented at the spring meeting of the American Geophysical Union, April 15.

Jones, Jeffrey M. 2008. Polluted Drinking Water Was No. 1 Concern before AP Report. Global Warming Way Down the List. Gallup Poll, conducted March 7–9. http://www.gallup.com/poll/104932/Polluted-Drinking-Water-No-Concern-Before-Report.aspx.

Kaplan, Stephen. 2000. Human Nature and Environmentally Responsible Behavior. *Journal of Social Issues* 56(3): 491–508.

Kellstedt, Paul M., Sammy Zahran, and Arnold Vedlitz. 2008. Personal Efficacy, the Information Environment, and Attitudes toward Global Warming and Climate Change in the United States. *Risk Analysis* 28(1): 113–126.

Klandermans, Bert, and Dirk Oegema. 1987. Potentials, Networks, Motivations, and Barriers: Steps Towards Participation in Social Movements. *American Sociological Review* 52(4): 519–531.

Lakoff, George. 2004. *Don't Think of an Elephant! Know Your Values and Frame the Debate*. White River Junction, VT: Chelsea Green Publishing.

Leighley, Jan E. 1995. Attitudes, Opportunities and Incentives: A Field Essay on Political Participation. *Political Research Quarterly* 48(1): 181–209.

Leiserowitz, Anthony. 2007. American Opinions on Global Warming: Survey Results. New Haven, CT: Yale University. http://environment.yale.edu/news/Research/5310/american-opinions-on-global-warming-summary/.

Leiserowitz, Anthony. 2008. International Public Opinion, Perception, and Understanding of Global Climate Change. New Haven, CT: Yale University.

Loeb, Paul R. 2004. Introduction to Part Four. In *The Impossible Will Take a Little While: A Citizen's Guide to Hope in a Time of Fear*, edited by P. R. Loeb. New York: Basic Books.

Macnaghten, Phil, and Michael Jacobs. 1997. Public Identification with Sustainable Development Investigating Cultural Barriers to Participation. *Global Environmental Change* 7(1): 5–24.

McCright, Aaron M., and Riley E. Dunlap. 2001. Challenging Global Warming as a Social Problem: An Analysis of the Conservative Movement's Counter-Claims. *Social Problems* 47(4): 499–522.

McCright, Aaron M., and Riley E. Dunlap. 2003. Defeating Kyoto: The Conservative Movement's Impact on U.S. Climate Change Policy. *Social Problems* 50(3): 348–373.

McPherson, Miller, Lynn Smith-Lovin, and James Cook. 2001. Birds of a Feather: Homophily in Social Networks. *Annual Review of Sociology* 27: 1–41.

Moser, Susanne C. 2007a. Communication Strategies to Mobilize the Climate Movement. In *Ignition: What You Can Do to Fight Global Warming and Spark a Movement*, edited by Jon Isham and Sissel Waage. Washington, DC: Island Press.

Moser, Susanne C. 2007b. In the Long Shadows of Inaction: The Quiet Building of a Climate Protection Movement in the United States. *Global Environmental Politics* 7(2): 124–144.

Moser, Susanne C. 2007c. More Bad News: The Risk of Neglecting Emotional Responses to Climate Change Information. In *Creating a Climate for Change: Communicating Climate Change and Facilitating Social Change*, edited by Susanne C. Moser and Lisa Dilling. Cambridge: Cambridge University Press.

Moser, Susanne C., and Lisa Dilling, eds. 2007a. *Creating a Climate for Change: Communicating Climate Change and Facilitating Social Change.* Cambridge: Cambridge University Press.

Moser, Susanne C., and Lisa Dilling, eds. 2007b. Toward the Social Tipping Point. In *Creating a Climate for Change: Communicating Climate Change and Facilitating Social Change*, edited by Susanne C. Moser and Lisa Dilling. Cambridge: Cambridge University Press.

Nisbet, Matthew C., and Teresa Myers. 2007. Twenty Years of Public Opinion about Global Warming. *Public Opinion Quarterly* 71(3): 444–470.

Norget, Kirstin. 1997. "The Politics of Liberation": The Popular Church, Indigenous Theology and Grassroots Mobilization in Oaxaca, Mexico. Paper presented at the 20th International Congress of the Latin American Studies Association, Guadalajara, Mexico, April 17–19.

O'Connor, Robert E., Robert J. Bord, and Ann Fisher. 1999. Risk Perceptions, General Environmental Beliefs, and Willingness to Address Climate Change. *Risk Analysis* 19(3): 461–471.

Olson, Robert L. 1995. Sustainability as a Social Vision. *Journal of Social Issues* 51(1): 15–35.

The Pew Research Center for the People and the Press. 2004. *News Audiences Increasingly Politicized*. Washington, DC: Pew Research Center Biennial News Consumption Survey. http://people-press.org/reports/pdf/215.pdf.

PIPA. 2005. *Americans on Climate Change: 2005*. Washington, DC: Program on International Policy Attitudes. http://65.109.167.118/pipa/pdf/jul05/ClimateChange05_Jul05_rpt.pdf.

Project for Excellence in Journalism. 2006. *The State of the News Media 2006. An Annual Report on American Journalism*. Washington, DC: Project for Excellence in Journalism.

Rabkin, Sarah, and Gershon, David. 2007. Changing the World One Household at a Time: Portland's 30–Day Program to Lose 5000 Pounds. In *Creating a Climate for Change: Communicating Climate Change and Facilitating Social Change*, edited by Susanne C. Moser and Lisa Dilling. Cambridge: Cambridge University Press.

Readership Institute. 2002. *Consumers, Media & U.S. Newspapers: Results from the Impact Study*. Evanston, IL: Readership Institute, Media Management Center at Northwestern University.

Rogers, Everett M. 2003. *Diffusion of Innovations*. New York, NY: Free Press.

Saad, Lydia. 2006. Americans Still Not Highly Concerned about Global Warming. Gallup News Service. http://www.gallup.com/poll/22291/Americans-Still-Highly-Concerned-About-Global-Warming.aspx.

Saad, Lydia. 2007. To Americans, the Risks of Global Warming Are Not Imminent: A Majority Worries about Climate Changes, But Thinks Problems Are a Decade or More Away. Gallup Poll, conducted February 22–25. http://www.gallup.com/poll/26842/Americans-Risks-Global-Warming-Imminent.aspx.

Tribbia, John. 2007. Stuck in the Slow Lane of Behavior Change? A Not-So-Superhuman Perspective on Getting Out of Our Cars. In *Creating a Climate for Change: Communicating Climate Change and Facilitating Social Change*, edited by Susanne C. Moser and Lisa Dilling. Cambridge: Cambridge University Press.

Warriors and Heroes: Twenty-Five Leaders Who Are Fighting to Stave Off the Planetwide Catastrophe—The Hawk: Jim Woolsey. 2005. *Rolling Stone*, November 3.

Conclusion

15

North American Climate Governance: Policymaking and Institutions in the Multilevel Greenhouse

Henrik Selin and Stacy D. VanDeveer

Introduction

Since the adoption of the Kyoto Protocol in 1997, conventional wisdom has held that Mexico is largely unconcerned by climate change, that Canada joined the group of European countries committed to meeting the challenge of reducing greenhouse gas (GHG) emissions, and that the United States refused to enact any meaningful climate change mitigation policies. This volume demonstrates that these commonplace views are incomplete—and quite often wrong. While North American federal governments are pursuing quite modest GHG mitigation efforts, a growing number of innovative and ambitious climate change policy initiatives are being developed in states, provinces, and municipalities and by private firms and universities. These initiatives are driven by a combination of factors, including acceptance of the science of human- induced climate change, concerns about vulnerabilities to a changing climate, efforts to protect the long-term viability of local economies, desires to act in the face of lagging federal climate policy, and the commitment and activities of climate change advocates networking across public, private, and civil society sectors.

Consequently, the state of climate change politics across North America is considerably more complicated and dynamic than the post-Kyoto conventional wisdom suggests. Concerns about climate change, energy security, and oil and gas prices have accelerated public- and private-sector initiatives since 2001. Cities, states, provinces, firms, and universities have pledged to cut hundreds of millions of tons of carbon emissions. One 2007 estimate of CO_2 emission reductions in U.S. states based on only three sets of policies—energy efficiency mandates, renewable portfolio standards, and impacts of the regional trading schemes under development in the Northeast and along the West Coast—suggests that such actions could cut 1.8 billion tons of CO_2 emissions by 2020 (Byrne, Hughes, Rickerson, et al. 2007). Another estimate suggests that if 17 of the states and 284 of the cities with explicit GHG reduction targets were to meet their goals by 2020, they would constitute almost 50 percent of the cuts needed for the United States to get back to 1990 emissions levels (Lutsey and Sperling 2008). In 2007 and 2008, British Columbia established itself as Canada's climate change policy leader, enacting ambitious GHG reduction goals, adopting a carbon tax, and joining California to develop a transborder emissions trading scheme.

Private sector membership in the Chicago Climate Exchange (CCX) has increased steadily since 2003, just one example of how a growing number of U.S. firms are responding to the climate change challenge. Furthermore, the Canadian Montreal Climate Exchange was established in collaboration with the CCX in 2006, with trading beginning in 2008, thereby expanding North American private sector efforts to reduce GHG emissions through trading schemes. The CCX furthermore responded to the expansion of California's GHG mitigation commitments by establishing a California-specific entity to develop a mechanism to trade GHG allowances pursuant to the states' laws and regulations. A host of reports and press releases demonstrate that these carbon exchanges in the United States and Canada, like their affiliates in Europe, are experiencing rapid year-on-year growth in the total amount of carbon traded and aggregate values of these trades. In addition, private-sector actors are leading through the creation of innovative policy mechanisms in areas such as emissions trading and renewable energy and by driving clean technology development.

Yet, the overall effectiveness of North American GHG mitigation efforts to date should not be overstated. There is continued growth in GHG emissions in all three North American countries (see chapter 1), and many emission trends show little sign of being immediately reversed (or even stabilized). For example, a recent U.S. government assessment of national GHG emissions suggest that they will grow by 11 percent by 2012 over 2002 levels (Revkin 2007). Many states, provinces, and municipalities are struggling to cut their GHG emissions, and most are unlikely to meet their self-imposed and relatively modest short-term reduction targets. As such, public- and private-sector actors must find much more efficient ways to reduce their GHG emissions if these are ever to be substantially brought down. Nevertheless, important political and technical precedents for future climate change actions are being set all over North America, and the trend toward more ambitious public- and private-sector initiatives continues apace.

The introductory chapter asks four critical questions about developing climate change action across the continent:

1. What are the new or emerging institutions, policies, and practices in the area of climate change governance under development in North America?
2. What roles do major public, private, and civil society actors play, and how do they interact to shape policy and governance?

3. Through which pathways are climate change policies and initiatives diffused across jurisdictions in North America?
4. To what extent can North American climate change action be characterized as existing or emerging multilevel governance, and are local and federal institutions across the continent facilitating or impeding such developments?

Answers to these four questions provide important insights into developing multilevel climate change politics across the continent. In the next four sections, we address these questions in turn as we draw on key insights and arguments from this volume's chapters. This is followed by a discussion about the future of continental climate change politics and governance as we identify four possible scenarios based on combinations of high and low federal and subnational involvement: federal inertia; federal resurgence; bottom-up expansion; and complex multilevel coordination. Aspects of all of these scenarios are possible, but some form of complex multilevel coordination seems most likely in the near term. The chapter—and this volume—ends with a few concluding remarks on continuing governance issues and challenges in the North American greenhouse.

Institution Building and Policy Innovation

What are the new or emerging institutions, policies, and practices in the area of climate change governance under development in North America? National governments and policymakers in Canada, the United States, and Mexico are engaged in building limited domestic and transnational institutions for GHG mitigation and climate change research, but much of the most significant institutional innovation in the post-Kyoto decade has taken place below federal organizations.

The three North American countries exhibit significant differences in their relationship to the Kyoto Protocol and in the views and strategies regarding how to bring down national GHG emissions. Canada ratified the Kyoto Protocol in 2002 and its leaders have repeatedly stated commitments to tackling climate change. However, Stoett argues that successive Canadian federal governments have failed to show necessary leadership to bring down national GHG emissions (or even stop annual increases). This argument is confirmed by the substantial increases in Canadian GHG emissions since 1990. Canadian governments have issued several action plans that are heavy on planning and rhetoric but low on action. The United States, under the George W. Bush administration, opposed any kind of mandatory GHG reduction targets, instead favoring voluntary measures and investments in scientific research and technological developments. While government-funded climate change science continues, there is little evidence that the federal voluntary GHG programs had any discernible impact on national emissions.

Canadian-U.S. national differences regarding climate change are also visible in Arctic political fora. The Arctic region has gained much symbolic value in international climate change debates and policymaking. It is also the geographic area where the earliest effects of a warming climate are visible, in the form of rapid melting of the Greenland ice sheet, thawing of permafrost, and reduced ice coverage in the Arctic Ocean. As Nilsson's chapter shows, the Bush administration's refusal to commit to mandatory GHG reductions often pitted the United States against the other seven Arctic countries and Arctic indigenous peoples organizations during the scientific work of the Arctic Climate Impact Assessment and its associated political efforts. Representatives of the Bush administration were also criticized domestically for interfering with scientific work and softening language in scientific reports for political purposes (Revkin and Wald 2007; Union of Concerned Scientists 2007).

Pulver argues that climate change is not yet seen as a major national political issue in Mexico, but that several public- and private-sector actors have focused on climate change–related issues since the 1990s. Mexican governments have taken few steps to develop major federal GHG reduction strategies, despite being active participants in the Kyoto negotiations and follow-up meetings. The federal government and private-sector entrepreneurs placed much faith in the Kyoto Protocol's clean development mechanism (CDM), hoping it would result in increased foreign investments in Mexico's energy sector and stimulate the development of GHG reduction projects in Mexico. The United States' rejection of the Kyoto Protocol dashed many of these hopes, since it meant that U.S. firms would not participate in the CDM. In other words, U.S. federal climate policy discouraged Mexican federal interest in climate change issues and policymaking. Mexico remains a strong supporter of the CDM and is partnering with Canada and other countries on a growing number of CDM projects. As a major developing country, Mexico is also facing pressure to accept mandatory GHG controls under a post-Kyoto agreement.

Given the diverging approaches to climate change and GHG mitigation policy by the three federal governments, it is hardly surprising that there have been few substantive efforts to create a more comprehensive North American climate change policy under the North American Commission for Environmental Cooperation, the environmental organ of the North American Free Trade Agreement (NAFTA), or in any other tri-national forum. Furthermore, Betsill argues that a NAFTA CO_2 trading system may not be desirable in the absence of more extensive and harmonized national policies in Canada, the United States, and Mexico. Thus, efforts to build GHG trading institutions, at least in the near future, are more likely to continue expanding via bottom-up initiatives among states and provinces. Transnational cooperation among cities is also flourishing in the absence of federal cooperation on the continent.

There are important continental energy issues at stake as a multitude of public- and private-sector actors seek to expand renewable energy generation and consumption (including through CDM projects in Mexico). However, actors express diverging opinions on what constitutes appropriate renewable energy sources. Rowlands's chapter identifies one controversial policy area, discussing politically significant differences between Canada and the United States over the use of large-scale hydropower facilities in renewable energy expansion efforts. These differences can have important ramifications for both future national energy investments and cross-border energy trade. Of course, nuclear power brings another set of controversial energy and environmental issues that are increasingly influenced by climate change debates and policymaking at various levels of governance in all three North American countries.

In contrast to the passivity of federal governments, several chapters demonstrate that the most innovative policies and practices for climate change action and GHG mitigation are developing at subnational political levels and (increasingly) in the private sector. Climate change policy advocates are creating new institutions and expanding existing ones. New public- and private-sector entities designed specifically or in part to address climate change issues include the Pew Center on Global Climate Change, the Regional Greenhouse Gas Initiative (RGGI), the Association for the Advancement of Sustainability in Higher Education (AASHE), the Climate Registry, and the CCX. In addition, a diverse set of actors has pushed to expand the scope of long-standing environmental and nonenvironmental organizations to engage issues of climate change mitigation and adaptation, including the Conference of the New England Governors and Eastern Canadian Premiers, the Western Governors Association, the International Council for Local Environmental Initiatives (ICLEI), the U.S. Conference of Mayors, and the Federation of Canadian Municipalities. Furthermore, the number of regional initiatives involving U.S. states and Canadian provinces continues to expand.

In the first years of the twenty-first century, U.S. state and local public-sector actors have essentially developed a set pattern in their climate change policy development efforts. The pattern consists of "inventorying their emissions, establishing climate change action plans, setting emission reduction targets similar to those of the Kyoto protocol, enacting state level regulations and standards explicitly targeting GHGs, and forging multi-government alliances to reinforce and support their actions" (Lutsey and Sperling 2008, 673). In other words, a concerted set of practices has developed from what had been a more ad hoc process of subnational climate change policy developments. The same appears to be happening among Canadian provinces, as more of them complete GHG inventories, develop climate change action plans, move toward setting GHG emissions targets, take a host of regulatory actions, and expand multijurisdictional cooperation.

State, provincial, and municipal authorities often seek to combine classical command-and-control policies and elements with more market-based and flexible approaches in their efforts to develop more aggressive climate change policy and action. Rabe, Selin and VanDeveer, and Farrell and Hanemann analyze many of the growing number of U.S. states (and to a lesser degree Canadian provinces) enacting mandatory GHG emission reduction policies for the energy sector and/or setting statewide renewable portfolio standards. Other states, including Hawaii, Illinois, New Jersey, and New Mexico followed California's lead, legislating aggressive statewide GHG reduction goals. Moreover, California leads efforts to impose CO_2 emissions standards on vehicles and to set lower carbon standards for vehicle fuels. By 2008, sixteen other states had committed to following California regulations for vehicle emissions and fuels. The latter efforts engendered legal challenges from the auto industry, which were supported in part by the Bush administration, in multiple jurisdictions.

States in the U.S. Northeast from Maryland to Maine are collaborating, with private and civil society sector actors, in a regional CO_2 trading system in the form of RGGI. Inspired in part by RGGI, California officials launched negotiations with their counterparts in U.S. states and Canadian provinces on the West Coast (and later Québec) to establish their own trading scheme. Related to efforts to create RGGI in the Northeast and California's leadership on the West Coast, states, provinces, and tribes also collaborated on a common GHG emissions reporting system, the Climate Registry. This effort seeks to harmonize important technical standards across a host of public-sector actors; 58 states, provinces, and tribes from all three North American countries (thirty-nine U.S. states, ten Canadian provinces, six Mexican states, and three Native American groups) and the District of Columbia were members as of late 2008. Data generated through the Climate Registry aid members' efforts to formulate targeted GHG emission reduction policies and also serve as a basis for national or even continental standardization of GHG estimation and reporting in the years to come.

Likewise, Gore and Robinson detail how a steadily growing number of U.S. and Canadian municipalities are joining climate change coalitions, formulating climate change policies, and setting GHG emission reduction standards. Cities such as Portland, Oregon, and Toronto, Ontario, demonstrate that much can be accomplished at the municipal level even when federal policy is not engendering national GHG emissions reductions. Many municipal efforts focus on energy efficiency programs and the development of green building codes, working with the Leadership in Energy and Environmental Design Green Building Rating System. This includes Mexico City, which launched a Green Plan that prioritizes climate change action and has joined the Clinton Foundation's Climate Initiative, a network of sixteen large cities around the globe pledging to reduce GHG emissions.[1] Furthermore, the Climate Registry is explicitly designed to complement and support municipal-level GHG reporting and reduction goals, such as those under ICLEI and the U.S. Conference of Mayors initiatives.

Jones and Levy's chapter discusses how a growing number of North American companies, which led much early private-sector opposition to GHG regulations, are taking a more proactive stand on climate change. This includes investments in clean energy and low-emissions technology and the development of new market-based approaches to GHG reductions, including through CCX. Firms are also increasingly lobbying policymakers to harmonize regulations across jurisdictions. More specifically, Haufler explores ways in which the insurance industry is responding to business challenges and opportunities stemming from climate change, as concerns about the potential costs of climate change impacts grow within the North American and global insurance and reinsurance industries.

Both the Haufler and the Jones and Levy chapters note the growing institutionalization of climate change concern within the private sector, as firms coordinate some of their climate change–related research and analysis through industrial and sectoral associations to which they belong. Levine furthermore details the many initiatives undertaken by a rapidly growing number of colleges and universities, which seek to institutionalize their climate change and energy initiatives within university administrative structures and to construct and join networks and organizations to enhance the effectiveness of their common efforts.

Actors and Their Interactions

What roles do major public, private, and civil society actors play, and how do they interact to shape policy and governance? Networked collaboration between private- and public-sector actors significantly influences North American policy developments. Virtually every chapter includes some discussion of actors' networking and institution-building activities. Several chapters demonstrate the importance of well-networked entrepreneurial leaders in developing subnational climate change action. Importantly, North American institution building and networking on climate change issues co-evolve over time. That is, the creation of climate change–focused institutions is driven by networked actors as new institutions help to form and maintain new and expanded networks. As network members collaborate over time, social interaction serves to identify and shape interests and preferences of climate change advocates across public, private, and civil society sectors. Actors' iterated interaction builds networks to diffuse information and shared norms and values.

Many public, private, and civil society actors have taken on crucial leadership roles in climate change governance and GHG mitigation. North American institution-building results from dynamic social processes and interactions shaped by the participants' interests and strategies, as well as by their behaviors, knowledge, and norms. National and local government officials are of course important actors in these social processes, but so are representatives of many private firms and advocacy groups. As such, analysis of changing climate change politics and policymaking across North America requires attention to the strategies and influence of networked actors across public, private, and civil society sectors. Simply put, the rapid growth in climate change initiatives across all of North America cannot be explained without attention to expanding networks of actors—or without attention to both material interests of engaged actors and changes in norms.

Many network members are motivated by shared environmental concerns, values, and beliefs. Networked actors, however, also seek to build coalitions with individuals and organizations who may gain from more aggressive climate change policy, but who may not necessarily share their enviromental concerns. Rabe, Selin and VanDeveer, and Farrell and Hanemann demonstrate that interactions among public, private, and civil society actors are at the core of much developing climate change policymaking in U.S. states. Gore and Robinson highlight the importance of national and transnational networks for North American municipalities as they seek to reduce their GHG emissions, similar to the networks that Jones and Levy discuss with respect to private-sector actors. Levine identifies a set of networks linking climate change and sustainability programs across public and private universities. In several of these chapters, it is clear that engaged actors have material interests at stake, and that changing understandings about climate change impacts, policies, and responsibilities are shaping the actions they take.

Many networks create and/or use organizations designed to facilitate interactions and lesson learning among a wide set of actors. Important organizational nodes include ICLEI's CCP program, the U.S. Conference of Mayors, the Federation of Canadian Municipalities, the PEW Center on Global Climate Change, NAFTA's North American Commission on Environmental Protection, AASHE, and the Arctic Council. In addition, public officials from all over North America (and the world) increasingly gather at major events such as the C40 Large Cities Climate Summit, which convened in New York City in 2007. State and provincial governors and civil servants also belong to a host of formal professional organizations (at national and regional levels) and informal networks of colleagues across jurisdictions. Stories about particular policy experiences and more systematic assessments of climate and energy policies and programs are diffused via professional conferences, newsletters, email list-serves, and personal exchanges.

Similarly, Moser focuses attention on individual actors—those functioning as communicators and audiences working to build loose bottom-up networks pushing for policy changes around climate change issues. She identifies a series of personal and societal barriers to more effective civic mobilization and engagement, but these barriers can be overcome in part through the design of better communication strategies by scientific and civil society leaders participating in climate change debates and policymaking efforts across Canada, the United States, and Mexico. As such, many public, private, and civil society participants—both individually and as network members—play important roles in fostering a broader civic engagement on climate change and contributing to behavioral changes throughout North American societies. To this end, individual and collective action can both address the climate change challenge and enrich democratic governance more broadly.

Pathways of Policy Change

Through which pathways are climate change policies and initiatives diffused across jurisdictions in North America? While expanding networks of public, private, and civil society actors are critical in North American climate change politics, the mere existence of a network does not explain the influence of either the network itself or its members. Selin and VanDeveer (2007) identified four overlapping pathways through which climate change networks influence policy developments at various levels of authority: (1) strategic demonstration of the feasibility of climate change action; (2) market creation and expansion; (3) policy diffusion and learning; and (4) norm creation and promulgation. Via these pathways, actors and institutions at one level of governance may influence others at the same level, or they may impact the behaviors and preferences of those at higher or lower levels of authority.

First, many public- and private-sector forerunners are explicit about their desire to lead by example in the face of lagging federal standard-setting and policymaking. They seek to demonstrate the feasibility and effectiveness of environmental policies that opponents claim cannot or should not be enacted. Furthermore, leaders frequently explicitly work to diffuse or disseminate their policy initiatives, horizontally and vertically, to other jurisdictions across North America and beyond. Thus many state, provincial, and municipal leaders seek to demonstrate that enacting and implementing more aggressive climate change policy is technically, economically, and politically possible and beneficial.

Rabe shows that the tradition of U.S. states acting as policy forerunners continues around climate change and energy issues, as a growing number of states across the United States develop climate change and cleaner energy policies that exceed, sometimes substantially, federal requirements. Farrell and Hanemann detail the ambition and complexity of California's growing set of climate change policy goals and measures, while Selin and VanDeveer show that states in the U.S. Northeast have attempted to play political leadership roles since the late 1990s on climate change issues, inspired by their earlier efforts to lead national policy in combating acid rain and mercury pollution. States in the United States have a long history of environmental leadership that subsequently influences federal policy.

Similarly, Gore and Robinson explore the activities and accomplishments of municipal leaders across Canada and the United States as they develop increasingly more ambitious municipal-based action in response to slow federal policymaking. In addition, Jones and Levy outline important changes in the private sector as a growing number of firms go beyond federal mandates and call for more organized federal policy expansion. Some public- and private-sector actors are discovering that GHG emission reductions can come at a positive economic net gain through energy savings and job creation. In part by demonstrating the feasibility and utility of more expansive efforts to address climate change, action on the municipal level and within the private sector engenders substantial horizontal diffusion of goals and practices by offering lessons for others to emulate.

Second, many climate change policy advocates in public and private sectors seek to promote collective and individual behavioral change by expanding existing markets for products and services deemed more climate friendly and by creating new markets designed to engender climate change mitigation. For example, statewide renewable portfolio standards in U.S. states are designed in part to drive additional investments in renewable energy generation capacity and to expand local and regional markets for renewable energy. In 2006, the Union of Concerned Scientists estimated that nearly 45,000 megawatts of renewable energy capacity will be developed by 2020 to satisfy state renewable portfolio standards (Byrne, Hughes, Rickerson, et al. 2007). Renewable energy generation capacity is steadily increasing to meet expanding renewable energy mandates across the continent (Rabe 2008).

Cap-and-trade schemes such as RGGI, as well as those under discussion on the West Coast and elsewhere, are also about market creation. Such schemes create marketable allowances and establish and regulate the markets in which these emission allowances are traded, thereby setting a market-based monetary value for CO_2 emissions. Similar dynamics play out in CDM and CCX and other trading schemes. In addition, policies that set energy efficiency standards for products or purchases expand markets for more energy-efficient technologies, consumer goods, and services. Over forty states have enacted such policies (Byrne, Hughes, Rickerson, et al. 2007). Meanwhile, Levine describes how many universities have pledged to buy renewable or cleaner energy. This is also true for a growing number of NGOs and firms. Advocates of such policies hope that increased incentives and competition, together with economies of scale dynamics, will help drive down costs for renewable energy generation and distribution and increase supplies.

Jones and Levy discuss changing attitudes and strategies among major businesses, but note that stronger government policies are critical to stimulate larger and more long-term private-sector commitments and investments. North American carbon markets, however, are growing rapidly, as is the amount of venture capital invested in more climate friendly technologies. There has also been a sharp growth in consultancy and accounting firms in North America that offer their services to private and public organizations that want to participate in credit and/or offset schemes for CO_2 reductions. Of course, some firms and sectors have more to gain or lose than others. Haufler's chapter assesses the many challenges faced by the insurance and reinsurance industries, for example. She argues that insurance and reinsurance firms on both sides of the Atlantic Ocean are beginning to develop new strategies to assess and reduce their risks associated with climate change, and provide new products to encourage individual property owners and firms to engage in climate change mitigation and adaptation.

A third pathway of change is the horizontal and vertical diffusion of policies and learning across jurisdictions. Networks move information about climate policies and management actions across public, private, and civil society sectors. Climate change policies developed among leader states, municipalities, and firms frequently serve as models for subsequent initiatives by other actors. For example, Rabe describes how U.S. states are drawing on each other's experiences in the design and implementation of renewable portfolio standards. Similarly, green building codes and energy reduction strategies are developed and diffused across North American municipalities through organizations such as ICLEI, the U.S. Conference of Mayors, and the Federation of Canadian Municipalities, while best practices are shared among universities via AASHE. Such learning is both "pulled" and "pushed" across jurisdictions; a growing number of public- and private-sector actors actively seek information on effective strategies and policies to emulate at the same time as many policy advocates seek new ways to communicate information and successful experiences.

A fourth pathway of influence is the creation and promulgation of norms related to expanding climate change action and policymaking. Normative change can be a powerful influence on policymaking. The chapters on state and municipal climate change actions demonstrate that a growing number of state and local officials have established CO_2 as a "pollutant" in need of regulation. In April 2007 the U.S. Supreme Court endorsed this view. Furthermore, the chapters focusing on private-sector actors suggest a growing acceptance of climate change among corporate leaders. These changing views and decisions are greatly influenced by recent scientific findings and the publication of the latest set of reports by the Intergovernmental

Panel on Climate Change in 2007 that link with increased confidence human activities that cause growing atmospheric GHG concentrations to rising global temperatures. However, changing views are also clearly linked to a growing sense that GHG mitigation is an appropriate, even necessary and expected, function of public- and private-sector organizations.

Through networks and meetings, actors develop and internalize shared norms on the importance of effective climate change action. Climate change policy advocates are generating broader political and public expectations that GHG emissions should decline. Over time, policies and behaviors that reduce GHG emissions are thus increasingly likely to be judged as more appropriate than those that engender increases. For example, changing norms on climate policy may affect how the public and policymakers view different forms of renewable energy. Similarly, normative changes can have significant impact on energy uses, building codes, fuel standards, and the design of goods if, over time, people come to expect that GHG emissions should decline. In this respect, Moser discusses the importance of well-designed communication strategies for effective civic mobilization and public engagement on climate change action and in democratic societies.

Norm development and promulgation also pertains to the use or choice of particular policies over others for GHG mitigation. As a growing number of public- and private-sector actors rally around particular policy instruments, these gain status as the most "appropriate" mechanisms. How particular policies are framed by political leaders and in the public imagination is critical (Rabe 2004, 2008). For example, even though many economists argue that the introduction of different kinds of carbon taxes would be an economically efficient way to price carbon emissions and bring emissions down over time, carbon taxes are ignored by many policy forerunners because they may be a hard sell politically. Instead, market-based policy instruments such as the creation of emissions trading schemes are increasingly being embraced by public- and private-sector representatives.

Multilevel Climate Change Governance

To what extent can North American climate change action be characterized as existing or emerging multilevel governance, and are local and federal institutions across the continent facilitating or impeding this? Based on a minimal definition of multilevel governance—actors operating across horizontal and vertical levels of social organization and jurisdictional authority around a particular issue—such governance is emerging in North America around climate change. These efforts over time are also getting more ambitious in terms of their mitigation goals. Yet, because federal governments have been so inactive, North American multilevel climate change governance consists of a multitude of uncoordinated efforts setting different goals within different time frames and applying a multitude of different political and technical means. There are, however, signs of a recent trend toward at least some increased technical standardization and policy harmonization across initiatives and jurisdictions.

States, provinces, municipalities, and firms use an increasingly standardized set of processes from GHG inventories to coordinate policy actions with other jurisdictions (Lutsey and Sperling 2008). Such coordination tends to occur within institutions created to achieve coordination, like RGGI or ICLEI, as greater numbers of actors learn from and emulate previous initiatives. Learning from others also reduces costs, compared to actors starting from square one every time. This kind of coordination is much different from more directed multilevel governance following from coherent federal and/or continentwide climate change leadership. For example, without federal action the regulatory distance and emissions levels between local jurisdictions taking action and those that refuse to do so is likely to grow, and policy laggards' emissions are likely to continue to increase. While many of the initiatives discussed in this volume are linked, at least partially, via some number of shared participants or institutions, the absence of committed federal authorities makes it more difficult to achieve significant, economywide GHG emissions reductions.

There are also important issues about actual and potential conflicts as multilevel governance efforts expand across North America. One looming challenge is the integration of the many developing public- and private-sector initiatives with expanded federal efforts, when these occur, into a coherent governance system. Stoett argues that Canadian federal-provincial relations on climate change have long been problematic and that this disharmony has impeded efforts to reduce GHG emissions. Rabe, Selin and VanDeveer, and Farrell and Hanemann show that many U.S. states have taken action in response to federal inaction, in many cases against the wishes of the federal government. Similarly, Gore and Robinson and Levine demonstrate that a multitude of municipalities and universities are taking action in the absence of federal leadership and also, in some cases, moving ahead of policies enacted by the states and provinces in which they are located. However, many issues of authority remain unsettled as federal, state, provincial, and municipal officials and organizations struggle over policymaking and leadership rights.

Whereas states and provinces have the capacity to regulate many activities critical to effective climate change governance, federal governments in Washington, DC and Ottawa have sometimes taken confrontational stands in their interactions with state and provincial officials. Federal opposition in the United States under the Bush administration included several years of refusing to recognize CO_2 as a pollutant under federal law, as well as working with the auto industry to prevent California and other states from regulating GHG emissions from vehicles (Eilperin 2007). While Canadian federal rhetoric has been much different than that in the United States, federal authorities have made few serious attempts to implement climate change policy or build enduring cooperation with the provinces around GHG mitigation. Furthermore, Canadian, U.S., and Mexican federal political systems do not divide decision-making authorities regarding energy, environmental, and product standards in the same ways, posing challenges for policy coordination across borders.

The United States and Canada have long histories of environmental federalism, as many laws and regulations are subject to shared or divided areas of authority between federal and state/provincial institutions. Yet, it remains unclear how far the many subnational policy efforts and private-sector initiatives can go and to what extent federal policymakers will attempt to supersede them when they choose to enact more ambitious climate change policy. Many analysts and activists worry that the U.S. government may preempt state climate change policy by imposing weaker policy than can be found already (or could be pursued) at the state level (Rabe 2008; Savacool and Barkenbus 2007). In the United States, the 1970s saw a great deal of federal preemption of state environmental policies as the federal government imposed more stringent policy on the states (Engel 2006). In contrast, more recent U.S. federal authorities have been more likely to use preemption to restrict states from exceeding standards (Savacool and Barkenbus 2007). In Canada, the provinces have a greater ability to stop federal attempts to limit their authority where environmental and natural resource management issues are concerned.

Where Next for Continental Climate Change Politics and Governance?

North American climate change governance is increasingly dynamic and multifaceted. All three federal governments face policymaking challenges and opportunities as the successor to the Kyoto Protocol is negotiated and implemented. These international negotiations offer federal governments opportunities to reassert domestic leadership. Yet, all three face a complex domestic terrain of subnational actors and institutions with divergent climate change–related commitments and preferences. Similarly, it is not clear what role continental institutions such as NAFTA will play in climate change governance. Betsill's chapter is rather pessimistic about the prospects of NAFTA serving as a venue for emissions trading. Furthermore, Rowlands argues that increased transborder conflicts around renewable energy generation and use are a distinct possibility; should such conflicts heat up, NAFTA bodies may be drawn into continental climate change politics alongside a host of federal and subnational actors.

Table 15.1
Four scenarios for North American multilevel climate change governance

Subnational policymaking	Federal policymaking: Low	Federal policymaking: High
Low	*1. Federal inertia* • Federal governments remain passive, or even obstructive, of subnational action • Subnational policymaking declines, due to a lack of federal support, active federal opposition, or a failure to realize GHG reduction goals	*2. Federal resurgence* • Federal governments enact policy ceilings, prohibiting subnational jurisdictions from exceeding federal policy • Subnational policymaking becomes more reactive due to federal limits, dependent on federal monetary support, or because federal actions are aggressive enough to make additional subnational policy efforts unlikely
High	*3. Bottom-up expansion* • Federal governments remain passive but are not overtly obstructive of local-level action • Subnational policymaking and implementation accelerates in response to a continued lack of federal leadership • Subnational authorities work to expand multijurisdictional collaboration and policy diffusion	*4. Complex multilevel coordination* • Federal governments set mandatory policy floors of minimum regulations and standards, allowing actors and jurisdictions to exceed federal policies in some areas • Subnational policymaking continues apace among leaders who exceed federal requirements • Continental climate change governance is characterized by debates about appropriate levels of policymaking and implementation

So, what does all this mean for the future of climate change governance in North America? One way to think about possible futures for North American multilevel governance is to explore combinations of federal and subnational climate change politics and policymaking efforts (Rabe 2004, 2008). Table 15.1 illustrates four combinations of high or low federal and subnational involvement in continued climate change governance. Within each combination of federal and subnational policymaking activity there are several possible sets of detailed policy developments. The four scenarios are not meant to be collectively exhaustive of all possible policymaking options on the continent. They are, however, intended to illustrate a host of broader issues (and some major political debates and controversies) about the future organization of multilevel governance around climate change that need to be addressed in one way or another across North America.

The first quadrant's scenario—federal inertia—is a worst-case scenario for climate change policy advocates wherein federal inaction (even obstruction) continues to contribute to a decline in subnational efforts. Such outcomes are possible if leading subnational jurisdictions find achieving their GHG reduction goals prohibitively expensive and/or technically or politically impossible in the face of lagging federal efforts. If, for example, leading efforts like those in California or the Northeast were to be undermined or banned by federal action or courts, it is possible that subnational efforts will decline in scope and quantity. Another possibility is that very few, if any, leading jurisdictions may be able to extract substantial GHG reductions from their activities and economies. Then a growing perception of failure might slow subnational policymaking and implementation substantially and have a chilling effect on voluntary private-sector initiatives. The first quadrant scenario seems most likely to produce continuing GHG emissions growth in some or all of the three North American economies.

The second quadrant's scenario—federal resurgence—includes a number of possibilities for more aggressive federal policymaking in conjunction with a decline of subnational policy efforts. Federal policy may override all, or large portions of, existing and ongoing subnational policymaking by effectively setting national policy "ceilings" (Rabe 2008; Savacool and Barkenbus 2007). In doing so, federal policy may limit subnational policy below the levels enacted by subnational leaders. However, federal resurgence may also lead to the setting of such aggressive GHG reduction goals and technical standards that few public or private entities would choose to exceed federal policies. If federal policymakers set relatively high standards, subnational policy implementation would accelerate, but desires to formulate additional policies may very well be much reduced, at least in the short term. Furthermore, two or three of the continent's federal governments could also work together to accelerate federal efforts in ways that exceed or are equal to current subnational leaders on the continent.

The third quadrant scenario—bottom-up expansion—basically amounts to a continuation of many aspects of the policymaking pattern apparent in the United States since the late 1990s and in Canada since the mid-2000s. If federal inaction continues, there may simply be a continuation of the kind of bottom-up climate change policymaking that has become commonplace among many states, provinces, municipalities, firms, and civil society organizations,. This scenario, however, relies on federal authorities—at a minimum—not to obstruct subnational policymaking as well as to refuse to enact more stringent standards. There may also be a continuing diffusion of policy initiative across jurisdictions and governance levels in combination with greater cooperation among states, provinces, municipalities, and other public- and private-sector organizations. In the private sector, for example, a growing number of voluntary GHG markets and publicly backed trading schemes could be developed as participants seek greater technical standardization and policy harmonization outside the realm of federal authorities.

Finally, in the case of the fourth quadrant scenario—complex multilevel coordination—subnational policymaking continues even as federal climate change action becomes more aggressive in attempts to bring down GHG emissions. Under this scenario, growing federal action, individually or in conjunction with other nation-states, establishes minimum standards with which subnational actors must comply, which could increase over time. Subnational authorities would be allowed to exceed many areas of federal policy. In other words, federal action would establish policy "floors" without enacting any restrictive policy "ceilings" (Rabe 2008; Savacool and Barkenbus 2007). For example, federal policymakers may set mandatory minimum goals for GHG mitigation or renewable energy generation while giving states the right to exceed these goals. In the United States, this approach is used in many social, economic, and environmental areas, including minimum wage legislation and civil rights issues as well as regulations on clean water, toxic substances, and brownfields (Rabe 2008; Savacool and Barkenbus 2007).

While all four scenarios are possible, complex multilevel coordination appears to be the most likely outcome in the near and medium terms. More aggressive federal climate change policy will be enacted as national policymakers come under increased pressure to act. Federal policy expansions may take place separately in each country, or they may involve different forms of transnational collaboration. There may also be substantial debate about which governance levels are the most appropriate for enacting and implementing specific policies—and about what roles are best played by public, private, or civil society actors. To this end, the European Union's greater experience with multilevel continental climate change governance, including debates about subsidiarity, may offer important lessons for North American governance efforts (Schreurs and Tiberghien 2007).

North American climate change governance characterized by complex multilevel coordination offers states and provinces continued opportunities to act as important laboratories for policy innovation that can shape subsequent federal and/or continental policymaking. Yet, public- and private-sector actors also share an interest in the harmonization of many climate- and energy-related standards across economies and societies, preferring regulatory uniformity over regulatory fragmentation. These shared interests in harmonization encourage federal policymakers to act, and the attending expansion of federal climate change policy creates greater regulatory uniformity. However, because subnational actors can continue to develop a wide range of climate change and energy policies within their jurisdictions, regulatory differences are likely to remain across North America.

Furthermore, local-level officials and private-sector representatives are likely to compete over which standards and policies do, and do not, get uploaded to federal and continental levels. Leader states and provinces and their representatives in federal legislatures hoping to reap early mover advantages have strong incentives to compete to upload their specific standards and programs into federal and continent-wide policy initiatives. Consequently, debates about differences and advantages between various policy approaches will increasingly be heard in Washington, DC, Ottawa, and Mexico City as pressures grow on federal policymakers to develop more aggressive national standards. At the same time, many state, provincial, and municipal officials will act forcefully to protect their authority over the federal government in many areas of climate change and energy policymaking.

Concluding Remarks: North American Governance in the Greenhouse

There is little indication that North American climate change forerunners will slow their efforts to reduce GHG emissions anytime soon. At the same time, climate change policy advocates readily acknowledge the importance of federal policy development that supports efforts in states, provinces, municipalities, and private firms for reaching medium- and long-term GHG emission reduction goals. Thus, local-level public and civil society actors seek to push federal regulatory standards upward, and subnational efforts are likely to have significant political and design impacts on future national, and perhaps international, policies (Selin and VanDeveer 2007). If federal policymakers decide to absorb many of the policies and institutions developed at subnational levels, then forerunner jurisdictions will be among the most influential in setting new national standards and rules (Selin and VanDeveer 2007).

North American citizens and policymakers, like those around the world, remain in the very early stages of developing and implementing effective climate change mitigation and adaptation policy. Myriad challenges confront efforts to decarbonize economies and decouple economic growth from GHG emissions. Clearly, much more effective national climate change policies in Canada, the United States, and Mexico are necessary to significantly reduce North American GHG emissions. Whether federal authorities seek to cooperate with subnational actors or restrict their ability to increase environmental standards and goals will be a central issue in future multilevel governance. Much has changed in the public, private, and civil society sectors in the post-Kyoto decade, and North America can now be said to be home to a host of increasingly ambitious and innovative climate change and energy-related policies and initiatives.

Much more extensive climate change–related cooperation between North American public- and private-sector actors and their counterparts outside the continent is likely in the future (Koehn 2008). Leading U.S. states and Canadian provinces have GHG emissions levels similar to those of many countries, and North American states and provinces face related challenges in substantially reducing GHG emissions from energy and transportation sectors as in many industrialized and developing countries. North American municipalities and firms moreover share many of the same problems in trying to develop policies and strategies for cutting GHG emissions as their foreign counterparts. Societies both inside and outside of North America would benefit from deepening international cooperation and lesson-learning outside the formal sphere of national governments.

Some of the most interesting developments in North American climate politics are found in the growing set of increasingly ambitious climate change and energy policies enacted in California. One of the largest economies on earth now has some of the most aggressive GHG mitigation goals and a set of policies designed to reduce emissions from transportation, the utilities sector, households, and service and industrial sectors. Public, private, and civil society actors across North America have taken notice and engaged California's efforts. This is true beyond North America as well. Many European and Asian actors, in particular, closely follow and assess policy developments in California and other North American jurisdictions to draw lessons and to get a sense of where U.S. and Canadian national climate change policy may be going. In this respect, they recognize that California, alongside other state and provincial forerunners, are setting the agenda for much future North American climate change politics.

Furthermore, the negotiation and content of a post–Kyoto Protocol agreement will impact North American politics. It may be tempting to argue that North American experiences suggest that it does not matter whether governments accept climate change treaties. Canada and Mexico ratified the Kyoto Protocol, yet their GHG emissions have gone up significantly since that treaty was adopted in 1997. The United States refused to ratify the Kyoto Protocol, even working against it in various international fora, and its emissions have risen slightly less quickly than those of Canada and Mexico. Yet, a multitude of states, provinces, cities, firms, and universities have declared their intentions to meet or exceed their countries' Kyoto goals, using the treaty as an important reference point. May subnational actors will pay close attention also to the post-Kyoto agreement as well as put increased pressure on the federal government to do better during the next international commitment period beyond 2012.

Climate change cooperation within North America and between North American actors and the rest of the world will not be seamless. The climate change challenge is enormous, and actors differ substantially in their views about the most appropriate ways to meet this challenge—as indicated by the many strongly held and differing opinions over how (and whether) to move forward with global climate change policymaking and institution building following the expiration of the Kyoto Protocol. Future North American climate change debates will be heated and diverse throughout the continent in part because time is running out to reduce GHG emissions in time to avert disastrous consequences of climate change (Moriarty and Honnery 2008; Timilsina 2008). However, many important policies and initiatives are developing across North America. While the continent's federal governments have been slow to act, other actors have not. Mainstream North American public debates are quickly shifting from the question of whether to act to a discussion on how to best combat climate change.

Note

1. Other Clinton Climate Initiative cities in North America include Chicago, Houston, New York City, Philadelphia, and Toronto.

References

Betsill, Michele, and Harriet Bulkeley. 2004. Transnational Networks and Global Climate Governance. *International Studies Quarterly* 48(2): 471–493.

Byrne, John, Kristen Hughes, Wilson Rickerson, and Lado Kurdgelashvili. 2007. American Policy Conflict in the Greenhouse: Divergent Trends in Federal, Regional, State and Local Green Energy and Climate Change Policy. *Energy Policy* 35(9): 4555–4573.

Eilperin, Juliet. 2007. U.S. Trying to Block Calif. on Emissions. *Washington Post*, September 25.

Engel, Kirsten H. 2006. Harnessing the Benefits of Dynamic Federalism in Environmental Law. *Emory Law Journal* 56(1): 159–188.

Koehn, Peter H. 2008. Underneath Kyoto: Emerging Subnational Government Initiatives and Incipient Issue-Bundling Opportunities in China and the United States. *Global Environmental Politics* 8(1): 53–77.

Lutsey, Nicholas, and Daniel Sperling. 2008. America's Bottom-Up Climate Change Mitigation Policy. *Energy Policy* 36(2): 673–685.

Moriarty, Patrick, and Damon Honnery. 2008. Mitigating Greenhouse: Limited Time, Limited Options. *Energy Policy* 36(4): 1251–1256.

Rabe, Barry G. 2004. *Statehouse and Greenhouse: The Emerging Politics of American Climate Change Policy*. Washington, DC: Brookings Institution Press.

Rabe, Barry G. 2008. States on Steroids: The Intergovernmental Odyssey of American Climate Policy. *Review of Policy Research* 25(2): 105–128.

Revkin, Andrew. 2007. U.S. Predicting Increase for Emissions. *New York Times*, March 3.

Revkin, Andrew C., and Matthew L. Wald. 2007. Material Shows Weakening of Climate Reports. *New York Times*, March 20.

Schreurs, Miranda A., and Yves Tiberghien. 2007. Multi-level Reinforcement: Explaining European Union Leadership in Climate Change Mitigation. *Global Environmental Politics* 7(3): 19–46.

Selin, Henrik, and Stacy D. VanDeveer. 2007. Political Science and Prediction: What's Next for U.S. Climate Change Policy? *Review of Policy Research* 24(1): 1–27.

Sovacool, Benjamin K., and Jack N. Barkenbus. 2007. Necessary but Insufficient: State Renewable Portfolio Standards and Climate Change Policies. *Environment* 49(6): 21–30.

Timilsina, Govinda. 2008. Atmospheric Stabilization of CO_2 Emissions: Near Term Reductions and Absolute Versus Intensity-Based Targets. *Energy Policy* 36(6): 1927–1936.

Union of Concerned Scientists. 2007. *Atmosphere of Pressure: Political Interference in Federal Climate Science*. Cambridge, MA: Union of Concerned Scientists.

About the Contributors

Michele M. Betsill is Associate Professor of Political Science at Colorado State University where she teaches courses in international relations, global environmental politics, and research methods. Her research focuses on the governance of global environmental problems, especially those related to climate change. She is author and coauthor of numerous book chapters and articles on climate change politics ranging from global to local levels. She is coauthor (with Harriet Bulkeley) of *Cities and Climate Change: Urban Sustainability and Global Environmental Governance* (Routledge, 2003). She is also a coeditor (with Kathryn Hochstetler and Dimitris Stevis) of *Palgrave Advances in International Environmental Politics* (Palgrave, 2006) and (with Elisabeth Corell) of *NGO Diplomacy: The Influence of Nongovernmental Organizations in International Environmental Negotiations* (MIT Press, 2008).

Alexander E. Farrell was an Associate Professor in the Energy Research Group at the University of California, Berkeley. His interdisciplinary research focused on technological and social aspects of energy and environmental issues. His primary research interest was improving our understanding of political, economic, social, and environmental impacts of energy production and transformation in California, across the United States, and around the world. He published widely on energy, environmental, and policy issues. Tragically, Dr. Farrell died in April 2008.

Christopher Gore is Assistant Professor in the Department of Politics and Public Administration and a faculty associate of the Environmental Science and Applied Management graduate program at Ryerson University, Toronto, Canada. His research focuses on the multilevel politics, policy, and administration of urban and environmental issues in North America and sub-Saharan Africa. Recent publications have focused on municipalities' climate change and citizen-government relations in Canada, and environmental policymaking and energy sector reform in East Africa. Chris is coeditor, with Peter Stoett, of the recent book *Environmental Challenges and Opportunities: Local-Global Perspectives on Canadian Issues* (Emond, 2009).

W. Michael Hanemann is Chancellor's Professor in the Department of Agriculture and Resource Economics at the University of California, Berkeley, where he is also on the faculty of the Goldman School of Public Policy. His research interests include nonmarket valuation, environmental economics and policy, water pricing and management, demand modeling for market research and policy design, the economics of irreversibility and adaptive management, and welfare economics. He has published well over a hundred articles, reports, book chapters, and monographs.

Virginia Haufler is Associate Professor of Government and Politics at the University of Maryland College Park. Her research and teaching focus on international political economy, global governance, private authority, and industry self-regulation. Her publications include *A Public Role for the Private Sector: Industry Self Regulation in the Global Political Economy; Private Authority in International Affairs* (with Claire Cutler and Tony Porter); *Dangerous Commerce: Insurance and the Management of International Risk*; "The Privatization of Diplomacy," *International Studies Perspectives* 2006; "Transnational Corporations in Conflict-prone Zones: Public Policy Responses and a Framework for Action" (with Jessica Banfield and Damian Lilly), *Oxford Development Studies* 2005; "Globalization and Industry Self-Regulation," in Miles Kahler and David Lake, editors, *Governance in a Global Economy* (Princeton University Press, 2004).

Charles A. Jones completed a Ph.D. in Public Policy and is a Lecturer in Management at the University of Massachusetts Boston, becoming a postdoctoral Research Fellow in Energy Technology Innovation Policy at Harvard's Kennedy School of Government. His fields of research include science and technology policy, industry growth, and the interactions between organizations and systems. He is a former submarine officer in the U.S. Navy. His dissertation examined the renewable energy industry in Massachusetts as a complex system.

Dovev Levine is a Ph.D. candidate in Natural Resources and Environmental Studies and a Lecturer in Climate Policy at the University of New Hampshire. His fields of research include campus-based climate actions and multilevel environmental governance. He is currently the academic counselor for the Graduate School of the University of New Hampshire.

David L. Levy is Professor and Chair of the Department of Management and Marketing at the University of Massachusetts, Boston. He received a DBA from Harvard Business School, and his research examines strategic contestation over the governance of controversial issues in the context of global production networks. He conducts research on corporate strategic responses to climate change and has also been examining the growth of the renewable energy business sector in the New England region. He has published widely on these topics, and his most recent book, coedited with Peter Newell, is *The Business of Global Environmental Governance* (MIT Press, 2005).

Susanne C. Moser is Director and Principal Scientist of Susanne Moser Research and Consulting in Santa Cruz, California. Her work focuses on interdisciplinary challenges such as the impacts of climate change and sea-level rise on coastal areas, community and state adaptations to such global change hazards, the interaction between science and policy/practice, and the communication of climate change risks in support of societal responses to climate change. Between 2003 and 2008, she was a Research Scientist at the National Center for Atmospheric Research in Boulder, Colorado. Prior to that, she was a postdoctoral fellow at Harvard University's John F. Kennedy School of Government and worked for the Heinz Center and the Union of Concerned Scientists. She was a contributing author to the IPCC Fourth Assessment Report (Working Group on Impacts, Vulnerability and Adaptation, Coastal chapter), and coeditor (with Lisa Dilling) of *Creating a Climate for Change* (Cambridge University Press, 2007).

Annika E. Nilsson is a Research Fellow at the Stockholm Environment Institute. Her work focuses on the interface between climate science and policy in the Arctic. Bringing together perspectives from international relations and science and technology studies, her research investigates how the structure of international cooperation affects society's capacity to learn about environmental issues. Her previous professional experience includes working as a science writer and editor covering biotechnology developments, global environmental change, and Arctic issues.

Simone Pulver is Assistant Professor of Environmental Studies at the University of California, Santa Barbara. From 2007 to 2009, she was the Joukowsky Family Assistant Professor (Research) at the Watson Institute for International Studies at Brown University. Her research focuses on the engagement of nonstate actors—particularly firms, nongovernmental organizations, and scientific experts—in climate change politics at international and national levels. She has guest-edited two special issues: "Where Next with Global Environmental Scenarios" for *Environmental Research Letters* (2008) and "Green Development: The Role of the Developing-Country Private Sector" for *Studies in Comparative International Development* (2007). Her research has also been published in *Organization & Environment, Global Environmental Politics*, and *Greener Management International.*

Barry G. Rabe is a Professor of Public Policy at the Gerald Ford School of Public Policy at the University of Michigan–Ann Arbor, where he also holds appointments in the School of Natural Resources and Environment and the Program in the Environment. Rabe is a senior fellow at the Brookings Institution, which has published his three most recent books. These include *Statehouse and Greenhouse: The Emerging Politics of American Climate Change Policy*, which won the 2005 Caldwell Award from the American Political Science Association. Rabe received a 2006 Climate Protection Award from the U.S. Environmental Protection Agency and the 2007 Elazar Award from the American Political Science Association for Distinguished Scholarship in Federalism and Intergovernmental Relations. In 2008–2009, he was a visiting professor of public affairs at the Miller Center of the University of Virginia.

Pamela Robinson is Assistant Professor, School of Urban and Regional Planning, Ryerson University, Toronto, Ontario, Canada. She is a professional urban planner whose research, teaching, and practice explore the application of urban sustainability principles to community design and civic engagement. She has written several articles on municipal response to climate change, served as an advisor on civic engagement to the City of Toronto, and is a coeditor for a forthcoming book on urban sustainability issues in Canada. In 2004, she won a Canadian Mortgage and Housing Corporation "Excellence in Education Award" for teaching urban sustainability. She is also on the Board of Directors for the Friends of the Greenbelt Foundation and an Advisory Committee member for Metrolinx.

Ian H. Rowlands is Professor in the Department of Environment and Resource Studies at the University of Waterloo. He is also the Associate Dean (Research) in the University's Faculty of Environment. His research and teaching interests lie in the areas of energy management strategies and policy, corporate environmentalism, and international environmental relations. He has published in many international journals, including *Energy Policy*, *Renewable Energy*, and *Business Strategy and the Environment.* He has served on a number of national and international committees, including those convened by the International Energy Agency and the Ontario Power Authority. Before joining the faculty of the University of Waterloo, he was a researcher at the United Nations Collaborating Centre on Energy and Environment in Denmark and a lecturer in International Relations and Development Studies at the London School of Economics and Political Science.

Henrik Selin is an Assistant Professor in the Department of International Relations at Boston University, where he conducts research and teaches classes on global and regional politics and policymaking on environment and sustainable development. On these issues, he has published numerous journal articles and book chapters. He is the coeditor of *Transatlantic Environment and Energy Politics* (Ashgate, 2009). Prior to his faculty position, he spent three years as a Wallenberg Postdoctoral Fellow in Environment and Sustainability at the Massachusetts Institute of Technology.

Peter J. Stoett is Associate Professor and Chair of the Department of Political Science at Concordia University in Montreal. His main areas of expertise include international relations and law, human rights and environmental issues, and Canadian foreign policy. Recent books (authored, coauthored, and coedited) include *Bilateral Ecopolitics: Canadian-American Environmental Relations* (Ashgate, 2006); *International Ecopolitical Theory: Critical Reflections* (UBC Press, 2006); *Global Politics: Origins, Currents, Directions*, third edition (ITP Nelson, 2005); and *Sustainable Development and Canada: National and International Perspectives* (Broadview Press, 2001). He has taught at a variety of Canadian universities, including the University of Guelph, Waterloo, British Columbia, McMaster, and Simon Fraser, before joining Concordia in 1998. He also teaches at the United Nations' University for Peace in San Jose, Costa Rica. He is currently working on a forthcoming book, *Global Biosecurity: The International Politics of Denial, Fear, and Injustice* (Broadview Press).

Stacy D. VanDeveer is an Associate Professor of Political Science at the University of New Hampshire. His research interests include international environmental policymaking and its domestic impacts, the connections between environmental and security issues, and the role of expertise in policymaking. He has received fellowships from the Belfer Center for Science and International Affairs at Harvard University's John F. Kennedy School of Government and the Watson Institute for International Studies at Brown University. In addition to authoring and coauthoring numerous articles, book chapters, working papers, and reports, he coedited several books, including *Comparative Environmental Politics* (MIT Press, forthcoming); *Transatlantic Environment and Energy Politics* (Ashgate, 2009); *EU Enlargement and the Environment: Institutional Change and Environmental Policy in Central and Eastern Europe* (Routledge, 2005); and *Saving the Seas: Values, Science and International Governance* (Maryland Sea Grant Press, 1997).

Index